D1083516

MONOGRAPHS ON
STATISTICS AND APPLIED PROBABILITY

General Editors

D.R. Cox, D.V. Hinkley, N. Reid, D.B. Rubin and B.W. Silverman

(Full details concerning this series are available from the Publishers.)

An Introduction to the Bootstrap

Bradley Efron
Department of Statistics
Stanford University
and
Robert J. Tibshirani
Department of Preventative Medicine and Biostatistics
and Department of Statistics, University of Toronto

CHAPMAN & HALL/CRC

Boca Raton London New York Washington, D.C.

Chapman & Hall/CRC
Taylor & Francis Group
6000 Broken Sound Parkway NW, Suite 300
Boca Raton, FL 33487-2742

Milton Park, Abingdon
Oxon OX 14 4RN

© 1994 by Taylor & Francis Group, LLC
Chapman & Hall/CRC is an imprint of Taylor & Francis Group

No claim to original U.S. Government works
Printed in the United States of America on acid-free paper
25 24 23 22 21 20 19 18 17 16 15 14 13 12 11
International Standard Book Number-13: 978-0-412-04231-7 (Hardcover)

Library of Congress Cataloging-in-Publication Data

Efron, Bradley.
An introduction to the bootstrap/Brad Efron, Rob Tibshirani.
    p. cm.
Includes bibliographical references and index.
ISBN 0-412-04231-2
1. Bootstrap (Statistics). I. Tibshirani, Robert. II. Title.
QA276.8.E3745 1993
519.5'44—dc20   93-4489

Visit the Taylor & Francis Web site at
http://www.taylorandfrancis.com

and the CRC Press Web site at
http://www.crcpress.com

TO

CHERYL, CHARLIE, RYAN AND JULIE

AND TO THE MEMORY OF

RUPERT G. MILLER, JR.

Contents

Preface

Dear friend, theory is all gray,
and the golden tree of life is green.
Goethe, from "Faust"

The ability to simplify means to eliminate the unnecessary so that
the necessary may speak.
Hans Hoffmann

Statistics is a subject of amazingly many uses and surprisingly few effective practitioners. The traditional road to statistical knowledge is blocked, for most, by a formidable wall of mathematics. Our approach here avoids that wall. The bootstrap is a computer-based method of statistical inference that can answer many real statistical questions without formulas. Our goal in this book is to arm scientists and engineers, as well as statisticians, with computational techniques that they can use to analyze and understand complicated data sets.

The word "understand" is an important one in the previous sentence. This is not a statistical cookbook. We aim to give the reader a good intuitive understanding of statistical inference.

One of the charms of the bootstrap is the direct appreciation it gives of variance, bias, coverage, and other probabilistic phenomena. What does it mean that a confidence interval contains the true value with probability .90? The usual textbook answer appears formidably abstract to most beginning students. Bootstrap confidence intervals are directly constructed from real data sets, using a simple computer algorithm. This doesn't necessarily make it easy to understand confidence intervals, but at least the difficulties are the appropriate conceptual ones, and not mathematical muddles.

Much of the exposition in our book is based on the analysis of real data sets. The mouse data, the stamp data, the tooth data, the hormone data, and other small but genuine examples, are an important part of the presentation. These are especially valuable if the reader can try his own computations on them. Personal computers are sufficient to handle most bootstrap computations for these small data sets.

This book does not give a rigorous technical treatment of the bootstrap, and we concentrate on the ideas rather than their mathematical justification. Many of these ideas are quite sophisticated, however, and this book is not just for beginners. The presentation starts off slowly but builds in both its scope and depth. More mathematically advanced accounts of the bootstrap may be found in papers and books by many researchers that are listed in the Bibliographic notes at the end of the chapters.

We would like to thank Andreas Buja, Anthony Davison, Peter Hall, Trevor Hastie, John Rice, Bernard Silverman, James Stafford and Sami Tibshirani for making very helpful comments and suggestions on the manuscript. We especially thank Timothy Hesterberg and Cliff Lunneborg for the great deal of time and effort that they spent on reading and preparing comments. Thanks to Maria-Luisa Gardner for providing expert advice on the "rules of punctuation." We would also like to thank numerous students at both Stanford University and the University of Toronto for pointing out errors in earlier drafts, and colleagues and staff at our universities for their support. Thanks to Tom Glinos of the University of Toronto for maintaining a healthy computing environment. Karola DeCleve typed much of the first draft of this book, and maintained vigilance against errors during its entire history. All of this was done cheerfully and in a most helpful manner, for which we are truly grateful. Trevor Hastie provided expert "S" and TeX advice, at crucial stages in the project.

We were lucky to have not one but two superb editors working on this project. Bea Schube got us going, before starting her retirement; Bea has done a great deal for the statistics profession and we wish her all the best. John Kimmel carried the ball after Bea left, and did an excellent job. We thank our copy-editor Jim Geronimo for his thorough correction of the manuscript, and take responsibility for any errors that remain.

The first author was supported by the National Institutes of Health and the National Science Foundation. Both groups have supported the development of statistical theory at Stanford, including much of the theory behind this book. The second author would like to thank his wife Cheryl for her understanding and support during this entire project, and his parents for a lifetime of encouragement. He gratefully acknowledges the support of the Natural Sciences and Engineering Research Council of Canada.

Palo Alto and Toronto   Bradley Efron
June 1993   Robert Tibshirani

CHAPTER 1

Introduction

Statistics is the science of learning from experience, especially experience that arrives a little bit at a time. The earliest information science was statistics, originating in about 1650. This century has seen statistical techniques become the analytic methods of choice in biomedical science, psychology, education, economics, communications theory, sociology, genetic studies, epidemiology, and other areas. Recently, traditional sciences like geology, physics, and astronomy have begun to make increasing use of statistical methods as they focus on areas that demand informational efficiency, such as the study of rare and exotic particles or extremely distant galaxies.

Most people are not natural-born statisticians. Left to our own devices we are not very good at picking out patterns from a sea of noisy data. To put it another way, we are all too good at picking out non-existent patterns that happen to suit our purposes. Statistical theory attacks the problem from both ends. It provides optimal methods for finding a real signal in a noisy background, and also provides strict checks against the overinterpretation of random patterns.

Statistical theory attempts to answer three basic questions:

(1) How should I collect my data?

(2) How should I analyze and summarize the data that I've collected?

(3) How accurate are my data summaries?

Question 3 constitutes part of the process known as statistical inference. The bootstrap is a recently developed technique for making certain kinds of statistical inferences. It is only recently developed because it requires modern computer power to simplify the often intricate calculations of traditional statistical theory.

The explanations that we will give for the bootstrap, and other

computer-based methods, involve explanations of traditional ideas
in statistical inference. The basic ideas of statistics haven't changed,
but their implementation has. The modern computer lets us apply these ideas flexibly, quickly, easily, and with a minimum of
mathematical assumptions. Our primary purpose in the book is to
explain when and why bootstrap methods work, and how they can
be applied in a wide variety of real data-analytic situations.

All three basic statistical concepts, data collection, summary and
inference, are illustrated in the New York Times excerpt of Figure
1.1. A study was done to see if small aspirin doses would prevent
heart attacks in healthy middle-aged men. The data for the aspirin study were collected in a particularly efficient way: by a controlled, randomized, double-blind study. One half of the subjects
received aspirin and the other half received a control substance, or
placebo, with no active ingredients. The subjects were randomly
assigned to the aspirin or placebo groups. Both the subjects and the
supervising physicians were blinded to the assignments, with the
statisticians keeping a secret code of who received which substance.
Scientists, like everyone else, want the project they are working on
to succeed. The elaborate precautions of a controlled, randomized,
blinded experiment guard against seeing benefits that don't exist,
while maximizing the chance of detecting a genuine positive effect.

The summary statistics in the newspaper article are very simple:

	heart attacks (fatal plus non-fatal)	subjects
aspirin group:	104	11037
placebo group:	189	11034

We will see examples of much more complicated summaries in later
chapters. One advantage of using a good experimental design is a
simplification of its results. What strikes the eye here is the lower
rate of heart attacks in the aspirin group. The ratio of the two
rates is

$$\widehat{\theta} = \frac{104/11037}{189/11034} = .55. \tag{1.1}$$

If this study can be believed, and its solid design makes it very
believable, the aspirin-takers only have 55% as many heart attacks
as placebo-takers.

Of course we are not really interested in $\widehat{\theta}$, the estimated ratio.
What we would like to know is θ, the true ratio, that is the ratio

HEART ATTACK RISK FOUND TO BE CUT BY TAKING ASPIRIN

LIFESAVING EFFECTS SEEN

Study Finds Benefit of Tablet Every Other Day Is Much Greater Than Expected

By HAROLD M. SCHMECK Jr.

A major nationwide study shows that a single aspirin tablet every other day can sharply reduce a man's risk of heart attack and death from heart attack.

The lifesaving effects were so dramatic that the study was halted in mid-December so that the results could be reported as soon as possible to the participants and to the medical profession in general.

The magnitude of the beneficial effect was far greater than expected, Dr. Charles H. Hennekens of Harvard, principal investigator in the research, said in a telephone interview. The risk of myocardial infarction, the technical name for heart attack, was cut almost in half.

'Extreme Beneficial Effect'

A special report said the results showed "a statistically extreme beneficial effect" from the use of aspirin. The report is to be published Thursday in The New England Journal of Medicine.

In recent years smaller studies have demonstrated that a person who has had one heart attack can reduce the risk of a second by taking aspirin, but there had been no proof that the beneficial effect would extend to the general male population.

Dr. Claude Lenfant, the director of the National Heart Lung and Blood Institute, said the findings were "extremely important," but he said the general public should not take the report as an indication that everyone should start taking aspirin.

Figure 1.1. *Front-page news from the New York Times of January 27, 1987. Reproduced by permission of the New York Times.*

we would see if we could treat all subjects, and not just a sample of them. The value $\widehat{\theta} = .55$ is only an estimate of θ. The sample seems large here, 22071 subjects in all, but the conclusion that aspirin works is really based on a smaller number, the 293 observed heart attacks. How do we know that $\widehat{\theta}$ might not come out much less favorably if the experiment were run again?

This is where statistical inference comes in. Statistical theory allows us to make the following inference: the true value of θ lies in the interval

$$.43 < \theta < .70 \tag{1.2}$$

with 95% confidence. Statement (1.2) is a classical confidence interval, of the type discussed in Chapters 12–14, and 22. It says that if we ran a much bigger experiment, with millions of subjects, the ratio of rates probably wouldn't be too much different than (1.1). We almost certainly wouldn't decide that θ exceeded 1, that is that aspirin was actually harmful. It is really rather amazing that the same data that give us an estimated value, $\widehat{\theta} = .55$ in this case, also can give us a good idea of the estimate's accuracy.

Statistical inference is serious business. A lot can ride on the decision of whether or not an observed effect is real. The aspirin study tracked strokes as well as heart attacks, with the following results:

	strokes	subjects
aspirin group:	119	11037
placebo group:	98	11034

$$\tag{1.3}$$

For strokes, the ratio of rates is

$$\widehat{\theta} = \frac{119/11037}{98/11034} = 1.21. \tag{1.4}$$

It now looks like taking aspirin is actually harmful. However the interval for the true stroke ratio θ turns out to be

$$.93 < \theta < 1.59 \tag{1.5}$$

with 95% confidence. This includes the neutral value $\theta = 1$, at which aspirin would be no better or worse than placebo vis-à-vis strokes. In the language of statistical hypothesis testing, aspirin was found to be significantly beneficial for preventing heart attacks, but not significantly harmful for causing strokes. The opposite conclusion had been reached in an older, smaller study concerning men who had experienced previous heart attacks. The aspirin treatment remains mildly controversial for such patients.

The bootstrap is a data-based simulation method for statistical inference, which can be used to produce inferences like (1.2) and (1.5). The use of the term bootstrap derives from the phrase *to pull oneself up by one's bootstrap*, widely thought to be based on one of the eighteenth century Adventures of Baron Munchausen, by Rudolph Erich Raspe. (The Baron had fallen to the bottom of a deep lake. Just when it looked like all was lost, he thought to pick himself up by his own bootstraps.) It is not the same as the term "bootstrap" used in computer science meaning to "boot" a computer from a set of core instructions, though the derivation is similar.

Here is how the bootstrap works in the stroke example. We create two populations: the first consisting of 119 ones and 11037-119=10918 zeroes, and the second consisting of 98 ones and 11034-98=10936 zeroes. We draw with replacement a sample of 11037 items from the first population, and a sample of 11034 items from the second population. Each of these is called a *bootstrap sample*. From these we derive the bootstrap replicate of $\hat{\theta}$:

$$\hat{\theta}^* = \frac{\text{Proportion of ones in bootstrap sample \#1}}{\text{Proportion of ones in bootstrap sample \#2}}. \tag{1.6}$$

We repeat this process a large number of times, say 1000 times, and obtain 1000 *bootstrap replicates* $\hat{\theta}^*$. This process is easy to implement on a computer, as we will see later. These 1000 replicates contain information that can be used to make inferences from our data. For example, the standard deviation turned out to be 0.17 in a batch of 1000 replicates that we generated. The value 0.17 is an estimate of the standard error of the ratio of rates $\hat{\theta}$. This indicates that the observed ratio $\hat{\theta} = 1.21$ is only a little more than one standard error larger than 1, and so the neutral value $\theta = 1$ cannot be ruled out. A rough 95% confidence interval like (1.5) can be derived by taking the 25th and 975th largest of the 1000 replicates, which in this case turned out to be (.93, 1.60).

In this simple example, the confidence interval derived from the bootstrap agrees very closely with the one derived from statistical theory. Bootstrap methods are intended to simplify the calculation of inferences like (1.2) and (1.5), producing them in an automatic way even in situations much more complicated than the aspirin study.

The terminology of statistical summaries and inferences, like regression, correlation, analysis of variance, discriminant analysis, standard error, significance level and confidence interval, has become the lingua franca of all disciplines that deal with noisy data. We will be examining what this language means and how it works in practice. The particular goal of bootstrap theory is a computer-based implementation of basic statistical concepts. In some ways it is easier to understand these concepts in computer-based contexts than through traditional mathematical exposition.

1.1 An overview of this book

This book describes the bootstrap and other methods for assessing statistical accuracy. The bootstrap does not work in isolation but rather is applied to a wide variety of statistical procedures. Part of the objective of this book is expose the reader to many exciting and useful statistical techniques through real-data examples. Some of the techniques described include nonparametric regression, density estimation, classification trees, and least median of squares regression.

Here is a chapter-by-chapter synopsis of the book. **Chapter 2** introduces the bootstrap estimate of standard error for a simple mean. **Chapters 3–5** contain some basic background material, and may be skimmed by readers eager to get to the details of the bootstrap in **Chapter 6**. Random samples, populations, and basic probability theory are reviewed in **Chapter 3**. **Chapter 4** defines the empirical distribution function estimate of the population, which simply estimates the probability of each of n data items to be $1/n$. **Chapter 4** also shows that many familiar statistics can be viewed as "plug-in" estimates, that is, estimates obtained by plugging in the empirical distribution function for the unknown distribution of the population. **Chapter 5** reviews standard error estimation for a mean, and shows how the usual textbook formula can be derived as a simple plug-in estimate.

The bootstrap is defined in **Chapter 6**, for estimating the standard error of a statistic from a single sample. The bootstrap standard error estimate is a plug-in estimate that rarely can be computed exactly; instead a simulation ("resampling") method is used for approximating it.

Chapter 7 describes the application of bootstrap standard errors in two complicated examples: a principal components analysis and a curve fitting problem.

Up to this point, only one-sample data problems have been discussed. The application of the bootstrap to more complicated data structures is discussed in **Chapter 8**. A two-sample problem and a time-series analysis are described.

Regression analysis and the bootstrap are discussed and illustrated in **Chapter 9**. The bootstrap estimate of standard error is applied in a number of different ways and the results are discussed in two examples.

The use of the bootstrap for estimation of bias is the topic of **Chapter 10**, and the pros and cons of bias correction are discussed. **Chapter 11** describes the jackknife method in some detail. We see that the jackknife is a simple closed-form approximation to the bootstrap, in the context of standard error and bias estimation.

The use of the bootstrap for construction of confidence intervals is described in **Chapters 12, 13** and **14**. There are a number of different approaches to this important topic and we devote quite a bit of space to them. In **Chapter 12** we discuss the bootstrap-t approach, which generalizes the usual Student's t method for constructing confidence intervals. The percentile method (**Chapter 13**) uses instead the percentiles of the bootstrap distribution to define confidence limits. The BC_a (bias-corrected accelerated interval) makes important corrections to the percentile interval and is described in **Chapter 14**.

Chapter 15 covers permutation tests, a time-honored and useful set of tools for hypothesis testing. Their close relationship with the bootstrap is discussed; **Chapter 16** shows how the bootstrap can be used in more general hypothesis testing problems.

Prediction error estimation arises in regression and classification problems, and we describe some approaches for it in **Chapter 17**. Cross-validation and bootstrap methods are described and illustrated. Extending this idea, **Chapter 18** shows how the bootstrap and cross-validation can be used to adapt estimators to a set of data.

Like any statistic, bootstrap estimates are random variables and so have inherent error associated with them. When using the bootstrap for making inferences, it is important to get an idea of the magnitude of this error. In **Chapter 19** we discuss the jackknife-after-bootstrap method for estimating the standard error of a bootstrap quantity.

Chapters 20–25 contain more advanced material on selected topics, and delve more deeply into some of the material introduced in the previous chapters. The relationship between the bootstrap and jackknife is studied via the "resampling picture" in **Chapter 20**. **Chapter 21** gives an overview of non-parametric and parametric inference, and relates the bootstrap to a number of other techniques for estimating standard errors. These include the delta method, Fisher information, infinitesimal jackknife, and the sandwich estimator.

Some advanced topics in bootstrap confidence intervals are discussed in **Chapter 22**, providing some of the underlying basis for the techniques introduced in Chapters 12–14. **Chapter 23** describes methods for efficient computation of bootstrap estimates including control variates and importance sampling. In **Chapter 24** the construction of approximate likelihoods is discussed. The bootstrap and other related methods are used to construct a "non-parametric" likelihood in situations where a parametric model is not specified.

Chapter 25 describes in detail a bioequivalence study in which the bootstrap is used to estimate power and sample size. In **Chapter 26** we discuss some general issues concerning the bootstrap and its role in statistical inference.

Finally, the **Appendix** contains a description of a number of different computer programs for the methods discussed in this book.

1.2 Information for instructors

We envision that this book can provide the basis for (at least) two different one semester courses. An upper-year undergraduate or first-year graduate course could be taught from some or all of the first 19 chapters, possibly covering Chapter 25 as well (both authors have done this). In addition, a more advanced graduate course could be taught from a selection of Chapters 6–19, and a selection of Chapters 20–26. For an advanced course, supplementary material might be used, such as Peter Hall's book *The Bootstrap and Edgeworth Expansion* or journal papers on selected technical topics. The Bibliographic notes in the book contain many suggestions for background reading.

We have provided numerous exercises at the end of each chapter. Some of these involve computing, since it is important for the student to get hands-on experience for learning the material. The bootstrap is most effectively used in a high-level language for data analysis and graphics. Our language of choice (at present) is "S" (or "S-PLUS"), and a number of S programs appear in the Appendix. Most of these programs could be easily translated into other languages such as Gauss, Lisp-Stat, or Matlab. Details on the availability of S and S-PLUS are given in the Appendix.

1.3 Some of the notation used in the book

Lower case bold letters such as $\mathbf{x}$ refer to vectors, that is, $\mathbf{x} = (x_1, x_2, \ldots x_n)$. Matrices are denoted by upper case bold letters such as $\mathbf{X}$, while a plain uppercase letter like X refers to a random variable. The transpose of a vector is written as $\mathbf{x}^T$. A superscript "*" indicates a bootstrap random variable: for example, $\mathbf{x}^*$ indicates a bootstrap data set generated from a data set $\mathbf{x}$. Parameters are denoted by Greek letters such as θ. A hat on a letter indicates an estimate, such as $\hat{\theta}$. The letters F and G refer to populations. In Chapter 21 the same symbols are used for the cumulative distribution function of a population. I_C is the indicator function equal to 1 if condition C is true and 0 otherwise. For example, $I_{\{x<2\}} = 1$ if $x < 2$ and 0 otherwise. The notation $\text{tr}(A)$ refers to the trace of the matrix A, that is, the sum of the diagonal elements. The derivatives of a function $g(x)$ are denoted by $g'(x), g''(x)$ and so on.

The notation

$$F \rightarrow (x_1, x_2, \ldots x_n)$$

indicates an independent and identically distributed sample drawn from F. Equivalently, we also write $x_i \overset{\text{i.i.d.}}{\sim} F$ for $i = 1, 2, \ldots n$.

Notation such as $\#\{x_i > 3\}$ means the number of x_is greater than 3. $\log x$ refers to the natural logarithm of x.

CHAPTER 2

The accuracy of a sample mean

The bootstrap is a computer-based method for assigning measures of accuracy to statistical estimates. The basic idea behind the bootstrap is very simple, and goes back at least two centuries. After reviewing some background material, this book describes the bootstrap method, its implementation on the computer, and its application to some real data analysis problems. First though, this chapter focuses on the one example of a statistical estimator where we really don't need a computer to assess accuracy: the sample mean. In addition to previewing the bootstrap, this gives us a chance to review some fundamental ideas from elementary statistics. We begin with a simple example concerning means and their estimated accuracies.

Table 2.1 shows the results of a small experiment, in which 7 out of 16 mice were randomly selected to receive a new medical treatment, while the remaining 9 were assigned to the non-treatment (control) group. The treatment was intended to prolong survival after a test surgery. The table shows the survival time following surgery, in days, for all 16 mice.

Did the treatment prolong survival? A comparison of the means for the two groups offers preliminary grounds for optimism. Let $x_1, x_2, \cdots, x_7$ indicate the lifetimes in the treatment group, so $x_1 = 94, x_2 = 197, \cdots, x_7 = 23$, and likewise let $y_1, y_2, \cdots, y_9$ indicate the control group lifetimes. The group means are

$$\bar{x} = \sum_{i=1}^{7} x_i/7 = 86.86 \quad \text{and} \quad \bar{y} = \sum_{i=1}^{9} y_i/9 = 56.22, \qquad (2.1)$$

so the difference $\bar{x} - \bar{y}$ equals 30.63, suggesting a considerable life-prolonging effect for the treatment.

But how accurate are these estimates? After all, the means (2.1) are based on small samples, only 7 and 9 mice, respectively. In

Table 2.1. *The mouse data. Sixteen mice were randomly assigned to a treatment group or a control group. Shown are their survival times, in days, following a test surgery. Did the treatment prolong survival?*

Group	Data			(Sample Size)	Mean	Estimated Standard Error
Treatment:	94	197	16			
	38	99	141			
	23			(7)	86.86	25.24
Control:	52	104	146			
	10	51	30			
	40	27	46	(9)	56.22	14.14
				Difference:	30.63	28.93

order to answer this question, we need an estimate of the accuracy of the sample means $\bar{x}$ and $\bar{y}$. For sample means, and essentially *only* for sample means, an accuracy formula is easy to obtain.

The *estimated standard error* of a mean $\bar{x}$ based on n independent data points $x_1, x_2, \cdots, x_n$, $\bar{x} = \sum_{i=1}^{n} x_i/n$, is given by the formula

$$\sqrt{\frac{s^2}{n}} \qquad (2.2)$$

where $s^2 = \sum_{i=1}^{n} (x_i - \bar{x})^2/(n - 1)$. (This formula, and standard errors in general, are discussed more carefully in Chapter 5.) The standard error of any estimator is defined to be the square root of its variance, that is, the estimator's root mean square variability around its expectation. This is the most common measure of an estimator's accuracy. Roughly speaking, an estimator will be less than one standard error away from its expectation about 68% of the time, and less than two standard errors away about 95% of the time.

If the estimated standard errors in the mouse experiment were very small, say less than 1, then we would know that $\bar{x}$ and $\bar{y}$ were close to their expected values, and that the observed difference of 30.63 was probably a good estimate of the true survival-prolonging capability of the treatment. On the other hand, if formula (2.2) gave big estimated standard errors, say 50, then the difference estimate would be too inaccurate to depend on.

The actual situation is shown at the right of Table 2.1. The estimated standard errors, calculated from (2.2), are 25.24 for $\bar{x}$ and 14.14 for $\bar{y}$. The standard error for the difference $\bar{x} - \bar{y}$ equals $28.93 = \sqrt{25.24^2 + 14.14^2}$ (since the variance of the difference of two independent quantities is the sum of their variances). We see that the observed difference 30.63 is only $30.63/28.93 = 1.05$ estimated standard errors greater than zero. Readers familiar with hypothesis testing theory will recognize this as an *insignificant* result, one that could easily arise by chance even if the treatment really had no effect at all.

There are more precise ways to verify this disappointing result, (e.g. the permutation test of Chapter 15), but usually, as in this case, estimated standard errors are an excellent first step toward thinking critically about statistical estimates. Unfortunately standard errors have a major disadvantage: for most statistical estimators other than the mean there is no formula like (2.2) to provide estimated standard errors. In other words, it is hard to assess the accuracy of an estimate other than the mean.

Suppose for example, we want to compare the two groups in Table 2.1 by their medians rather than their means. The two medians are 94 for treatment and 46 for control, giving an estimated difference of 48, considerably more than the difference of the means. But how accurate are these medians? Answering such questions is where the bootstrap, and other computer-based techniques, come in. The remainder of this chapter gives a brief preview of the bootstrap estimate of standard error, a method which will be fully discussed in succeeding chapters.

Suppose we observe independent data points $x_1, x_2, \cdots, x_n$, for convenience denoted by the vector $\mathbf{x} = (x_1, x_2, \cdots, x_n)$, from which we compute a statistic of interest $s(\mathbf{x})$. For example the data might be the $n = 9$ control group observations in Table 2.1, and $s(\mathbf{x})$ might be the sample mean.

The bootstrap estimate of standard error, invented by Efron in 1979, looks completely different than (2.2), but in fact it is closely related, as we shall see. A *bootstrap sample* $\mathbf{x}^* = (x_1^*, x_2^*, \cdots, x_n^*)$ is obtained by randomly sampling n times, with replacement, from the original data points $x_1, x_2, \cdots, x_n$. For instance, with $n = 7$ we might obtain $\mathbf{x}^* = (x_5, x_7, x_5, x_4, x_7, x_3, x_1)$.

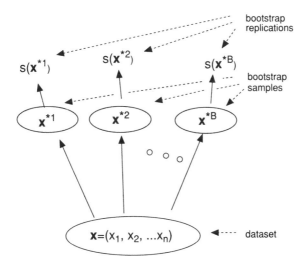

Figure 2.1. *Schematic of the bootstrap process for estimating the standard error of a statistic $s(\mathbf{x})$. B bootstrap sampleᵤ are generated from the original data set. Each bootstrap sample has n elements, generated by sampling with replacement n times from the original data set. Bootstrap replicates $s(\mathbf{x}^{*1}), s(\mathbf{x}^{*2}), \ldots s(\mathbf{x}^{*B})$ are obtained by calculating the value of the statistic $s(\mathbf{x})$ on each bootstrap sample. Finally, the standard deviation of the values $s(\mathbf{x}^{*1}), s(\mathbf{x}^{*2}), \ldots s(\mathbf{x}^{*B})$ is our estimate of the standard error of $s(\mathbf{x})$.*

Figure 2.1 is a schematic of the bootstrap process. The bootstrap algorithm begins by generating a large number of independent bootstrap samples $\mathbf{x}^{*1}, \mathbf{x}^{*2}, \cdots, \mathbf{x}^{*B}$, each of size n. Typical values for B, the number of bootstrap samples, range from 50 to 200 for standard error estimation. Corresponding to each bootstrap sample is a *bootstrap replication* of s, namely $s(\mathbf{x}^{*b})$, the value of the statistic s evaluated for $\mathbf{x}^{*b}$. If $s(\mathbf{x})$ is the sample median, for instance, then $s(\mathbf{x}^{*})$ is the median of the bootstrap sample. The bootstrap estimate of standard error is the standard deviation of the bootstrap replications,

$$\widehat{se}_{\text{boot}} = \left\{ \sum_{b=1}^{B} [s(\mathbf{x}^{*b}) - s(\cdot)]^2 / (B-1) \right\}^{\frac{1}{2}}, \qquad (2.3)$$

where $s(\cdot) = \sum_{b=1}^{B} s(\mathbf{x}^{*b})/B$. Suppose $s(\mathbf{x})$ is the mean $\bar{x}$. In this

Table 2.2. *Bootstrap estimates of standard error for the mean and me-dian; treatment group, mouse data, Table 2.1. The median is less accu-rate (has larger standard error) than the mean for this data set.*

B:	50	100	250	500	1000	∞
mean:	19.72	23.63	22.32	23.79	23.02	23.36
median:	32.21	36.35	34.46	36.72	36.48	37.83

case, standard probability theory tells us (Problem 2.5) that as B gets very large, formula (2.3) approaches

$$\{\sum_{i=1}^{n}(x_i - \bar{x})^2/n^2\}^{\frac{1}{2}}. \tag{2.4}$$

This is almost the same as formula (2.2). We could make it ex-actly the same by multiplying definition (2.3) by the factor $[n/(n-1)]^{\frac{1}{2}}$, but there is no real advantage in doing so.

Table 2.2 shows bootstrap estimated standard errors for the mean and the median, for the treatment group mouse data of Ta-ble 2.1. The estimated standard errors settle down to limiting val-ues as the number of bootstrap samples B increases. The limiting value 23.36 for the mean is obtained from (2.4). The formula for the limiting value 37.83 for the standard error of the median is quite complicated: see Problem 2.4 for a derivation.

We are now in a position to assess the precision of the differ-ence in medians between the two groups. The bootstrap procedure described above was applied to the control group, producing a stan-dard error estimate of 11.54 based on $B = 100$ replications ($B = \infty$ gave 9.73). Therefore, using $B = 100$, the observed difference of 48 has an estimated standard error of $\sqrt{36.35^2 + 11.54^2} = 38.14$, and hence is $48/38.14 = 1.26$ standard errors greater than zero. This is larger than the observed difference in means, but is still insignifi-cant.

For most statistics we don't have a formula for the limiting value of the standard error, but in fact no formula is needed. Instead we use the numerical output of the bootstrap program, for some convenient value of B. We will see in Chapters 6 and 19, that B in the range 50 to 200 usually makes $\widehat{se}_{boot}$ a good standard error estimator, even for estimators like the median. It is easy to write a bootstrap program that works for any computable statistic $s(\mathbf{x})$, as shown in Chapters 6 and the Appendix. With these programs in place, the data analyst is free to use any estimator, no matter how complicated, with the assurance that he or she will also have a reasonable idea of the estimator's accuracy. The price, a factor of perhaps 100 in increased computation, has become affordable as computers have grown faster and cheaper.

Standard errors are the simplest measures of statistical accuracy. Later chapters show how bootstrap methods can assess more complicated accuracy measures, like biases, prediction errors, and confidence intervals. Bootstrap confidence intervals add another factor of 10 to the computational burden. The payoff for all this computation is an increase in the statistical problems that can be analyzed, a reduction in the assumptions of the analysis, and the elimination of the routine but tedious theoretical calculations usually associated with accuracy assessment.

2.1 Problems

2.1 [†] Suppose that the mouse survival times were expressed in weeks instead of days, so that the entries in Table 2.1 were all divided by 7.

    (a)   What effect would this have on $\bar{x}$ and on its estimated standard error (2.2)? Why does this make sense?

    (b)   What effect would this have on the ratio of the difference $\bar{x} - \bar{y}$ to its estimated standard error?

2.2 Imagine the treatment group in Table 2.1 consisted of R repetitions of the data actually shown, where R is a positive integer. That is, the treatment data consisted of R 94's, R 197's, etc. What effect would this have on the estimated standard error (2.2)?

2.3 It is usually true that the error of a statistical estimator decreases at a rate of about 1 over the square root of the sample size. Does this agree with the result of Problem 2.2?

2.4 Let $x_{(1)} < x_{(2)} < x_{(3)} < x_{(4)} < x_{(5)} < x_{(6)} < x_{(7)}$ be an ordered sample of size $n = 7$. Let $\mathbf{x}^*$ be a bootstrap sample, and $s(\mathbf{x}^*)$ be the corresponding bootstrap replication of the median. Show that

(a) $s(\mathbf{x}^*)$ equals one of the original data values $x_{(i)}$, $i = 1, 2, \cdots, 7$.

(b) † $s(\mathbf{x}^*)$ equals $x_{(i)}$ with probability

$$p(i) = \sum_{j=0}^{3} \{\mathrm{Bi}(j; n, \frac{i-1}{n}) - \mathrm{Bi}(j; n, \frac{i}{n})\}, \qquad (2.5)$$

where $\mathrm{Bi}(j; n, p)$ is the binomial probability $\binom{n}{j}p^j(1-p)^{n-j}$. [The numerical values of $p(i)$ are .0102, .0981, .2386, .3062, .2386, .0981, .0102. These values were used to compute $\widehat{\mathrm{se}}_{\mathrm{boot}}\{\mathrm{median}\} = 37.83$, for $B = \infty$, Table 2.2.]

2.5 Apply the weak law of large numbers to show that expression (2.3) approaches expression (2.4) as n goes to infinity.

$\dagger$ Indicates a difficult or more advanced problem.

## CHAPTER 3

Random samples and probabilities

3.1 Introduction

Statistics is the theory of accumulating information, especially information that arrives a little bit at a time. A typical statistical situation was illustrated by the mouse data of Table 2.1. No one mouse provides much information, since the individual results are so variable, but seven, or nine mice considered together begin to be quite informative. Statistical theory concerns the best ways of extracting this information. Probability theory provides the mathematical framework for statistical inference. This chapter reviews the simplest probabilistic model used to model random data: the case where the observations are a random sample from a single unknown population, whose properties we are trying to learn from the observed data.

3.2 Random samples

It is easiest to visualize random samples in terms of a finite population or "universe" $\mathcal{U}$ of individual units $U_1, U_2, \cdots, U_N$, any one of which is equally likely to be selected in a single random draw. The population of units might be all the registered voters in an area undergoing a political survey, all the men that might conceivably be selected for a medical experiment, all the high schools in the United States, etc. The individual units have properties we would like to learn, like a political opinion, a medical survival time, or a graduation rate. It is too difficult and expensive to examine every unit in $\mathcal{U}$, so we select for observation a random sample of manageable size.

A *random sample of size n* is defined to be a collection of n

units $u_1, u_2, \cdots, u_n$ selected at random from $\mathcal{U}$. In principle the sampling process goes as follows: a random number device independently selects integers $j_1, j_2, \cdots, j_n$, each of which equals any value between 1 and N with probability $1/N$. These integers determine which members of $\mathcal{U}$ are selected to be in the random sample, $u_1 = U_{j_1}, u_2 = U_{j_2}, \cdots, u_n = U_{j_n}$. In practice the selection process is seldom this neat, and the population $\mathcal{U}$ may be poorly defined, but the conceptual framework of random sampling is still useful for understanding statistical inference. (The methodology of good experimental design, for example the random assignment of selected units to Treatment or Control groups as was done in the mouse experiment, helps make random sampling theory more applicable to real situations like that of Table 2.1.)

Our definition of random sampling allows a single unit U_i to appear more than once in the sample. We could avoid this by insisting that the integers $j_1, j_2, \cdots, j_n$ be distinct, called "sampling without replacement." It is a little simpler to allow repetitions, that is to "sample with replacement", as in the previous paragraph. If the size n of the random sample is much smaller than the population size N, as is usually the case, the probability of sample repetitions will be small anyway. See Problem 3.1. Random sampling always means sampling *with* replacement in what follows, unless otherwise stated.

Having selected a random sample $u_1, u_2, \cdots, u_n$, we obtain one or more measurements of interest for each unit. Let x_i indicate the measurements for unit u_i. The *observed data* are the collection of measurements $x_1, x_2, \cdots, x_n$. Sometimes we will denote the observed data $(x_1, x_2, \cdots, x_n)$ by the single symbol $\mathbf{x}$.

We can imagine making the measurements of interest on every member $U_1, U_2, \cdots, U_N$ of $\mathcal{U}$, obtaining values $X_1, X_2, \cdots, X_N$. This would be called a census of U.

The symbol $\mathcal{X}$ will denote the census of measurements $(X_1, X_2, \cdots, X_N)$. We will also refer to $\mathcal{X}$ as the population of measurements, or simply the population, and call $\mathbf{x}$ a random sample of size n from $\mathcal{X}$. In fact, we usually can't afford to conduct a census, which is why we have taken a random sample. The goal of statistical inference is to say what we have learned about the population $\mathcal{X}$ from the observed data $\mathbf{x}$. In particular, we will use the bootstrap to say how accurately a statistic calculated from $x_1, x_2, \cdots, x_n$ (for instance the sample median) estimates the corresponding quantity for the whole population.

Table 3.1. *The law school data. A random sample of size $n = 15$ was taken from the collection of $N = 82$ American law schools participating in a large study of admission practices. Two measurements were made on the entering classes of each school in 1973: LSAT, the average score for the class on a national law test, and GPA, the average undergraduate grade-point average for the class.*

School	LSAT	GPA	School	LSAT	GPA
1	576	3.39	9	651	3.36
2	635	3.30	10	605	3.13
3	558	2.81	11	653	3.12
4	578	3.03	12	575	2.74
5	666	3.44	13	545	2.76
6	580	3.07	14	572	2.88
7	555	3.00	15	594	2.96
8	661	3.43			

Table 3.1 shows a random sample of size $n = 15$ drawn from a population of $N = 82$ American law schools. What is actually shown are two measurements made on the entering classes of 1973 for each school in the sample: LSAT, the average score of the class on a national law test, and GPA, the average undergraduate grade point average achieved by the members of the class. In this case the measurement x_i on u_i, the ith member of the sample, is the pair

$$x_i = (\text{LSAT}_i, \text{GPA}_i) \qquad i = 1, 2, \cdots, 15.$$

The observed data $x_1, x_2, \cdots, x_n$ is the collection of 15 pairs of numbers shown in Table 3.1.

This example is an artificial one because the census of data $X_1, X_2, \cdots, X_{82}$ was actually made. In other words, LSAT and GPA are available for the entire population of $N = 82$ schools. Figure 3.1 shows the census data and the sample data. Table 3.2 gives the entire population of N measurements.

In a real statistical problem, like that of Table 3.1, we would see only the sample data, from which we would be trying to infer the properties of the population. For example, consider the 15 LSAT scores in the observed sample. These have mean 600.27 with estimated standard error 10.79, based on the data in Table 3.1 and formula (2.2). There is about a 68% chance that the true LSAT

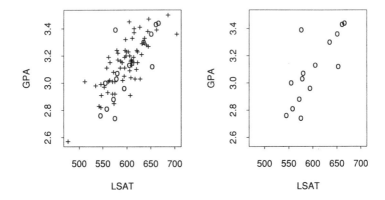

Figure 3.1. *The left panel is a scatterplot of the (LSAT, GPA) data for all $N = 82$ law schools; circles indicate the $n = 15$ data points comprising the "observed sample" of Table 3.1. The right panel shows only the observed sample. In problems of statistical inference, we are trying to infer the situation on the left from the picture on the right.*

mean, the mean for the entire population from which the observed data was sampled, lies in the interval 600.27 ± 10.79.

We can check this result, since we are dealing with an artificial example for which the complete population data are known. The mean of all 82 LSAT values is 597.55, lying nicely within the predicted interval 600.27 ± 10.79.

3.3 Probability theory

Statistical inference concerns learning from experience: we observe a random sample $\mathbf{x} = (x_1, x_2, \cdots, x_n)$ and wish to infer properties of the complete population $\mathcal{X} = (X_1, X_2, \cdots, X_N)$ that yielded the sample. Probability theory goes in the opposite direction: from the composition of a population $\mathcal{X}$ we deduce the properties of a random sample $\mathbf{x}$, and of statistics calculated from $\mathbf{x}$. Statistical inference as a mathematical science has been developed almost exclusively in terms of probability theory. Here we will review briefly

Table 3.2. *The population of measurements (LSAT,GPA), for the universe of 82 law schools. The data in Table 3.1 was sampled from this population. The +'s indicate the sampled schools.*

school	LSAT	GPA	school	LSAT	GPA	school	LSAT	GPA
1	622	3.23	28	632	3.29	56	641	3.28
2	542	2.83	29	587	3.16	57	512	3.01
3	579	3.24	30	581	3.17	58	631	3.21
4+	653	3.12	31+	605	3.13	59	597	3.32
5	606	3.09	32	704	3.36	60	621	3.24
6+	576	3.39	33	477	2.57	61	617	3.03
7	620	3.10	34	591	3.02	62	637	3.33
8	615	3.40	35+	578	3.03	62	572	3.08
9	553	2.97	36+	572	2.88	64	610	3.13
10	607	2.91	37	615	3.37	65	562	3.01
11	558	3.11	38	606	3.20	66	635	3.30
12	596	3.24	39	603	3.23	67	614	3.15
13+	635	3.30	40	535	2.98	68	546	2.82
14	581	3.22	41	595	3.11	69	598	3.20
15+	661	3.43	42	575	2.92	70+	666	3.44
16	547	2.91	43	573	2.85	71	570	3.01
17	599	3.23	44	644	3.38	72	570	2.92
18	646	3.47	45+	545	2.76	73	605	3.45
19	622	3.15	46	645	3.27	74	565	3.15
20	611	3.33	47+	651	3.36	75	686	3.50
21	546	2.99	48	562	3.19	76	608	3.16
22	614	3.19	49	609	3.17	77	595	3.19
23	628	3.03	50+	555	3.00	78	590	3.15
24	575	3.01	51	586	3.11	79+	558	2.81
25	662	3.39	52+	580	3.07	80	611	3.16
26	627	3.41	53+	594	2.96	81	564	3.02
27	608	3.04	54	594	3.05	82+	575	2.74
			55	560	2.93			

some fundamental concepts of probability, including probability distributions, expectations, and independence.

As a first example, let x represent the outcome of rolling a fair die so x is equally likely to be $1, 2, 3, 4, 5$, or 6. We write this in probability notation as

$$\text{Prob}\{x = k\} = 1/6 \qquad \text{for} \quad k = 1, 2, 3, 4, 5, 6. \qquad (3.1)$$

A random quantity like x is often called a *random variable*.

Probabilities are idealized or theoretical proportions. We can imagine a universe $\mathcal{U} = \{U_1, U_2, \cdots, U_N\}$ of possible rolls of the die, where U_j completely describes the physical act of the jth roll, with corresponding results $\mathcal{X} = (X_1, X_2, \cdots, X_N)$. Here N might be very large, or even infinite. The statement $\text{Prob}\{x = 5\} = 1/6$ means that a randomly selected member of $\mathcal{X}$ has a $1/6$ chance of equaling 5, or more simply that $1/6$ of the members of $\mathcal{X}$ equal 5. Notice that probabilities, like proportions, can never be less than 0 or greater than 1.

For convenient notation define the *frequencies* f_k,

$$f_k = \text{Prob}\{x = k\}, \tag{3.2}$$

so the fair die has $f_k = 1/6$ for $k = 1, 2, \cdots, 6$. The *probability distribution of a random variable* x, which we will denote by F, is any complete description of the probabilistic behavior of x. F is also called the probability distribution of the population $\mathcal{X}$. Here we can take F to be the vector of frequencies

$$F = (f_1, f_2, \cdots, f_6) = (1/6, 1/6, \cdots, 1/6). \tag{3.3}$$

An *unfair* die would be one for which F did not equal $(1/6, 1/6, \cdots, 1/6)$.

Note: In many books, the symbol F is used for the cumulative probability distribution function $F(x_0) = \text{Prob}\{x \leq x_0\}$ for $-\infty < x_0 < \infty$. This is an equally valid description of the probabilistic behavior of x, but it is only convenient for the case where x is a real number. We will also be interested in cases where x is a vector, as in Table 3.1, or an even more general object. This is the reason for defining F as *any* description of x's probabilities, rather than the specific description in terms of the cumulative probabilities. When no confusion can arise, in later chapters we use symbols like F and G to represent cumulative distribution functions.

Some probability distributions arise so frequently that they have received special names. A random variable x is said to have the *binomial distribution* with size n and probability of success p, denoted

$$x \sim \text{Bi}(n, p), \tag{3.4}$$

if its frequencies are

$$f_k = \binom{n}{k} p^k (1 - p)^{n-k} \quad \text{for} \quad k = 0, 1, 2, \cdots, n. \tag{3.5}$$

Here n is a positive integer, p is a number between 0 and 1, and $\binom{n}{k}$ is the binomial coefficient $n!/[k!(n-k)!]$. Figure 3.2 shows the distribution $F = (f_0, f_1, \cdots, f_n)$ for $x \sim \text{Bi}(n, p)$, with $n = 25$ and $p = .25, .50$, and $.90$. We also write $F = \text{Bi}(n, p)$ to indicate situation (3.4).

Let A be a set of integers. Then the probability that x takes a value in A, or more simply the *probability of A*, is

$$\text{Prob}\{x \in A\} = \text{Prob}\{A\} = \sum_{k \in A} f_k. \tag{3.6}$$

For example if $A = \{1, 3, 5, \cdots, 25\}$ and $x \sim \text{Bi}(25, p)$, then $\text{Prob}\{A\}$ is the probability that a binomial random variable of size 25 and probability of success p equals an odd integer. Notice that since f_k is the theoretical proportion of times x equals k, the sum $\sum_{k \in A} f_k = \text{Prob}\{A\}$ is the theoretical proportion of times x takes its value in A.

The *sample space* of x, denoted $\mathcal{S}_x$, is the collection of possible values x can have. For a fair die, $\mathcal{S}_x = \{1, 2, \cdots, 6\}$, while $\mathcal{S}_x = \{0, 1, 2, \cdots, n\}$ for a $\text{Bi}(n, p)$ distribution. By definition, x occurs in $\mathcal{S}_x$ every time, that is, with theoretical proportion 1, so

$$\text{Prob}\{\mathcal{S}_x\} = \sum_{k \in \mathcal{S}_x} f_k = 1. \tag{3.7}$$

For any probability distribution on the integers the frequencies f_j are nonnegative numbers summing to 1.

In our examples so far, the sample space $\mathcal{S}_x$ has been a subset of the integers. One of the convenient things about probability distributions is that they can be defined on quite general spaces. Consider the law school data of Figure 3.1. We might take $\mathcal{S}_x$ to be the positive quadrant of the plane,

$$\mathcal{S}_x = \mathcal{R}^{2+} = \{(y, z), y > 0, z > 0\}. \tag{3.8}$$

(This includes values like $x = (10^6, 10^9)$, but it doesn't hurt to let $\mathcal{S}_x$ be too big.) For a subset A of $\mathcal{S}_x$, we would still write $\text{Prob}\{A\}$ to indicate the probability that x occurs in A.

For example, we could take

$$A = \{(y, z) : 0 < y < 600, 0 < z < 3.0\}. \tag{3.9}$$

A law school $x \in A$ if its 1973 entering class had LSAT less than 600 and GPA less than 3.0. In this case we happen to know the complete population $\mathcal{X}$; it is the 82 points indicated on the left panel of Figure 3.1 and in Table 3.2. Of these, 16 are in A, so

$$\text{Prob}\{A\} = 16/82 = .195. \tag{3.10}$$

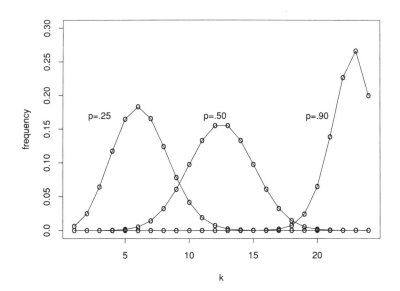

Figure 3.2. *The frequencies* $f_0, f_1, \cdots, f_n$ *for the binomial distributions* $Bi(n,p)$, $n = 25$ *and* $p = .25, .50,$ *and* $.90$. *The points have been connected by lines to enhance visibility.*

Here the idealized proportion $\text{Prob}\{A\}$ is an actual proportion. Only in cases where we have a complete census of the population is it possible to directly evaluate probabilities as proportions.

The probability distribution F of x is still defined to be any complete description of x's probabilities. In the law school example, F can be described as follows: for any subset A of $\mathcal{S}_x = \mathcal{R}^{2+}$,

$$\text{Prob}\{x \in A\} = \#\{X_j \in A\}/82, \qquad (3.11)$$

where $\#\{X_j \in A\}$ is the number of the 82 points in the left panel of Figure 3.1 that lie in A. Another way to say the same thing is that F is a *discrete distribution* putting probability (or frequency) $1/82$ on each of the indicated 82 points.

Probabilities can be defined continuously, rather than discretely as in (3.6) or (3.11). The most famous example is the *normal* (or *Gaussian*, or *bell-shaped*) distribution. A real-valued random variable x is defined to have the normal distribution with mean μ and variance σ^2, written

$$x \sim N(\mu, \sigma^2) \quad \text{or} \quad F = N(\mu, \sigma^2), \tag{3.12}$$

if

$$\text{Prob}\{x \in A\} = \int_A \frac{1}{\sqrt{2\pi\sigma^2}} e^{-\frac{1}{2}\left(\frac{x-\mu}{\sigma}\right)^2} dx \tag{3.13}$$

for any subset A of the real line $\mathcal{R}^1$. The integral in (3.13) is over the values of $x \in A$.

There are higher dimensional versions of the normal distribution, which involve taking integrals similar to (3.13) over multidimensional sets A. We won't need continuous distributions for development of the bootstrap (though they will appear later in some of the applications) and will avoid mathematical derivations based on calculus. As we shall see, one of the main incentives for the development of the bootstrap is the desire to substitute computer power for theoretical calculations involving special distributions.

The *expectation* of a real-valued random variable x, written $\text{E}(x)$, is its average value, where the average is taken over the possible outcomes of x weighted according to its probability distribution F. Thus

$$\text{E}(x) = \sum_{x=0}^{n} x \binom{n}{x} p^x (1-p)^x \quad \text{for} \quad x \sim \text{Bi}(n, p), \tag{3.14}$$

and

$$\text{E}(x) = \int_{-\infty}^{\infty} x \frac{1}{\sqrt{2\pi\sigma^2}} e^{-\frac{1}{2}\left(\frac{x-\mu}{\sigma}\right)^2} dx \quad \text{for} \quad x \sim N(\mu, \sigma^2). \tag{3.15}$$

It is not difficult to show that $\text{E}(x) = np$ for $x \sim \text{Bi}(n, p)$, and $\text{E}(x) = \mu$ for $x \sim N(\mu, \sigma^2)$. (See Problems 3.6 and 3.7.)

We sometimes write the expectation as $\text{E}_F(x)$, to indicate that the average is taken with respect to the distribution F.

Suppose $r = g(x)$ is some function of the random variable x. Then $\text{E}(r)$, the expectation of r, is the theoretical average of $g(x)$ weighted according to the probability distribution of x. For example if $x \sim N(\mu, \sigma^2)$ and $r = x^3$, then

$$\text{E}(r) = \int_{-\infty}^{\infty} x^3 \frac{1}{\sqrt{2\pi\sigma^2}} e^{-\frac{1}{2}\left(\frac{x-\mu}{\sigma}\right)^2} dx. \tag{3.16}$$

Probabilities are a special case of expectations. Let A be a subset of $\mathcal{S}_x$, and take $r = I_{\{x \in A\}}$ where $I_{\{x \in A\}}$ is the *indicator function*

$$I_{\{x \in A\}} = \begin{cases} 1 & \text{if} \quad x \in A \\ 0 & \text{if} \quad x \notin A \end{cases}. \tag{3.17}$$

Then $E(r)$ equals $\text{Prob}\{x \in A\}$, or equivalently

$$E(I_{\{x \in A\}}) = \text{Prob}\{x \in A\}. \tag{3.18}$$

For example if $x \sim N(\mu, \sigma^2)$, then

$$\begin{aligned} E(r) &= \int_{-\infty}^{\infty} I_{\{x \in A\}} \frac{1}{\sqrt{2\pi\sigma^2}} e^{-\frac{1}{2}(\frac{x-\mu}{\sigma})^2} dx \\ &= \int_A \frac{1}{\sqrt{2\pi\sigma^2}} e^{-\frac{1}{2}(\frac{x-\mu}{\sigma})^2} dx, \end{aligned} \tag{3.19}$$

which is $\text{Prob}\{x \in A\}$ according to (3.13).

The notion of an expectation as a theoretical average is very general, and includes cases where the random variable x is not real-valued. In the law school situation, for instance, we might be interested in the expectation of the ratio of LSAT and GPA. Writing $x = (y, z)$ as in (3.8), then $r = y/z$, and the expectation of r is

$$E(\text{LSAT}/\text{GPA}) = \frac{1}{82} \sum_{j=1}^{82} (y_j/z_j) \tag{3.20}$$

where $x_j = (y_j, z_j)$ is the jth point in Table 3.2. Numerical evaluation of (3.20) gives $E(\text{LSAT}/\text{GPA}) = 190.8$.

Let $\mu_x = E_F(x)$, for x a real-valued random variable with distribution F. The *variance* of x, indicated by σ_x^2 or just σ^2, is defined to be the expected value of $y = (x - \mu)^2$. In other words, σ^2 is the theoretical average squared distance of a random variable x from its expectation μ_x,

$$\sigma_x^2 = E_F(x - \mu_x)^2. \tag{3.21}$$

The variance of $x \sim N(\mu, \sigma^2)$ equals σ^2; the variance of $x \sim \text{Bi}(n, p)$ equals $np(1 - p)$, see Problem 3.9. The *standard deviation* of a random variable is defined to be the square root of its variance.

Two random variables y and z are said to be *independent* if

$$E[g(y)h(z)] = E[g(y)]E[h(z)] \tag{3.22}$$

for all functions $g(y)$ and $h(z)$. Independence is well named: (3.22) implies that the random outcome of y doesn't affect the random outcome of z, and vice-versa.

To see this, let B and C be subsets of $\mathcal{S}_y$ and $\mathcal{S}_z$ respectively, the sample spaces of y and z, and take g and h to be the indicator functions $g(y) = I_{\{y \in B\}}$ and $h(z) = I_{\{z \in C\}}$. Notice that

$$I_{\{y \in B\}} I_{\{z \in C\}} = \begin{cases} 1 & \text{if } y \in B \text{ and } z \in C \\ 0 & \text{otherwise.} \end{cases} \tag{3.23}$$

So $I_{\{y \in B\}} I_{\{z \in C\}}$ is the indicator function of the intersection $\{y \in B\} \cap \{z \in C\}$. Then by (3.18) and the independence definition (3.22),

$$\begin{aligned} \text{Prob}\{(y, z) \in B \cap C\} &= \text{E}(I_{\{y \in B\}} I_{\{z \in C\}}) = \text{E}(I_{\{y \in B\}}) \text{E}(I_{\{z \in C\}}) \\ &= \text{Prob}\{y \in B\} \text{Prob}\{z \in C\}. \end{aligned} \tag{3.24}$$

Looking at Figure 3.1, we can see that (3.24) does *not* hold for the law school example, see Problem 3.10, so LSAT and GPA are not independent.

Whether or not y and z are independent, expectations follow the simple addition rule

$$\text{E}[g(y) + h(z)] = \text{E}[g(y)] + \text{E}[h(z)]. \tag{3.25}$$

In general,

$$\text{E}[\sum_{i=1}^{n} g_i(x_i)] = \sum_{i=1}^{n} \text{E}[g_i(x_i)] \tag{3.26}$$

for any functions g_i of any n random variables $x_1, x_2, \cdots, x_n$.

Random sampling with replacement guarantees independence: if $\mathbf{x} = (x_1, x_2, \cdots, x_n)$ is a random sample of size n from a population $\mathcal{X}$, then all n observations x_i are identically distributed and mutually independent of each other. In other words, all of the x_i have the same probability distribution F, and

$$\begin{aligned} \text{E}_F[g_1(x_1) g_2(x_2), \cdots, g_n(x_n)] = \\ \text{E}_F[g_1(x_1)] \text{E}_F[g_2(x_2)] \cdots \text{E}_F[g_n(x_n)] \end{aligned} \tag{3.27}$$

for any functions $g_1, g_2, \cdots, g_n$. (This is almost a definition of what random sampling means.) We will write

$$F \rightarrow (x_1, x_2, \cdots, x_n) \tag{3.28}$$

to indicate that $\mathbf{x} = (x_1, x_2, \cdots, x_n)$ is a random sample of size n from a population with probability distribution F. This is sometimes written as

$$x_i \overset{\text{i.i.d.}}{\sim} F \qquad i = 1, 2, \cdots, n, \tag{3.29}$$

where i.i.d. stands for independent and identically distributed.

3.4 Problems

3.1 A random sample of size n is taken *with* replacement from a population of size N. Show that the probability of having no repetitions in the sample is given by the product

$$\prod_{j=0}^{n-1} (1 - \frac{j}{N}).$$

3.2 Why might you suspect that the sample of 15 law schools in Table (3.1) was obtained by sampling without replacement, rather than with replacement?

3.3 The mean GPA for all 82 law schools is 3.13. How does this compare with the mean GPA for the observed sample of 15 law schools in Table 3.1? Is this difference compatible with the estimated standard error (2.2)?

3.4 Denote the mean and standard deviation of a set of numbers $X_1, X_2, \cdots, X_N$ by $\overline{X}$ and S respectively, where

$$\overline{X} = \sum_{j=1}^{N} X_j/N \qquad S = \{\sum_{j=1}^{N} (X_j - \overline{X})^2/N\}^{1/2}.$$

(a)   A sample $x_1, x_2, \cdots, x_n$ is selected from $X_1, X_2, \cdots, X_N$ by random sampling with replacement. Denote the standard deviation of the sample average $\bar{x} = \sum_{i=1}^{n} x_i/n$, usually called *the standard error* of $\bar{x}$, by $se(\bar{x})$. Use a basic result of probability theory to show that

$$se(\bar{x}) = \frac{S}{\sqrt{n}}.$$

(b) [†] Suppose instead that $x_1, x_2, \cdots, x_n$ is selected by random sampling *without* replacement (so we must have

$n \leq N$), show that

$$\mathrm{se}(\bar{x}) = \frac{S}{\sqrt{n}} \left[\frac{N-n}{N-1} \right]^{\frac{1}{2}}$$

(c) We see that sampling without replacement gives a smaller standard error for $\bar{x}$. Proportionally how much smaller will it be in the case of the law school data?

3.5 Given a random sample $x_1, x_2, \cdots, x_n$, the *empirical probability* of a set A is defined to be the proportion of the sample in A, written

$$\widehat{\mathrm{Prob}}\{A\} = \#\{x_i \in A\}/n. \tag{3.30}$$

(a) Find $\widehat{\mathrm{Prob}}\{A\}$ for the data in Table 3.1, with A as given in (3.9).
(b) The standard error of an empirical probability is $[\mathrm{Prob}\{A\} \cdot (1 - \mathrm{Prob}\{A\})/n]^{1/2}$. How many standard errors is $\widehat{\mathrm{Prob}}\{A\}$ from $\mathrm{Prob}\{A\}$, given in (3.10)?

3.6 A very simple probability distribution F puts probability on only two outcomes, 0 or 1, with frequencies

$$f_0 = 1 - p, \quad f_1 = p. \tag{3.31}$$

This is called the *Bernoulli* distribution. Here p is a number between 0 and 1. If $x_1, \cdots, x_n$ is a random sample from F, then elementary probability theory tells us that the sum

$$s = x_1 + x_2 + \cdots + x_n \tag{3.32}$$

has the binomial distribution (3.5),

$$s \sim \mathrm{Bi}(n, p). \tag{3.33}$$

(a) Show that the empirical probability (3.30) satisfies

$$n \cdot \widehat{\mathrm{Prob}}\{A\} \sim \mathrm{Bi}(n, \mathrm{Prob}\{A\}). \tag{3.34}$$

Expression (3.34) can also be written as $\widehat{\mathrm{Prob}}\{A\} \sim \mathrm{Bi}(n, \mathrm{Prob}\{A\})/n.$)
(b) Prove that if $x \sim \mathrm{Bi}(n, p)$, then $\mathrm{E}(x) = np$.

3.7 Without using calculus, give a symmetry argument to show that $\mathrm{E}(x) = \mu$ for $x \sim N(\mu, \sigma^2)$.

3.8 Suppose that y and z are independent random variables, with variances σ_y^2 and σ_z^2.

    (a)   Show that the variance of $y + z$ is the sum of the variances

$$\sigma_{y+z}^2 = \sigma_y^2 + \sigma_z^2 \ . \tag{3.35}$$

    (In general, the variance of the sum is the sum of the variances for independent random variables $x_1, x_2, \cdots, x_n$.)

    (b)   Suppose $F \rightarrow (x_1, x_2, \cdots, x_n)$ where the probability distribution F has expectation μ and variance σ^2. Show that $\bar{x}$ has expectation μ and variance σ^2/n.

3.9 Use the results in Problems (3.6) and (3.8) to show that $\sigma_x^2 = np(1 - p)$ for $x \sim \text{Bi}(n, p)$.

3.10 Forty-three of the 82 points in Table 3.1 have LSAT < 600; 17 of the 82 points have GPA < 3.0. Why do we know that LSAT and GPA are not independent?

3.11 In the discussion of random sampling, $j_1, j_2, \cdots, j_n$ were taken to be independent integers having a uniform distribution on the numbers $1, 2, \cdots, N$. That is, $j_1, j_2, \cdots, j_n$ is itself a random sample, say

$$F_{1:N} \rightarrow (j_1, j_2, \cdots, j_n), \tag{3.36}$$

where $F_{1:N}$ is the discrete distribution having frequencies $f_j = 1/N$, for $j = 1, 2, \cdots, N$. In practice, we depend on our computer's random number generator to give us (3.36). If (3.36) holds, then a random sample as defined in this chapter has the "i.i.d." property defined in (3.29). Give a brief argument why this is so.

† Indicates a difficult or more advanced problem.

## CHAPTER 4

The empirical distribution function and the plug-in principle

4.1 Introduction

Problems of statistical inference often involve estimating some aspect of a probability distribution F on the basis of a random sample drawn from F. The empirical distribution function, which we will call $\hat{F}$, is a simple estimate of the entire distribution F. An obvious way to estimate some interesting aspect of F, like its mean or median or correlation, is to use the corresponding aspect of $\hat{F}$. This is the "plug-in principle." The bootstrap method is a direct application of the plug-in principle, as we shall see in Chapter 6.

4.2 The empirical distribution function

Having observed a random sample of size n from a probability distribution F,

$$F \rightarrow (x_1, x_2, \cdots, x_n), \tag{4.1}$$

the *empirical distribution function* $\hat{F}$ is defined to be the discrete distribution that puts probability $1/n$ on each value x_i, $i = 1, 2, \cdots, n$. In other words, $\hat{F}$ assigns to a set A in the sample space of x its empirical probability

$$\widehat{\text{Prob}}\{A\} = \#\{x_i \in A\}/n, \tag{4.2}$$

the proportion of the observed sample $\mathbf{x} = (x_1, x_2, \cdots, x_n)$ occurring in A. We will also write $\text{Prob}_{\hat{F}}\{A\}$ to indicate (4.2). The hat symbol "∧" always indicates quantities calculated from the observed data.

Table 4.1. *A random sample of 100 rolls of the die. The outcomes* 1, 2, 3, 4, 5, 6 *occurred* 13, 19, 10, 17, 14, 27 *times, respectively, so the empirical distribution is* (.13, .19, .10, .17, .14, .27).

6	3	2	4	6	6	6	5	3	6	2	2	6	2	3	1	5	1
6	6	4	1	5	3	6	6	4	1	4	2	5	6	6	5	5	3
6	2	6	6	1	4	1	5	6	1	6	3	3	2	2	2	5	2
2	4	1	4	5	6	6	6	2	2	4	6	1	2	2	2	5	1
5	3	5	4	2	1	4	6	6	5	6	4	6	4	3	6	4	1
4	5	4	4	2	3	2	1	4	6								

Consider the law school sample of size $n = 15$, shown in Table 3.1 and in the right panel of Figure 3.1. The empirical distribution $\hat{F}$ puts probability $1/15$ on each of the 15 data points. Five of the 15 points lie in the set $A = \{(y, z) : 0 < y < 600, 0 < z < 3.00\}$, so $\widehat{\text{Prob}}\{A\} = 5/15 \doteq .333$. Notice that we get a different empirical probability for the set $\{0 < y < 600, 0 < z \leq 3.00\}$, since one of the 15 data points has GPA = 3.00, LSAT < 600.

Table 4.1 shows a random sample of $n = 100$ rolls of a die: $x_1 = 6, x_2 = 3, x_3 = 2, \cdots, x_{100} = 6$. The empirical distribution $\hat{F}$ puts probability $1/100$ on each of the 100 outcomes. In cases like this, where there are repeated values, we can express $\hat{F}$ more economically as the vector of *observed frequencies* $\hat{f}_k$, $k = 1, 2, \cdots, 6$,

$$\hat{f}_k = \#\{x_i = k\}/n. \tag{4.3}$$

For the data in Table 4.1, $\hat{F} = (.13, .19, .10, .17, .14, .27)$.

An empirical distribution is a list of the values taken on by the sample $\mathbf{x} = (x_1, x_2, \cdots, x_n)$, along with the proportion of times each value occurs. Often each value occurring in the sample appears only once, as with the law data. Repetitions, as with the die of Table 4.1, allow the list to be shortened. In either case each of the n data points x_i is assigned probability $1/n$ by the empirical distribution.

Is it obvious that we have not lost information in going from the full data set $(x_1, x_2, \cdots, x_{100})$ in Table 4.1 to the reduced representation in terms of the frequencies? No, but it is true. It can be proved that the vector of observed frequencies $\hat{F} = (\hat{f}_1, \hat{f}_2, \cdots)$ is a *sufficient statistic* for the true distribution $F = (f_1, f_2, \cdots)$. This means that all of the information about F contained in $\mathbf{x}$ is also contained in $\hat{F}$.

Table 4.2. *Rainfall data. The yearly rainfall, in inches, in Nevada City, California, 1873 through 1978. An example of time series data.*

	0	1	2	3	4	5	6	7	8	9
1870:				80	40	65	46	68	32	58
1880:	60	61	60	45	48	63	44	66	39	35
1890:	44	104	36	45	69	50	72	57	53	30
1900:	40	56	55	46	46	72	50	68	71	37
1910:	64	46	69	31	33	61	56	55	40	37
1920:	40	34	60	54	52	20	49	43	62	44
1930:	33	45	30	53	32	38	56	63	52	79
1940:	30	62	75	70	60	34	54	51	35	53
1950:	44	53	73	80	54	52	40	77	52	75
1960:	42	43	39	54	70	40	73	41	75	43
1970:	80	60	59	41	67	83	56	29	21	

The sufficiency theorem assumes that the data have been generated by random sampling from some distribution F. This is certainly *not* always true. For example the mouse data of Table 2.1 involve two probability distributions, one for Treatment and one for Control. Table 4.2 shows a *time-series* of 106 numbers: the annual rainfall in Nevada City, California from 1873 through 1978. We could calculate the empirical distribution $\hat{F}$ for this data set, but it would not include any of time series information, for example, if high numbers follow high numbers. Later, in Chapter 8, we will see how to apply bootstrap methods to situations like the rainfall data. For now we are restricting attention to data obtained by random sampling from a single distribution, the so-called *one-sample situation*. This is not as restrictive as it sounds. In the mouse data example, for instance, we can apply one-sample results separately to the Treatment and Control populations.

In applying statistical theory to real problems, the answers to questions of interest are usually phrased in terms of probability distributions. We might ask if the die giving the data in Table 4.1 is fair. This is equivalent to asking if the die's probability distribution F equals $(1/6, 1/6, 1/6, 1/6, 1/6, 1/6)$. In the law school example, the question might be how correlated are LSAT and GPA. In terms of F, the distribution of $x = (y, z) = (\text{LSAT}, \text{GPA})$, this is a question about the value of the *population correlation coefficient*

$$\text{corr}(y, z) = \frac{\sum_{j=1}^{82}(Y_j - \mu_y)(Z_j - \mu_z)}{[\sum_{j=1}^{82}(Y_j - \mu_y)^2 \sum_{j=1}^{82}(Z_j - \mu_z)^2]^{1/2}}, \qquad (4.4)$$

where (Y_j, Z_j) is the jth point in the law school population $\mathcal{X}$, and $\mu_y = \sum_{j=1}^{82} Y_j/82$, $\mu_z = \sum_{j=1}^{82} Z_j/82$.

When the probability distribution F is known (i.e. when we have a complete census of the population $\mathcal{X}$), answering such questions involves no more than arithmetic. For the law school population, the census in Table 3.2 gives $\mu_y = 597.5$, $\mu_z = 3.13$, and

$$\text{corr}(y, z) = .761. \qquad (4.5)$$

This is the original definition of "statistics." Usually we don't have a census. Then we need statistical inference, the more modern statistical theory for inferring properties of F from a random sample **x**.

If we had available only the law school sample of size 15, Table 3.1, we could estimate $\text{corr}(y, z)$ by the *sample correlation coefficient*

$$\widehat{\text{corr}}(y, z) = \frac{\sum_{i=1}^{15}(y_i - \hat{\mu}_y)(z_i - \hat{\mu}_z)}{[\sum_{i=1}^{15}(y_i - \hat{\mu}_y)^2 \sum_{i=1}^{15}(z_i - \hat{\mu}_z)^2]^{1/2}} \qquad (4.6)$$

where (y_i, z_i) is the ith point in Table 3.1, $i = 1, 2, \cdots, 15$, and $\hat{\mu}_y = \sum_{i=1}^{15} y_i/15$, $\hat{\mu}_z = \sum_{i=1}^{15} z_i/15$. Table 3.1 gives $\hat{\mu}_y = 600.3$, $\hat{\mu}_z = 3.09$, and

$$\widehat{\text{corr}}(y, z) = .776. \qquad (4.7)$$

Here is another example of a plug-in estimate. Suppose we are interested in estimating the probability of a LSAT score greater than 600, that is

$$\theta = \frac{1}{82} \sum_{1}^{82} I_{\{Y_i > 600\}}. \qquad (4.8)$$

Since 39 of the 82 LSAT scores exceed 600, $\theta = 39/82 \doteq 0.48$. The plug estimate of θ is

$$\hat{\theta} = \frac{1}{15} \sum_{1}^{15} I_{\{y_i > 600\}} \qquad (4.9)$$

the sample proportion of LSAT scores above 600. Six of the 15
LSAT scores exceed 600, so $\hat{\theta} = 6/15 = 0.4$.

For the die of Table 4.1, we don't have census data but only the
sample $\mathbf{x}$, so any questions about the fairness of the die must be
answered by inference from the empirical frequencies

$$\hat{F} = (\hat{f}_1, \hat{f}_2, \cdots, \hat{f}_6) = (.13, .19, .10, .17, .14, .27). \qquad (4.10)$$

Discussions of statistical inference are phrased in terms of *parameters* and *statistics*. A parameter is a function of the probability distribution F. A statistic is a function of the sample $\mathbf{x}$. Thus
$\mathrm{corr}(y, z)$, (4.4), is a parameter of F, while $\widehat{\mathrm{corr}}(y, z)$, (4.6), is a
statistic based on $\mathbf{x}$. Similarly f_k is a parameter of F in the die
example, while $\hat{f}_k$ is a statistic, $k = 1, 2, 3, \cdots, 6$.

We will sometimes write parameters directly as functions of F,
say

$$\theta = t(F). \qquad (4.11)$$

This notation emphasizes that the value θ of the parameter is obtained by applying some numerical evaluation procedure $t(\cdot)$ to the
distribution function F. For example if F is a probability distribution in the real line, the expectation can be thought of as the
parameter

$$\theta = t(F) = \mathrm{E}_F(x). \qquad (4.12)$$

Here $t(F)$ gives θ by the expectation process, that is, the average
value of x weighted according to F. For a given distribution F such
as $F = \mathrm{Bi}(n, p)$ we can evaluate $t(F) = np$. Even if F is unknown,
the form of $t(F)$ tells us the functional mapping that inputs F and
outputs θ.

4.3 The plug-in principle

The plug-in principle is a simple method of estimating parameters
from samples. The *plug-in estimate* of a parameter $\theta = t(F)$ is
defined to be

$$\hat{\theta} = t(\hat{F}). \qquad (4.13)$$

In other words, we estimate the function $\theta = t(F)$ of the probability
distribution F by the same function of the empirical distribution
$\hat{F}$, $\hat{\theta} = t(\hat{F})$. (Statistics like (4.13) that are used to estimate parameters are sometimes called *summary statistics*, as well as *estimates*

and *estimators.*)

We have already used the plug-in principle in estimating f_k by $\hat{f}_k$, and in estimating $\mathrm{corr}(y, z)$ by $\widehat{\mathrm{corr}}(y, z)$. To see this, note that our law school population F can be written as $F = (f_1, f_2, \ldots f_{82})$ where each f_j, the probability of the jth law school, has value $1/82$. This is the probability distribution on $\mathcal{X}$, the 82 law school pairs. The population correlation coefficient can be written as

$$\mathrm{corr}(y, z) = \frac{\sum_{j=1}^{82} f_j (Y_j - \mu_y)(Z_j - \mu_z)}{[\sum_{j=1}^{82} f_j (Y_j - \mu_y)^2 \sum_{j=1}^{82} f_j (Z_j - \mu_z)^2]^{1/2}}, \quad (4.14)$$

where

$$\mu_y = \sum_{j=1}^{82} f_j Y_j, \quad \mu_z = \sum_{j=1}^{82} f_j Z_j. \quad (4.15)$$

Setting each $f_j = 1/82$ gives expression (4.4). Now for our sample $(x_1, x_2, \ldots x_{15})$, the sample frequency $\hat{f}_j$ is the proportion of sample points equal to X_j:

$$\hat{f}_j = \#\{x_i = X_j\}/15, \quad j = 1, 2, \ldots 82. \quad (4.16)$$

For the sample of Table 3.1, $\hat{f}_1 = 0, \hat{f}_2 = 0, \hat{f}_3 = 0, \hat{f}_4 = 1/15$ etc. Now plugging these values $\hat{f}_j$ into expressions (4.15) and (4.14) gives $\hat{\mu}_y, \hat{\mu}_z$ and $\widehat{\mathrm{corr}}(y, z)$ respectively. That is, $\hat{\mu}_y, \hat{\mu}_z$ and $\widehat{\mathrm{corr}}(y, z)$ are *plug-in* estimates of μ_y, μ_z and $\mathrm{corr}(y, z)$.

In general, the plug-in estimate of an expectation $\theta = \mathrm{E}_F(x)$ is

$$\hat{\theta} = \mathrm{E}_{\hat{F}}(x) = \frac{1}{n} \sum_{i=1}^{n} x_i = \bar{x}. \quad (4.17)$$

How good is the plug-in principle? It is usually quite good, if the only available information about F comes from the sample $\mathbf{x}$. Under this circumstance $\hat{\theta} = t(\hat{F})$ cannot be improved upon as an estimator of $\theta = t(F)$, at least not in the usual asymptotic $(n \to \infty)$ sense of statistical theory. For example if $\hat{f}_k$ is the plug-in frequency estimate $\#\{x_i = k\}/n$, then

$$\hat{f}_k \sim \mathrm{Bi}(n, f_k)/n \quad (4.18)$$

as in Problem 3.6. In this case the estimator $\hat{f}_k$ is *unbiased* for f_k, $\mathrm{E}(\hat{f}_k) = f_k$, with variance $f_k(1 - f_k)/n$. This is the smallest possible variance for an unbiased estimator of f_k.

We will use the bootstrap to study the bias and standard error of the plug-in estimate $\hat{\theta} = t(\hat{F})$. The bootstrap's virtue is that it produces biases and standard errors in an automatic way, no matter how complicated the functional mapping $\theta = t(F)$ may be. We will see that the bootstrap itself is an application of the plug-in principle.

The plug-in principle is less good in situations where there is information about F other than that provided by the sample $\mathbf{x}$. We might know, or assume, that F is a member of a *parametric family*, like the family of multivariate normal distributions. Or we might be in a *regression* situation, where we have available a collection of random samples $\mathbf{x}(z)$ depending on a predictor variable z. Then even if we are only interested in F_{z_0}, the distribution function for some specific value z_0 of z, there may be information about F_{z_0} in the other samples $\mathbf{x}(z)$, especially those for which z is near z_0. Regression models are discussed in Chapters 7 and 9.

The plug-in principle and the bootstrap can be adopted to parametric families and to regression models. See Section 6.5 of Chapter 6 and Chapter 9. For the next few chapters we assume that we are in the situation where we have only the one random sample $\mathbf{x}$ from a completely unknown distribution F. This is called the one-sample *nonparametric* setup.

4.4 Problems

4.1 Say carefully why the plug-in estimate of the expectation of a real-valued random variable is $\bar{x}$, the sample average.

4.2 We would like to estimate the variance σ_x^2 of a real-valued random variable x, having observed a random sample $x_1, x_2, \cdots, x_n$. What is the plug-in estimate of σ_x^2?

4.3 (a) Show that the standard error of an empirical frequency $\hat{f}_k$ is $\sqrt{\hat{f}_k(1 - \hat{f}_k)/n}$. (You can use the result in problem 3.5b.)

(b) Do you believe that the die used to generate Table 4.1 is fair?

4.4 Suppose a random variable x has possible values $1, 2, 3, \cdots$. Let A be a subset of the positive integers.

(a) Show that $\widehat{\text{Prob}}\{A\} = \sum_{k \in A} \hat{f}_k$.

(b)   Compare problems 4.3a and 3.5b, and conclude that the observed frequencies $\hat{f}_k$ are not independent of each other.

(c)   Say in words why the observed frequencies aren't independent.

## CHAPTER 5

Standard errors and estimated standard errors

5.1 Introduction

Summary statistics such as $\hat{\theta} = t(\hat{F})$ are often the first outputs of a data analysis. The next thing we want to know is the accuracy of $\hat{\theta}$. The bootstrap provides accuracy estimates by using the plug-in principle to estimate the standard error of a summary statistic. This is the subject of Chapter 6. First we will discuss estimation of the standard error of a mean, where the plug-in principle can be carried out explicitly.

5.2 The standard error of a mean

Suppose that x is a real-valued random variable with probability distribution F. Let us denote the expectation and variance of F by the symbols μ_F and σ_F^2 respectively,

$$\mu_F = \mathrm{E}_F(x), \qquad \sigma_F^2 = \mathrm{var}_F(x) = \mathrm{E}_F[(x - \mu_F)^2]. \qquad (5.1)$$

These are the quantities called μ_x and σ_x^2 in Chapter 3. Here we are emphasizing the dependence on F. The alternative notation "$\mathrm{var}_F(x)$" for the variance, sometimes abbreviated to $\mathrm{var}(x)$, means the same thing as σ_F^2. In what follows we will sometimes write

$$x \sim (\mu_F, \sigma_F^2) \qquad (5.2)$$

to indicate concisely the expectation and variance of x.

Now let $(x_1, \cdots, x_n)$ be a random sample of size n from the distribution F. The mean of the sample $\bar{x} = \sum_{i=1}^{n} x_i/n$ has expectation μ_F and variance σ_F^2/n,

$$\bar{x} \sim (\mu_F, \sigma_F^2/n). \qquad (5.3)$$

In other words, the expectation of $\bar{x}$ is the same as the expectation of a single x, but the variance of $\bar{x}$ is $1/n$ times the variance of x. See Problem 3.8b. This is the reason for taking averages: the larger n is, the smaller $\mathrm{var}(\bar{x})$ is, so bigger n means a better estimate of μ_F.

The *standard error* of the mean $\bar{x}$, written $\mathrm{se}_F(\bar{x})$ or $\mathrm{se}(\bar{x})$, is the square root of the variance of $\bar{x}$,

$$\mathrm{se}_F(\bar{x}) = [\mathrm{var}_F(\bar{x})]^{1/2} = \sigma_F/\sqrt{n}. \tag{5.4}$$

Standard error is a general term for the standard deviation of a summary statistic.[1] They are the most common way of indicating statistical accuracy. Roughly speaking, we expect $\bar{x}$ to be less than one standard error away from μ_F about 68% of the time, and less than two standard errors away from μ_F about 95% of the time.

These percentages are based on the *central limit theorem*. Under quite general conditions on F, the distribution of $\bar{x}$ will be approximately normal as n gets large, which we can write as

$$\bar{x} \overset{.}{\sim} N(\mu_F, \sigma_F^2/n). \tag{5.5}$$

The expectation μ_F and variance σ_F^2/n in (5.5) are exact, only the normality being approximate. Using (5.5), a table of the normal distribution gives

$$\mathrm{Prob}\{|\bar{x} - \mu_F| < \frac{\sigma_F}{\sqrt{n}}\} \overset{.}{=} .683, \qquad \mathrm{Prob}\{|\bar{x} - \mu_F| < \frac{2\sigma_F}{\sqrt{n}}\} \overset{.}{=} .954, \tag{5.6}$$

as illustrated in Figure 5.1. One of the advantages of the bootstrap is that we do not have to rely entirely on the central limit theorem. Later we will see how to get accuracy statements like (5.6) directly from the data (see Chapters 12–14 on bootstrap confidence intervals). It will then be clear that (5.6), which is correct for large values of n, can sometimes be quite inaccurate for the sample size actually available. Keeping this in mind, it is still true that the standard error of an estimate usually gives a good idea of its accuracy.

A simple example shows the limitations of the central limit theorem approximation. Suppose that F is a distribution that puts

---

[1] In some books, the term "standard error" is used to denote an estimated standard deviation, that is, an estimate of σ_F based on the data. That differs from our usage of the term.

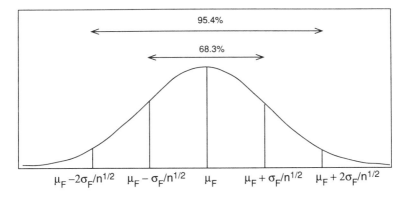

Figure 5.1. *For large values of n, the mean $\bar{x}$ of a random sample from F will have an approximate normal distribution with mean μ_F and variance σ_F^2/n.*

probability on only two outcomes, 0 or 1, as in problem 3.6, say

$$\text{Prob}_F\{x = 1\} = p \qquad \text{and} \qquad \text{Prob}_F\{x = 0\} = 1 - p. \qquad (5.7)$$

Here p is a parameter of F, often called the probability of success, having a value between 0 and 1. A random sample $F \rightarrow (x_1, x_2, \cdots, x_n)$ can be thought of as n independent flips of a coin having probability of success (or of "heads", or of $x = 1$) equaling p. Then the sum $s = \sum_{i=1}^n x_i$ is the number of successes in n independent flips of the coin; s has the binomial distribution (3.3),

$$s \sim \text{Bi}(n, p). \qquad (5.8)$$

The average $\bar{x} = s/n$ equals $\hat{p}$, the plug-in estimate of p. Distribution (5.7) has $\mu_F = p$, $\sigma_F^2 = p(1 - p)$, so (5.3) gives

$$\hat{p} \sim (p, p(1 - p)/n) \qquad (5.9)$$

for the mean and variance of $\hat{p}$. In other words, $\hat{p}$ is an unbiased estimate of p, $\text{E}(\hat{p}) = p$, with standard error

$$\text{se}(\hat{p}) = \left[\frac{p(1 - p)}{n}\right]^{1/2}. \qquad (5.10)$$

Figure 5.2 shows the central limit theorem working for the binomial distribution with $n = 25$, $p = .25$ and $p = .90$. (Problem 5.3 says what is actually plotted in Figure 5.2.) The central limit theorem gives a good approximation to the binomial distribution

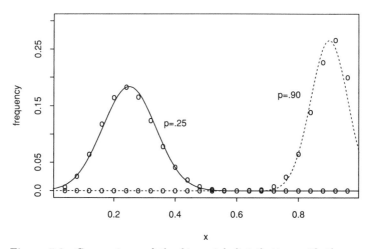

Figure 5.2. *Comparison of the binomial distribution with the normal distribution suggested by the central limit theorem; n = 25, p = .25 and p = .90. The smooth curves are the normal densities, see problem 5.3; circles indicate the binomial probabilities (3.5). The approximation is good for p = .25, but is somewhat off for p = .90.*

for $n = 25, p = .25$, but is somewhat less good for $n = 25, p = .9$.

5.3 Estimating the standard error of the mean

Suppose that we have in hand a random sample of numbers $F \rightarrow x_1, x_2, \cdots, x_n$, such as the $n = 9$ Control measurements for the mouse data of Table 2.1. We compute the estimate $\bar{x}$ for the expectation μ_F, equaling 56.22 for the mouse data, and want to know the standard error of $\bar{x}$. Formula (5.4), $\mathrm{se}_F(\bar{x}) = \sigma_F/\sqrt{n}$, involves the unknown distribution F and so cannot be directly used.

At this point we can use the plug-in principle: we substitute $\hat{F}$ for F in the formula $\mathrm{se}_F(\bar{x}) = \sigma_F/\sqrt{n}$. The plug-in estimate of $\sigma_F = [\mathrm{E}_F(x - \mu_F)^2]^{1/2}$ is

$$\hat{\sigma} = \sigma_{\hat{F}} = \{\frac{1}{n}\sum_{i=1}^{n}(x_i - \bar{x})^2\}^{1/2}, \tag{5.11}$$

since $\mu_{\hat{F}} = \bar{x}$ and $E_{\hat{F}} g(x) = \frac{1}{n} \sum_{i=1}^{n} g(x_i)$ for any function g. This gives the *estimated standard error* $\widehat{se}(\bar{x}) = se_{\hat{F}}(\bar{x})$,

$$\widehat{se}(\bar{x}) = \sigma_{\hat{F}}/\sqrt{n} = \{\sum_{i=1}^{n} (x_i - \bar{x})^2/n^2\}^{1/2}. \tag{5.12}$$

For the mouse Control group data, $\widehat{se}(\bar{x}) = 13.33$.

Formula (5.12) is slightly different than the usual estimated standard error (2.2). That is because σ_F is usually estimated by $\bar{\sigma} = \{\sum (x_i - \bar{x})^2/(n-1)\}^{1/2}$ rather than by $\hat{\sigma}$, (5.11). Dividing by $n-1$ rather than n makes $\bar{\sigma}^2$ unbiased for σ_F^2. For most purposes $\hat{\sigma}$ is just as good as $\bar{\sigma}$ for estimating σ_F.

Notice that we have used the plug-in principle twice: first to estimate the expectation μ_F by $\mu_{\hat{F}} = \bar{x}$, and then to estimate the standard error $se_F(\bar{x})$ by $se_{\hat{F}}(\bar{x})$. The bootstrap estimate of standard error, which is the subject of Chapter 6, amounts to using the plug-in principle to estimate the standard error of an arbitrary statistic $\hat{\theta}$. Here we have seen that if $\hat{\theta} = \bar{x}$, then this approach leads to (almost) the usual estimate of standard error. As we will see, the advantage of the bootstrap is that it can be applied to virtually any statistic $\hat{\theta}$, not just the mean $\bar{x}$.

5.4 Problems

5.1 Formula (5.4) exemplifies a general statistical truth: most estimates of unknown quantities improve at a rate proportional to the square root of the sample size. Suppose that it were necessary to know μ_F for the mouse Control group with a standard error of no more than 3 days. How many more Control mice should be sampled?

5.2 State clearly why $\hat{p} = s/n$ is the plug-in estimate of p for the binomial situation (5.8).

5.3 Figure 5.2 compares the function

$$\binom{n}{nx} p^{nx} (1-p)^{n(1-x)} \qquad \text{for} \qquad x = 0, 1/25, 2/25, \cdots, 1$$

with

$$\frac{1}{n} \frac{1}{\sqrt{2\pi p(1-p)/n}} \exp\left\{-\frac{1}{2}\left[\frac{x - np}{\sqrt{np(1-p)}}\right]^2\right\} \qquad \text{for} \quad x \in [0, 1].$$

Why is this the correct comparison?

5.4 In the binomial case there seems to be two plug-in estimates for $\text{se}_F(\hat{p}) = \sigma_F/\sqrt{n} = [p(1-p)/n]^{1/2}$, one based on (5.12) and the other equal to $[\hat{p} \cdot (1-\hat{p})/n]^{1/2}$. Show that they are the same. [It helps to write the variance in the form $\sigma_F^2 = \text{E}_F(x^2) - \mu_F^2$.]

5.5 The *coefficient of variation* of a random variable x is defined to be the ratio of its standard deviation to the absolute value of its mean, say

$$\text{cv}_F(x) = \sigma_F/|\mu_F|. \tag{5.13}$$

(cv_F measures the randomness in x relative to the magnitude of its deterministic part μ_F.)

(a)   Show that $\text{cv}_F(\bar{x}) = \text{cv}_F(x)/\sqrt{n}$.

(b)   Suppose $x \sim \text{Bi}(n,p)$. How large must n be in order that $\text{cv}(x) = .10$? $\text{cv}(x) = .05$? $\text{cv}(x) = .01$? Give a formula for n as a function of p, and give specific values for $p = .5, .25,$ and $.1$.

## CHAPTER 6

The bootstrap estimate of standard error

6.1 Introduction

Suppose we find ourselves in the following common data-analytic situation: a random sample $\mathbf{x} = (x_1, x_2, \cdots, x_n)$ from an unknown probability distribution F has been observed and we wish to estimate a parameter of interest $\theta = t(F)$ on the basis of $\mathbf{x}$. For this purpose, we calculate an estimate $\hat{\theta} = s(\mathbf{x})$ from $\mathbf{x}$. [Note that $s(\mathbf{x})$ may be the plug-in estimate $t(\hat{F})$, but doesn't have to be.] How accurate is $\hat{\theta}$? The *bootstrap* was introduced in 1979 as a computer-based method for estimating the standard error of $\hat{\theta}$. It enjoys the advantage of being completely automatic. The bootstrap estimate of standard error requires no theoretical calculations, and is available no matter how mathematically complicated the estimator $\hat{\theta} = s(\mathbf{x})$ may be. It is described and illustrated in this chapter.

6.2 The bootstrap estimate of standard error

Bootstrap methods depend on the notion of a *bootstrap sample*. Let $\hat{F}$ be the empirical distribution, putting probability $1/n$ on each of the observed values x_i, $i = 1, 2, \cdots, n$, as described in Chapter 4. A bootstrap sample is defined to be a random sample of size n drawn from $\hat{F}$, say $\mathbf{x}^* = (x_1^*, x_2^*, \cdots, x_n^*)$,

$$\hat{F} \rightarrow (x_1^*, x_2^*, \cdots, x_n^*). \tag{6.1}$$

The star notation indicates that $\mathbf{x}^*$ is not the actual data set $\mathbf{x}$, but rather a randomized, or *resampled*, version of $\mathbf{x}$.

There is another way to say (6.1): the bootstrap data points $x_1^*, x_2^*, \cdots, x_n^*$ are a random sample of size n drawn *with* replacement from the population of n objects $(x_1, x_2, \cdots, x_n)$. Thus we might have $x_1^* = x_7, x_2^* = x_3, x_3^* = x_3, x_4^* = x_{22}, \cdots, x_n^* = x_7$. The bootstrap data set $(x_1^*, x_2^*, \cdots, x_n^*)$ consists of members of the original data set $(x_1, x_2, \cdots, x_n)$, some appearing zero times, some appearing once, some appearing twice, etc.

Corresponding to a bootstrap data set $\mathbf{x}^*$ is a *bootstrap replication* of $\hat\theta$,

$$\hat\theta^* = s(\mathbf{x}^*). \tag{6.2}$$

The quantity $s(\mathbf{x}^*)$ is the result of applying the same function $s(\cdot)$ to $\mathbf{x}^*$ as was applied to $\mathbf{x}$. For example if $s(\mathbf{x})$ is the sample mean $\bar x$ then $s(\mathbf{x}^*)$ is the mean of the bootstrap data set, $\bar x^* = \sum_{i=1}^n x_i^*/n$.

The bootstrap estimate of $\mathrm{se}_F(\hat\theta)$, the standard error of a statistic $\hat\theta$, is a plug-in estimate that uses the empirical distribution function $\hat F$ in place of the unknown distribution F. Specifically, the bootstrap estimate of $\mathrm{se}_F(\hat\theta)$ is defined by

$$\mathrm{se}_{\hat F}(\hat\theta^*). \tag{6.3}$$

In other words, the bootstrap estimate of $\mathrm{se}_F(\hat\theta)$ is the standard error of $\hat\theta$ for data sets of size n randomly sampled from $\hat F$.

Formula (6.3) is called the *ideal bootstrap estimate of standard error* of $\hat\theta$. Unfortunately, for virtually any estimate $\hat\theta$ other than the mean, there is no neat formula like (5.4) on page 40 that enables us to compute the numerical value of the ideal estimate exactly. The bootstrap algorithm, described next, is a computational way of obtaining a good approximation to the numerical value of $\mathrm{se}_{\hat F}(\hat\theta^*)$.

It is easy to implement bootstrap sampling on the computer. A random number device selects integers $i_1, i_2, \cdots, i_n$, each of which equals any value between 1 and n with probability $1/n$. The bootstrap sample consists of the corresponding members of $\mathbf{x}$,

$$x_1^* = x_{i_1}, x_2^* = x_{i_2}, \cdots, x_n^* = x_{i_n}. \tag{6.4}$$

The bootstrap algorithm works by drawing many independent bootstrap samples, evaluating the corresponding bootstrap replications, and estimating the standard error of $\hat\theta$ by the empirical standard deviation of the replications. The result is called the bootstrap estimate of standard error, denoted by $\widehat{\mathrm{se}}_B$, where B is the number of bootstrap samples used.

Algorithm 6.1 is a more explicit description of the bootstrap procedure for estimating the standard error of $\hat\theta = s(\mathbf{x})$ from the observed data $\mathbf{x}$.

Algorithm 6.1

The bootstrap algorithm for estimating standard errors

1. Select B independent bootstrap samples $\mathbf{x}^{*1}, \mathbf{x}^{*2}, \cdots, \mathbf{x}^{*B}$, each consisting of n data values drawn with replacement from $\mathbf{x}$, as in (6.1) or (6.4). [For estimating a standard error, the number B will ordinarily be in the range $25 - 200$, see Table 6.1.]

2. Evaluate the bootstrap replication corresponding to each bootstrap sample,

$$\hat{\theta}^*(b) = s(\mathbf{x}^{*b}) \qquad b = 1, 2, \cdots, B. \qquad (6.5)$$

3. Estimate the standard error $\mathrm{se}_F(\hat{\theta})$ by the sample standard deviation of the B replications

$$\widehat{\mathrm{se}}_B = \left\{ \sum_{b=1}^{B} [\hat{\theta}^*(b) - \hat{\theta}^*(\cdot)]^2 / (B-1) \right\}^{1/2}, \qquad (6.6)$$

where $\hat{\theta}^*(\cdot) = \sum_{b=1}^{B} \hat{\theta}^*(b)/B$.

Figure 6.1 is a schematic diagram of the bootstrap standard error algorithm. The Appendix gives programs for computing $\widehat{\mathrm{se}}_B$, written in the S language.

The limit of $\widehat{\mathrm{se}}_B$ as B goes to infinity is the ideal bootstrap estimate of $\mathrm{se}_F(\hat{\theta})$,

$$\lim_{B \to \infty} \widehat{\mathrm{se}}_B = \mathrm{se}_{\hat{F}} = \mathrm{se}_{\hat{F}}(\hat{\theta}^*). \qquad (6.7)$$

The fact that $\widehat{\mathrm{se}}_B$ approaches $\mathrm{se}_{\hat{F}}$ as B goes to infinity amounts to saying that an empirical standard deviation approaches the population standard deviation as the number of replications grows large. The "population" in this case is the population of values $\hat{\theta}^* = s(\mathbf{x}^*)$, where $\hat{F} \to (x_1^*, x_2^*, \cdots, x_n^*) = \mathbf{x}^*$.

The ideal bootstrap estimate $\mathrm{se}_{\hat{F}} \hat{\theta}^*$ and its approximation $\widehat{\mathrm{se}}_B$ are sometimes called *nonparametric bootstrap* estimates because they are based on $\hat{F}$, the nonparametric estimate of the population F. In Section 6.5 we discuss the *parametric bootstrap*, which uses a different estimate of F.

Figure 6.1. *The bootstrap algorithm for estimating the standard error of a statistic $\hat{\theta} = s(\mathbf{x})$; each bootstrap sample is an independent random sample of size n from $\hat{F}$. The number of bootstrap replications B for estimating a standard error is usually between 25 and 200. As $B \to \infty$, $\widehat{se}_B$ approaches the plug-in estimate of $se_F(\hat{\theta})$.*

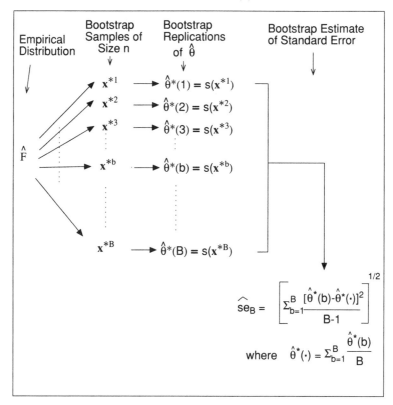

A word about notation: in (6.7) we write $se_{\hat{F}}(\hat{\theta}^*)$ rather than $se_{\hat{F}}(\hat{\theta})$ to avoid confusion between $\hat{\theta}$, the value of $s(\mathbf{x})$ based on the observed data, and $\hat{\theta}^* = s(\mathbf{x}^*)$ thought of as a random variable based on the bootstrap sample. The fuller notation $se_{\hat{F}}(\hat{\theta}(\mathbf{x}^*))$ emphasizes that $se_{\hat{F}}$ is a bootstrap standard error: the actual data $\mathbf{x}$ is held fixed in (6.7); the randomness in the calculation comes from the variability of the bootstrap samples $\mathbf{x}^*$, *given* $\mathbf{x}$. Similarly we will write $E_{\hat{F}}g(\mathbf{x}^*)$ to indicate the *bootstrap expectation* of a function $g(\mathbf{x}^*)$, the expectation with $\mathbf{x}$ (and $\hat{F}$) fixed and $\mathbf{x}^*$ varying according to (6.1).

The reader is asked in Problem 6.5 to show that there is a total of $\binom{2n-1}{n}$ distinct bootstrap samples. Denote these by $\mathbf{z}^1, \mathbf{z}^2, \ldots \mathbf{z}^m$ where $m = \binom{2n-1}{n}$. For example, if $n = 2$, the distinct samples are $(x_1, x_1), (x_2, x_2)$ and (x_1, x_2); since the order doesn't matter, (x_2, x_1) is the same as (x_1, x_2). The probability of obtaining one of these samples under sampling with replacement can be obtained from the multinomial distribution: details are in Problem 6.7. Denote the probability of the j*th* distinct sample by $w_j, j = 1, 2, \ldots \binom{2n-1}{n}$. Then a direct way to calculate the ideal bootstrap estimate of standard error would be to use the population standard deviation of the m bootstrap values $s(\mathbf{z}^j)$:

$$\mathrm{se}_{\hat{F}}(\hat{\theta}^*) = [\sum_{j=1}^{m} w_j \{s(\mathbf{z}^j) - s(\cdot)\}^2]^{1/2} \qquad (6.8)$$

where $s(\cdot) = \sum_{j=1}^{m} w_j s(\mathbf{z}^j)$. The difficulty with this approach is that unless n is quite small (≤ 5), the number $\binom{2n-1}{n}$ is very large, making computation of (6.8) impractical. Hence the need for bootstrap sampling as described above.

6.3 Example: the correlation coefficient

We have already seen two examples of the bootstrap standard error estimate, for the mean and the median of the Treatment group of the mouse data, Table 2.1. As a second example consider the sample correlation coefficient between $y = \mathrm{LSAT}$ and $z = \mathrm{GPA}$ for the $n = 15$ law school data points, Table 3.1, $\widehat{\mathrm{corr}}(y, z) = .776$. How accurate is the estimate .776? Table 6.1 shows the bootstrap estimate of standard error $\widehat{\mathrm{se}}_B$ for B ranging from 25 to 3200. The last value, $\widehat{\mathrm{se}}_{3200} = .132$, is our estimate for $\mathrm{se}_F(\widehat{\mathrm{corr}})$. Later we will see that $\widehat{\mathrm{se}}_{200}$ is nearly as good an estimate of se_F as is $\widehat{\mathrm{se}}_{3200}$.

Looking at the right side of Figure 3.1, the reader can imagine the bootstrap sampling process at work. The sample correlation of the $n = 15$ actual data points is $\widehat{\mathrm{corr}} = .776$. A bootstrap sample consists of 15 points selected at random and with replacement from the actual 15. The sample correlation of the bootstrap sample is a bootstrap replication $\widehat{\mathrm{corr}}^*$, which may be either bigger or smaller than $\widehat{\mathrm{corr}}$. Independent repetitions of the bootstrap sampling process give bootstrap replications $\widehat{\mathrm{corr}}^*(1), \widehat{\mathrm{corr}}^*(2), \cdots, \widehat{\mathrm{corr}}^*(B)$. Fi-

Table 6.1. *The bootstrap estimate of standard error for* $\widehat{corr}(y, z) = .776$, *the law school data of Table 3.1,* $n = 15$; *a run of 3200 bootstrap replications gave the tabled values of* $\widehat{se}_B$ *as* B *increased from 25 to 3200.*

B:	25	50	100	200	400	800	1600	3200
$\widehat{se}_B$:	.140	.142	.151	.143	.141	.137	.133	.132

nally, $\widehat{se}_B$ is the sample standard deviation of the $\widehat{corr}^*(b)$ values.

The left panel of Figure 6.2 is a histogram of the 3200 bootstrap replications $\widehat{corr}^*(b)$. It is always a good idea to look at the bootstrap data graphically, rather than relying entirely on a single summary statistic like $\widehat{se}_B$. In the correlation example it may turn out that a few outlying values of $\widehat{corr}^*(b)$ are greatly inflating $\widehat{se}_B$, in which case it pays to use a more robust measure of standard deviation; see Problem 6.6. In this case the histogram is noticeably non-normal, having a long tail toward the left. Inferences based on the normal curve, as in (5.6) and Figure 5.1, are suspect when the bootstrap histogram is markedly non-normal. Chapters 12–14, discuss bootstrap confidence intervals, which use more of the information in the bootstrap histogram than just its standard deviation $\widehat{se}_B$.

In the law school situation we happen to have the complete population $\mathcal{X}$ of $N = 82$ points, Table 3.2. The right side of Figure 6.2 shows the histogram of $\widehat{corr}(y, z)$ for 3200 samples of size $n = 15$ drawn from $\mathcal{X}$. In other words, 3200 random samples $\mathbf{x} = (x_1, x_2, \cdots, x_{15})$ were drawn with replacement from the 82 points in $\mathcal{X}$, and $\widehat{corr}(\mathbf{x})$ evaluated for each one. The standard deviation of the 3200 $\widehat{corr}(\mathbf{x})$ values was .131, so $\widehat{se}_B$ is a good estimate of the population standard error in this case. More impressively, the bootstrap histogram on the left strongly resembles the population histogram on the right. Remember, in a real problem we would only have the information on the left, from which we would be trying to infer the situation on the right.

6.4 The number of bootstrap replications B

How large should we take B, the number of bootstrap replications used to evaluate $\widehat{se}_B$? The *ideal bootstrap estimate* "$\widehat{se}_\infty$" takes $B = \infty$, in which case $\widehat{se}_\infty$ equals the plug-in estimate $se_{\hat{F}}(\hat{\theta}^*)$. Formula (5.12) gives $\widehat{se}_\infty$ for $\hat{\theta} = \bar{x}$, the mean, but for most other

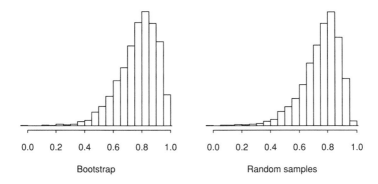

Bootstrap   Random samples

Figure 6.2. *Left panel: histogram of 3200 bootstrap replications of* $\widehat{corr}(\mathbf{x}^*)$, *from the law school data,* $n = 15$, *Table 3.1. Right panel: histogram of 3200 replications* $\widehat{corr}(\mathbf{x})$, *where* $\mathbf{x}$ *is a random sample of size* n *from the* $N = 82$ *points in the law school population, Table 3.2. The bootstrap histogram strongly resembles the population histogram. Both are notably non-normal.*

statistics we must actually do the bootstrap sampling. The amount of computer time, which depends mainly on how long it takes to evaluate the bootstrap replications (6.5), increases linearly with B. Time constraints may dictate a small value of B if $\hat{\theta} = s(\mathbf{x})$ is a very complicated function of $\mathbf{x}$, as in the examples of Chapter 7.

We want the same good behavior from a standard error estimate as from an estimate of any other quantity of interest: small bias and small standard deviation. The bootstrap estimate of standard error usually has relatively little bias. The ideal bootstrap estimate $\widehat{se}_\infty$ has the smallest possible standard deviation among nearly unbiased estimates of $se_F(\hat{\theta})$, at least in an asymptotic $(n \to \infty)$ sense. These good properties follow from the fact that $\widehat{se}_\infty$ is the plug-in estimate $se_{\hat{F}}(\hat{\theta}^*)$. It is not hard to show that $\widehat{se}_B$ always has greater standard deviation than $\widehat{se}_\infty$; see Problem 6.3. The practical question is "how much greater?"

An approximate, but quite satisfactory answer can be phrased in terms of the coefficient of variation of $\widehat{se}_B$, the ratio of $\widehat{se}_B$'s standard deviation to its expectation, see Problem 5.5. The increased variability due to stopping after B bootstrap replications, rather than going on to infinity, is reflected in an increased coefficient of variation,

$$\mathrm{cv}(\widehat{\mathrm{se}}_B) \doteq \left\{ \mathrm{cv}(\widehat{\mathrm{se}}_\infty)^2 + \frac{\mathrm{E}(\Delta) + 2}{4B} \right\}^{1/2}. \qquad (6.9)$$

This formula is derived in Chapter 19. Here Δ is a parameter that measures how long-tailed the distribution of $\hat{\theta}^*$ is: Δ is zero for the normal distribution, it ranges from -2 for the shortest-tailed distributions to arbitrarily large values when F is long-tailed. [1] In practice, Δ is usually no larger than 10. The coefficient of variation in equation (6.9) refers to variation both at the resampling (bootstrap) level and at the population sampling level. The ideal estimate $\widehat{\mathrm{se}}_\infty = \mathrm{se}_{\hat{F}}(\hat{\theta}^*)$ isn't perfect. It can still have considerable variability as an estimate of $\mathrm{se}_F(\hat{\theta})$, due to the variability of $\hat{F}$ as an estimate of F. For example if $x_1, x_2, \cdots, x_n$ is a random sample from a normal distribution and $\hat{\theta} = \bar{x}$, then $\mathrm{cv}(\widehat{\mathrm{se}}_\infty) \doteq 1/\sqrt{2n}$, equaling .22 for $n = 10$. Formula (6.9) has an important practical consequence: for the values of $\mathrm{cv}(\widehat{\mathrm{se}}_\infty)$ and Δ likely to arise in practice, $\mathrm{cv}(\widehat{\mathrm{se}}_B)$ is not much greater than $\mathrm{cv}(\widehat{\mathrm{se}}_\infty)$ for $B \geq 200$.

Table 6.2 compares $\mathrm{cv}(\widehat{\mathrm{se}}_B)$ with $\mathrm{cv}(\widehat{\mathrm{se}}_\infty)$ for various choices of B, assuming $\Delta = 0$. Very often we can expect to have $\mathrm{cv}(\widehat{\mathrm{se}}_\infty)$ no smaller than .10, in which case $B = 100$ gives quite satisfactory results.

Here are two rules of thumb, gathered from the authors' experience:

(1) Even a small number of bootstrap replications, say $B = 25$, is usually informative. $B = 50$ is often enough to give a good estimate of $\mathrm{se}_F(\hat{\theta})$.

(2) Very seldom are more than $B = 200$ replications needed for estimating a standard error. (Much bigger values of B are required for bootstrap confidence intervals; see Chapters 12–14 and 19.)

Approximations obtained by random sampling or simulation are called *Monte Carlo* estimates. We will see in Chapter 23 that computational methods other than straightforward Monte Carlo simulation can sometimes reduce manyfold the number of replications

---

[1] Let $\delta_{\hat{F}}$ be the *kurtosis* of $\hat{\theta}^* = s(\mathbf{x}^*)$, i.e. $\delta_{\hat{F}} = \mathrm{E}_{\hat{F}}(\hat{\theta}^* - \hat{\mu})^4 / (\mathrm{E}_{\hat{F}}(\hat{\theta}^* - \hat{\mu})^2)^2 - 3$, where $\hat{\mu} = \mathrm{E}_{\hat{F}}(\hat{\theta}^*)$. Then Δ is the expected value of $\delta_{\hat{F}}$, where $\hat{F}$ is the empirical distribution based on a random sample of size n from F. If $\hat{\theta} = \bar{x}$, then Δ equals about $1/n$ times the kurtosis of F itself. See Section 9 of Efron and Tibshirani (1986).

Table 6.2. *The coefficient of variation of $\widehat{se}_B$ as a function of the coefficient of variation of the ideal bootstrap estimate $\widehat{se}_\infty$ and the number of bootstrap samples B; from formula (6.9) assuming $\Delta = 0$.*

		$B \rightarrow$				
		25	50	100	200	∞
$\text{cv}(\widehat{se}_\infty)$	.25	.29	.27	.26	.25	.25
$\downarrow$	.20	.24	.22	.21	.21	.20
	.15	.21	.18	.17	.16	.15
	.10	.17	.14	.12	.11	.10
	.05	.15	.11	.09	.07	.05
	.00	.14	.10	.07	.05	.00

B needed to attain a prespecified accuracy. Meanwhile it pays to remember that bootstrap data, like real data, deserves a close look. In particular, it is almost never a waste of time to display the histogram of the bootstrap replications.

6.5 The parametric bootstrap

It might seem strange to use a resampling algorithm to estimate standard errors, when a textbook formula could be used. In fact, bootstrap sampling can be carried out *parametrically* and when it is used in that way, the results are closely related to textbook standard error formulae.

The parametric bootstrap estimate of standard error is defined as

$$\text{se}_{\hat{F}_{\text{par}}}(\hat{\theta}^*), \tag{6.10}$$

where $\hat{F}_{\text{par}}$ is an estimate of F derived from a *parametric model* for the data. Parametric models are discussed in Chapter 21: here we will give a simple example to illustrate the idea. For the law school data, instead of estimating F by the empirical distribution $\hat{F}$, we could assume that the population has a bivariate normal distribution. Reasonable estimates of the mean and covariance of this population are given by $(\bar{y}, \bar{z})$ and

$$\frac{1}{14} \begin{pmatrix} \sum(y_i - \bar{y})^2 & \sum(y_i - \bar{y})(z_i - \bar{z}) \\ \sum(y_i - \bar{y})(z_i - \bar{z}) & \sum(z_i - \bar{z})^2 \end{pmatrix}. \tag{6.11}$$

Denote the bivariate normal population with this mean and co-variance by $\hat{F}_{\text{norm}}$; it is an example of a *parametric* estimate of the population F. Using this, the parametric bootstrap estimate of standard error of the correlation $\hat{\theta}$ is $\text{se}_{\hat{F}_{\text{norm}}}(\hat{\theta}^*)$. As in the non-parametric case, the ideal parametric bootstrap estimate cannot be easily evaluated except when $\hat{\theta}$ is the mean. Therefore we approximate the ideal bootstrap estimate by bootstrap sampling, but in a different manner than before. Instead of sampling with replacement from the data, we draw B samples of size n from the parametric estimate of the population $\hat{F}_{\text{par}}$:

$$\hat{F}_{\text{par}} \rightarrow (x_1^*, x_2^*, \ldots x_n^*)$$

After generating the bootstrap samples, we proceed exactly as in steps 2 and 3 of the bootstrap algorithm of Section 6.2: we evaluate our statistic on each bootstrap sample, and then compute the standard deviation of the B bootstrap replications.

In the correlation coefficient example, assuming a bivariate normal population, we draw B samples of size 15 from $\hat{F}_{\text{norm}}$ and compute the correlation coefficient for each bootstrap sample. (Problem 6.8 shows how to generate bivariate normal random variables.) The left panel of Figure 6.3 shows the histogram of $B = 3200$ bootstrap replicates obtained in this way. It looks quite similar to the histograms of Figure 6.2. The parametric bootstrap estimate of standard error from these replicates was .124, close to the value of .131 obtained from nonparametric bootstrap sampling.

The textbook formula for the standard error of the correlation coefficient is $(1 - \hat{\theta}^2)/\sqrt{n - 3}$. Substituting $\hat{\theta} = .776$, this gives a value of .115 for the law school data.

We can make a further comparison to our parametric bootstrap result. Textbook results also state that Fisher's transformation of $\hat{\theta}$

$$\hat{\zeta} = .5 \cdot \log\left(\frac{1 + \hat{\theta}}{1 - \hat{\theta}}\right) \tag{6.12}$$

is approximately normally distributed with mean $\zeta = .5 \cdot \log\left(\frac{1+\theta}{1-\theta}\right)$ and standard deviation $1/\sqrt{n - 3}$, θ being the population correlation coefficient. From this, one typically carries out inference for ζ and then transforms back to make an inference about the correlation coefficient. To compare this with our parametric bootstrap analysis, we calculated $\hat{\zeta}$ rather than $\hat{\theta}$ for each of our 3200 boot-

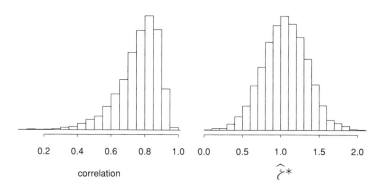

Figure 6.3. *Left panel: histogram of 3200 parametric bootstrap replications of* $\widehat{corr}(\mathbf{x}^*)$, *from the law school data, n = 15. Right panel: histogram of 3200 replications of* $\hat{\zeta}$, *Fisher's transformation of the correlation coefficient, defined in (6.12). The left histogram looks much like the histograms of (6.2), while the right histogram looks quite normal as predicted by statistical theory.*

strap samples. A histogram of the $\hat{\zeta}^*$ values is shown in the right panel of Figure 6.3, and looks quite normal. Furthermore, the standard deviation of the 3200 $\hat{\zeta}^*$ values was .290, very close to the value $1/\sqrt{15-3} = .289$.

This agreement holds quite generally. Most textbook formulae for standard errors are approximations based on normal theory, and will typically gives answers close to the parametric bootstrap that draws samples from a normal distribution. The relationship between the bootstrap and traditional statistical theory is a more advanced topic mathematically, and is explored in Chapter 21.

The bootstrap has two somewhat different advantages over traditional textbook methods: 1) when used in nonparametric mode, it relieves the analyst from having to make parametric assumptions about the form of the underlying population, and 2) when used in parametric mode, it provides more accurate answers than textbook formulas, and can provide answers in problems for which no textbook formulae exist.

Most of this book concentrates on the nonparametric application of the bootstrap, with some exceptions being Chapter 21 and examples in Chapters 14 and 25. The parametric bootstrap is useful in problems where some knowledge about the form of the underlying population is available, and for comparison to nonparametric analyses. However, a main reason for making parametric assumptions in traditional statistical analysis is to facilitate the derivation of textbook formulas for standard errors. Since we don't need formulas in the bootstrap approach, we can avoid restrictive parametric assumptions.

Finally, we mention that in Chapters 13 and 14 we describe bootstrap methods for construction of confidence intervals in which transformations such as (6.12) are incorporated in an automatic way.

6.6 Bibliographic notes

The bootstrap was introduced by Efron (1979a), with further general developments given in Efron (1981a, 1981b). The monograph of Efron (1982) expands on many of the topics in the 1979 paper and discusses some new ones. Expositions of the bootstrap for a statistical audience include Efron and Gong (1983), Efron and Tibshirani (1986) and Hinkley (1988). Efron (1992a) outlines some statistical questions that arose from bootstrap research. The lecture notes of Beran and Ducharme(1991) and Hall's (1992) monograph give a mathematically sophisticated treatment of the bootstrap. Non-technical descriptions may be found in Diaconis and Efron (1983), Lunneborg (1985), Rasmussen (1987), and Efron and Tibshirani (1991). A general discussion of computers and statistics may be found in Efron (1979b). Young (1988a) studies bootstrapping of the correlation coefficient.

While Efron's 1979 paper formally introduced and studied the bootstrap, similar ideas had been suggested in different contexts. These include the Monte Carlo hypothesis testing methods of Barnard (1963), Hope (1968) and Marriott (1979). Particularly notable contributions were made by Hartigan (1969, 1971, 1975) in his typical value theory for constructing confidence intervals. J.L. Simon discussed computational methods very similar to the bootstrap in a sociometrics textbook of the 1960's; see Simon and Bruce (1991).

The jackknife and cross-validation techniques predate the bootstrap and are closely related to it. References to these methods are given in the bibliographic notes in Chapters 11 and 17.

6.7 Problems

6.1 We might have divided by B instead of $B - 1$ in definition (6.6) of the bootstrap standard error estimate. How would that change Table 6.1?

6.2 With $\widehat{se}_B$ defined as in (6.6), show that

$$E_{\hat{F}}(\widehat{se}_B^2) = \widehat{se}_\infty^2, \tag{6.13}$$

where $\widehat{se}_\infty^2$ equals the ideal bootstrap estimate $se_{\hat{F}}(\hat{\theta}^*)$. In other words, the variance estimate $\widehat{se}_B^2$ based on B bootstrap replications has bootstrap expectation equal to the ideal bootstrap variance $\widehat{se}_\infty^2$.

6.3 † Show that $E_F(\widehat{se}_B^2) = E_F(\widehat{se}_\infty^2)$, but $var_F(\widehat{se}_B^2) \geq var_F(\widehat{se}_\infty^2)$. In other words $\widehat{se}_B^2$ has the same expectation as $\widehat{se}_\infty^2$, but larger variance. (Notice that these results involve the usual expectation and variance E_F and var_F, not the bootstrap quantities $E_{\hat{F}}$ and $var_{\hat{F}}$.)

6.4 The data in Table 3.2 allow us to compute the quantities $cv(\widehat{se}_\infty)$ and Δ in formula (6.9) for the law school data: $cv(\widehat{se}_\infty) = .41$, $\Delta = 4$. What value of B makes $cv(\widehat{se}_B)$ only 10% larger than $cv(\widehat{se}_\infty)$? 5%? 1%?

6.5 † Given a data set of n distinct values, show that the number of distinct bootstrap samples is

$$\binom{2n-1}{n}. \tag{6.14}$$

How many are there for $n = 15$?

6.6 A biased but more robust estimate of the bootstrap standard error is

$$\widetilde{se}_{B,\alpha} = \frac{\hat{\theta}^{*(\alpha)} - \hat{\theta}^{*(1-\alpha)}}{2z^{(\alpha)}}, \tag{6.15}$$

where $\hat{\theta}^{*(\alpha)}$ is the 100αth quantile of the bootstrap replications (i.e. the 100αth largest value in an ordered list of the $\hat{\theta}^*(b)$), and $z^{(\alpha)}$ is the 100αth percentile of a standard normal distribution, $z^{(.95)} = 1.645$ etc. Here is a table of the quantiles for the 3200 bootstrap replications of $\hat{\theta}^*$ in Table 6.1 and the left panel of Figure 6.2:

α:	.05	.10	.16	.50	.84	.90	.95
	.524	.596	.647	.793	.906	.927	.948

(a)   Compute $\widetilde{se}_{B,\alpha}$ for $\alpha = .95, .90$, and $.84$.

(b)   Suppose that a transcription error caused one of the $\hat{\theta}^*(b)$ values to change from .42 to -4200. *Approximately* how much would this change $\hat{se}_B$? $\widetilde{se}_{B,\alpha}$?

6.7 Suppose a bootstrap sample of size n, drawn with replacement from $x_1, x_2, \ldots x_n$, contains j_1 copies of x_1, j_2 copies of x_2, and so on, up to j_n copies of x_n, with $j_1 + j_2 \ldots + j_n = n$. Show that the probability of obtaining this sample is the multinomial probability

$$\binom{n}{j_1 j_2 \cdots j_n} \prod_{i=1}^{n} \left(\frac{1}{n}\right)^{j_i}, \tag{6.16}$$

where

$$\binom{n}{j_1 j_2 \cdots j_n} = \frac{n!}{j_1! j_2! \cdots j_n!}. \tag{6.17}$$

6.8 *Generation of bivariate normal random variables.* Suppose we have a random number generator that produces independent standard normal variates[2] r_1 and r_2 and we wish to generate bivariate random variables y and z with means μ_y, μ_z and covariance matrix

$$\begin{pmatrix} \sigma_y^2 & \sigma_{yz} \\ \sigma_{yz} & \sigma_z^2 \end{pmatrix}.$$

Let $\rho = \sigma_{yz}/(\sigma_y \sigma_z)$ and define

$$y = \mu_y + \sigma_y r_1; \quad z = \mu_z + \frac{\sigma_z}{\sqrt{1 + c^2}}(r_1 + c \cdot r_2)$$

where $c = \sqrt{(1/\rho^2) - 1}$. Show that y and z have the required bivariate normal distribution.

6.9 Generate 100 bootstrap replicates of the correlation coefficient for the law school data. From these, compute the

---

[2] Most statistical packages have the facility for generating independent standard normal variates. For a comprehensive reference on the subject, see Devroye (1986).

bootstrap estimate of standard error for the correlation coefficient. Compare your results to those in Table 6.1 and
Figure 6.2.

6.10 † Consider an artificial data set consisting of the 8 numbers

$$1, 2, 3.5, 4, 7, 7.3, 8.6, 12.4, 13.8, 18.1.$$

Let $\hat{\theta}$ be the 25% trimmed mean, computed by deleting the
smallest two numbers and largest two numbers, and then
taking the average of the remaining four numbers.

(a) Calculate $\widehat{se}_B$ for $B = 25, 100, 200, 500, 1000, 2000$. From
these results estimate the ideal bootstrap estimate $\widehat{se}_\infty$.

(b) Repeat part (a) using ten different random number
seeds and hence assess the variability in the estimates.
How large should we take B to provide satisfactory accuracy?

(c) Calculate the ideal bootstrap estimate $\widehat{se}_\infty$ directly using formula (6.8). Compare the answer to that obtained
in part (a).

† Indicates a difficult or more advanced problem.

## CHAPTER 7

Bootstrap standard errors: some examples

7.1 Introduction

Before the computer age statisticians calculated standard errors using a combination of mathematical analysis, distributional assumptions, and, often, a lot of hard work on mechanical calculators. One classical result was given in Section 6.5: it concerns the sample correlation coefficient $\widehat{\text{corr}}(y, z)$ defined in (4.6). If we are willing to assume that the probability distribution F giving the n data points (y_i, z_i) is bivariate normal, then a reasonable estimate for the standard error of $\widehat{\text{corr}}$ is

$$\widehat{\text{se}}_{\text{normal}} = (1 - \widehat{\text{corr}}^2)/\sqrt{n - 3}. \tag{7.1}$$

An obvious objection to $\widehat{\text{se}}_{\text{normal}}$ concerns the use of the bivariate normal distribution. What right do we have to assume that F is normal? To the trained eye, the data plotted in the right panel of Figure 3.1 look suspiciously non-normal – the point at $(576, 3.39)$ is too far removed from the other 14 points. The real reason for considering bivariate normal distributions is mathematical tractability. No other distributional form leads to a simple approximation for $\text{se}(\widehat{\text{corr}})$.

There is a second important objection to $\widehat{\text{se}}_{\text{normal}}$: it requires a lot of mathematical work to derive formulas like (7.1). If we choose a statistic more complicated than $\widehat{\text{corr}}$, or a distribution less tractable than the bivariate normal, then no amount of mathematical cleverness will yield a simple formula. Because of such limitations, pre-computer statistical theory focused on a small set of distributions and a limited class of statistics. Computer-based methods like the bootstrap free the statistician from these constraints. Standard errors, and other measures of statistical accuracy, are produced automatically, without regard to mathematical complexity. [1]

Bootstrap methods come into their own in complicated estimation problems. This chapter discusses standard errors for two such problems, one concerning the eigenvalues and eigenvectors of a covariance matrix, the other a computer-based curve-fitting algorithm called "loess." Describing these problems requires some matrix terminology that may be unfamiliar to the reader. However, matrix-theoretic calculations will be avoided, and in any case the theory isn't necessary to understand the main point being made here, that the simple bootstrap algorithm of Chapter 6 can provide standard errors for very complicated situations.

At the end of this chapter, we discuss a simple problem in which the bootstrap fails and look at the reason for the failure.

7.2 Example 1: test score data

Table 7.1 shows the *score data*, from Mardia, Kent and Bibby (1979); $n = 88$ students each took 5 tests, in mechanics, vectors, algebra, analysis, and statistics.

The first two tests were closed book, the last three open book. It is convenient to think of the score data as an 88×5 data matrix $\mathbf{X}$, the ith row of $\mathbf{X}$ being

$$\mathbf{x}_i = (x_{i1}, x_{i2}, x_{i3}, x_{i4}, x_{i5}), \qquad (7.2)$$

the 5 scores for student i, $i = 1, 2, \cdots, 88$.

The mean vector $\bar{\mathbf{x}} = \sum_{i=1}^{88} \mathbf{x}/88$ is the vector of column means,

$$
\begin{aligned}
\bar{\mathbf{x}} &= (\bar{x}_1, \bar{x}_2, \bar{x}_3, \bar{x}_4, \bar{x}_5) \\
&= \left(\sum_{i=1}^{88} x_{i1}/88, \sum_{i=1}^{88} x_{i2}/88, \cdots, \sum_{i=1}^{88} x_{i5}/88 \right) \\
&= (38.95, 50.59, 50.60, 46.68, 42.31).
\end{aligned}
\qquad (7.3)
$$

The *empirical covariance* matrix $\mathbf{G}$ is the 5×5 matrix with (j, k)th

---

[1] This is not all pure gain. Theoretical formulas like (7.1) can help us understand a situation in a different way than the numerical output of a bootstrap program. (Later, in Chapter 21, we will examine the close connections between formulas like (7.1) and the bootstrap.) It pays to remember that methods like the bootstrap free the statistician to look *more* closely at the data, without fear of mathematical difficulties, not *less* closely.

Table 7.1. *The score data, from Mardia, Kent and Bibby (1979); $n = 88$ students each took five tests, in mechanics, vectors, algebra, analysis, and statistics; "c" and "o" indicate closed and open book, respectively.*

#	mec (c)	vec (c)	alg (o)	ana (o)	sta (o)	#	mec (c)	vec (c)	alg (o)	ana (o)	sta (o)
1	77	82	67	67	81	45	46	61	46	38	41
2	63	78	80	70	81	46	40	57	51	52	31
3	75	73	71	66	81	47	49	49	45	48	39
4	55	72	63	70	68	48	22	58	53	56	41
5	63	63	65	70	63	49	35	60	47	54	33
6	53	61	72	64	73	50	48	56	49	42	32
7	51	67	65	65	68	51	31	57	50	54	34
8	59	70	68	62	56	52	17	53	57	43	51
9	62	60	58	62	70	53	49	57	47	39	26
10	64	72	60	62	45	54	59	50	47	15	46
11	52	64	60	63	54	55	37	56	49	28	45
12	55	67	59	62	44	56	40	43	48	21	61
13	50	50	64	55	63	57	35	35	41	51	50
14	65	63	58	56	37	58	38	44	54	47	24
15	31	55	60	57	73	59	43	43	38	34	49
16	60	64	56	54	40	60	39	46	46	32	43
17	44	69	53	53	53	61	62	44	36	22	42
18	42	69	61	55	45	62	48	38	41	44	33
19	62	46	61	57	45	63	34	42	50	47	29
20	31	49	62	63	62	64	18	51	40	56	30
21	44	61	52	62	46	65	35	36	46	48	29
22	49	41	61	49	64	66	59	53	37	22	19
23	12	58	61	63	67	67	41	41	43	30	33
24	49	53	49	62	47	68	31	52	37	27	40
25	54	49	56	47	53	69	17	51	52	35	31
26	54	53	46	59	44	70	34	30	50	47	36
27	44	56	55	61	36	71	46	40	47	29	17
28	18	44	50	57	81	72	10	46	36	47	39
29	46	52	65	50	35	73	46	37	45	15	30
30	32	45	49	57	64	74	30	34	43	46	18
31	30	69	50	52	45	75	13	51	50	25	31
32	46	49	53	59	37	76	49	50	38	23	9
33	40	27	54	61	61	77	18	32	31	45	40
34	31	42	48	54	68	78	8	42	48	26	40
35	36	59	51	45	51	79	23	38	36	48	15
36	56	40	56	54	35	80	30	24	43	33	25
37	46	56	57	49	32	81	3	9	51	47	40
38	45	42	55	56	40	82	7	51	43	17	22
39	42	60	54	49	33	83	15	40	43	23	18
40	40	63	53	54	25	84	15	38	39	28	17
41	23	55	59	53	44	85	5	30	44	36	18
42	48	48	49	51	37	86	12	30	32	35	21
43	41	63	49	46	34	87	5	26	15	20	20
44	46	52	53	41	40	88	0	40	21	9	14

element

$$G_{jk} = \frac{1}{88} \sum_{i=1}^{88} (x_{ij} - \bar{x}_j)(x_{ik} - \bar{x}_k) \qquad j, k = 1, 2, 3, 4, 5 . \quad (7.4)$$

Notice that the diagonal element G_{jk} is the plug-in estimate (5.11) for the variance of the scores on test j. We compute

$$\mathbf{G} = \begin{pmatrix} 302.3 & 125.8 & 100.4 & 105.1 & 116.1 \\ 125.8 & 170.9 & 84.2 & 93.6 & 97.9 \\ 100.4 & 84.2 & 111.6 & 110.8 & 120.5 \\ 105.1 & 93.6 & 110.8 & 217.9 & 153.8 \\ 116.1 & 97.9 & 120.5 & 153.8 & 294.4 \end{pmatrix} . \quad (7.5)$$

Educational testing theory is often concerned with the *eigenvalues* and *eigenvectors* of the covariance matrix $\mathbf{G}$. A 5×5 covariance matrix has 5 positive eigenvalues, labeled in decreasing order $\hat{\lambda}_1 \geq \hat{\lambda}_2 \geq \hat{\lambda}_3 \geq \hat{\lambda}_4 \geq \hat{\lambda}_5$. Corresponding to each $\hat{\lambda}_i$ is a 5 dimensional eigenvector $\hat{\mathbf{v}}_i = (\hat{v}_{i1}, \hat{v}_{i2}, \hat{v}_{i3}, \hat{v}_{i4}, \hat{v}_{i5})$. Readers not familiar with eigenvalues and vectors may prefer to think of a function "eigen", a black box [2] which inputs the matrix $\mathbf{G}$ and outputs the $\hat{\lambda}_i$ and corresponding $\hat{\mathbf{v}}_i$. Here are the eigenvectors and values for matrix (7.5):

$$\hat{\lambda}_1 = 679.2 \qquad \hat{\mathbf{v}}_1 = (.505, .368, .346, .451, .535)$$
$$\hat{\lambda}_2 = 199.8 \qquad \hat{\mathbf{v}}_2 = (-.749, -.207, .076, .301, .548)$$
$$\hat{\lambda}_3 = 102.6 \qquad \hat{\mathbf{v}}_3 = (-.300, .416, .145, .597, -.600)$$
$$\hat{\lambda}_4 = 83.7 \qquad \hat{\mathbf{v}}_4 = (.296, -.783, -.003, .518, -.176)$$
$$\hat{\lambda}_5 = 31.8 \qquad \hat{\mathbf{v}}_5 = (.079, .189, -.924, .286, .151).$$
$$(7.6)$$

Of what interest are the eigenvalues and eigenvectors of a covariance matrix? They help explain the structure of *multivariate* data like that in Table 7.1, data for which we have many independent units, the $n = 88$ students in this case, but correlated measurements within each unit. Notice that the 5 test scores are highly correlated with each other. A student who did well on the mechanics test is likely to have done well on vectors, etc. A very

---

[2] The eigenvalues and eigenvectors of a matrix are actually computed by a complicated series of algebraic manipulations requiring on the order of p^3 calculations when $\mathbf{G}$ is a $p \times p$ matrix. Chapter 8 of Golub and Van Loan, 1983, describes the algorithm.

simple model for correlated scores is

$$\mathbf{x}_i = Q_i \mathbf{v} \qquad\qquad i = 1, 2, \cdots, 88 \ . \tag{7.7}$$

Here Q_i is a single number representing the capability of student i, while $\mathbf{v} = (v_1, v_2, v_3, v_4, v_5)$ is a fixed vector of 5 numbers, applying to all students. Q_i can be thought of as student i's scientific Intelligence Quotient (IQ). IQs were originally motivated by a model just slightly more complicated than (7.7).

If model (7.7) were true, then we would find this out from the eigenvalues: only $\hat{\lambda}_1$ would be positive, $\hat{\lambda}_2 = \hat{\lambda}_3 = \hat{\lambda}_4 = \hat{\lambda}_5 = 0$; also the first eigenvector $\hat{\mathbf{v}}_1$ would equal $\mathbf{v}$. Let $\hat{\theta}$ be the ratio of the largest eigenvalue to the total,

$$\hat{\theta} = \hat{\lambda}_1 / \sum_{i=1}^{5} \hat{\lambda}_i \ . \tag{7.8}$$

Model (7.7) is equivalent to $\hat{\theta} = 1$. Of course we don't expect (7.7) to be exactly true for noisy data like test scores, even if the model is basically correct.

Figure 7.1 gives a stylized illustration. We have taken just two of the scores, and on the left depicted what their scatterplot would look like if a single number Q_i captured both scores. The scores lie exactly on a line; Q_i could be defined as the distance along the line of each point from the origin. The right panel shows a more realistic situation. The points do not lie exactly on a line, but are fairly collinear. The line shown in the plot points in the direction given by the first eigenvector of the covariance matrix. It is sometimes called the *first principal component* line, and has the property that it minimizes the sum of squared orthogonal distances from the points to the line (in contrast to the least-squares line which minimizes the sum of vertical distances from the points to the line). The orthogonal distances are shown by the short line segments in the right panel. It is difficult to make such a graph for the score data: the principal component line would be a line in five dimensional space lying closest to the data. If we consider the projection of each data point onto the line, the principal component line also maximizes the sample variance of the collection of projected points.

For the score data

$$\hat{\theta} = \frac{679.2}{679.2 + 199.8 + \cdots + 31.8} = .619 \ . \tag{7.9}$$

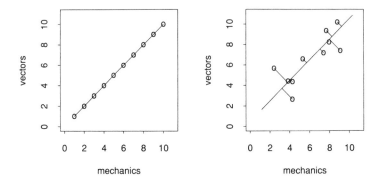

Figure 7.1. *Hypothetical plot of mechanics and vector scores. On the left, the pairs line exactly on a straight line (that is, have correlation 1) and hence a single measure captures the two scores. On the right, the scores have correlation less than one. The principal component line minimizes the sum of orthogonal distances to the line and has direction given by the largest eigenvector of the covariance matrix.*

In many situations this would be considered an interestingly large value of $\hat{\theta}$, indicating a high degree of explanatory power for model (7.7). The value of $\hat{\theta}$ measures the percentage of the variance explained by the first principal component. The closer the points lie to the principal component line, the higher the value of $\hat{\theta}$.

How accurate is $\hat{\theta}$? This is the kind of question that the bootstrap was designed to answer. The mathematical complexity going into the computation of $\hat{\theta}$ is irrelevant, as long as we can compute $\hat{\theta}^*$ for any bootstrap data set. In this case a bootstrap data set is an 88×5 matrix $\mathbf{X}^*$. The rows $\mathbf{x}_i^*$ of $\mathbf{X}^*$ are a random sample of size 88 from the rows of the actual data matrix $\mathbf{X}$,

$$\mathbf{x}_1^* = \mathbf{x}_{i_1}, \mathbf{x}_2^* = \mathbf{x}_{i_2}, \cdots, \mathbf{x}_{88}^* = \mathbf{x}_{i_{88}}, \tag{7.10}$$

as in (6.4). Some of the rows of $\mathbf{X}$ appear zero times as rows of $\mathbf{X}^*$, some once, some twice, etc., for a total of 88 rows.

Having generated $\mathbf{X}^*$, we calculate its covariance matrix $\mathbf{G}^*$ as

Table 7.2. *Quantiles of the bootstrap distribution of $\hat{\theta}^*$ defined in (7.12)*

α	.05	.10	.16	.50	.84	.90	.95
quantile	.545	.557	.576	.629	.670	.678	.693

in (7.4)

$$G_{jk}^* = \frac{1}{88} \sum_{i=1}^{88} (x_{ij}^* - \bar{x}_j^*)(x_{ik}^* - \bar{x}_k^*) \qquad j, k = 1, 2, 3, 4, 5.$$
(7.11)

We then compute the eigenvalues of $\mathbf{G}^*$, namely $\hat{\lambda}_1^*, \hat{\lambda}_2^*, \cdots, \hat{\lambda}_5^*$, and finally

$$\hat{\theta}^* = \hat{\lambda}_1^* / \sum_{j=1}^{5} \hat{\lambda}_j^*,$$
(7.12)

the bootstrap replication of $\hat{\theta}$.

Figure 7.2 is a histogram of $B = 200$ bootstrap replications $\hat{\theta}^*$. These gave estimated standard error $\widehat{se}_{200} = .047$ for $\hat{\theta}$. The mean of the 200 replications was .625, only slightly larger than $\hat{\theta} = .619$. This indicates that $\hat{\theta}$ is close to unbiased. The histogram looks reasonably normal, but $B = 200$ is not enough replications to see the distributional shape clearly. Some quantiles of the empirical distribution of the $\hat{\theta}^*$ values are shown in Table 7.2. [The αth *quantile* is the number $q(\alpha)$ such that $100\alpha\%$ of the $\hat{\theta}^*$'s are less than $q(\alpha)$. The .50 quantile is the median.]

The *standard confidence interval* for the true value of θ, (the value of $\hat{\theta}$ we would see if $n \to \infty$) is

$$\theta \in \hat{\theta} \pm z^{(1-\alpha)} \cdot \widehat{se} \quad \text{(with probability } 1 - 2\alpha\text{)}$$
(7.13)

where $z^{(1-\alpha)}$ is the $100(1 - \alpha)$th percentile of a standard normal distribution $z^{(.975)} = 1.960$, $z^{(.95)} = 1.645$, $z^{(.841)} = 1.000$, etc. This is based on an asymptotic theory which extends (5.6) to general summary statistics $\hat{\theta}$. In our case

$$\theta \in .619 \pm .047 = [.572, .666] \quad \text{with probability } .683$$
$$\theta \in .619 \pm 1.645, .047 = [.542, .696] \quad \text{with probability } .900.$$

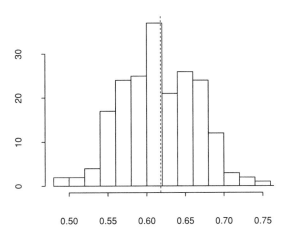

Figure 7.2. *200 bootstrap replications of the statistic* $\hat{\theta} = \hat{\lambda}_1 / \sum_1^5 \hat{\lambda}_i$. *The bootstrap standard error is .047. The dashed line indicates the observed value* $\hat{\theta} = .619$.

Chapters 12–14 discuss improved bootstrap confidence intervals that are less reliant on asymptotic normal distribution theory.

The eigenvector $\hat{\mathbf{v}}_1$ corresponding to the largest eigenvalue is called the *first principal component* of **G**. Suppose we wanted to summarize each student's performance by a single number, rather than 5 numbers, perhaps for grading purposes. It can be shown that the best single linear combination of the scores is

$$y_i = \sum_{k=1}^{5} \hat{v}_{1k} x_{ik}, \tag{7.14}$$

that is, the linear combination that uses the components of $\hat{\mathbf{v}}_1$ as weights. This linear combination is "best" in the sense that it captures the largest amount of the variation in the original five scores among all possible choices of **v**. If we want a two-number summary for each student, say (y_i, z_i), the second linear combination should be

$$z_i = \sum_{k=1}^{5} \hat{v}_{2k} x_{ik}, \qquad (7.15)$$

with weights given by the *second* principal component $\hat{\mathbf{v}}_2$, the second eigenvector of $\mathbf{G}$.

The weights assigned by the principal components often give insight into the structure of a multivariate data set. For the score data the interpretation might go as follows: the first principal component $\hat{\mathbf{v}}_1 = (.51, .37, .35, .45, .54)$ puts positive weights of approximately equal size on each test score, so y_i is roughly equivalent to taking student i's total (or average) score. The second principal component $\hat{\mathbf{v}}_2 = (-.75, -.21, .08, .30, .55)$ puts negative weights on the two closed-book tests and positive weights on the three open-book tests so z_i is a *contrast* between a student's open and closed book performances. (A student with a high z score did much better on the open book tests than the closed book tests.)

The principal component vectors $\hat{\mathbf{v}}_1$ and $\hat{\mathbf{v}}_2$ are summary statistics, just like $\hat{\theta}$, even though they have several components each. We can use a bootstrap analysis to learn how variable they are. The same 200 bootstrap samples that gave the $\hat{\theta}^*$'s also gave bootstrap replications $\hat{\mathbf{v}}_1^*$ and $\hat{\mathbf{v}}_2^*$. These are calculated as the first two eigenvectors of $\mathbf{G}^*$, (7.11).

Table 7.3 shows $\hat{\mathrm{se}}_{200}$, for each component of $\hat{\mathbf{v}}_1$ and $\hat{\mathbf{v}}_2$. The first thing we notice is the greater accuracy of $\hat{\mathbf{v}}_1$; the bootstrap standard error for the components of $\hat{\mathbf{v}}_1$ are less than half those of $\hat{\mathbf{v}}_2$. Table 7.3 also gives the robust percentile-based bootstrap standard errors $\tilde{\mathrm{se}}_{200,\alpha}$ of Problem 6.6 calculated for $\alpha = .84, .90,$ and .95. For the components of $\hat{\mathbf{v}}_1$, $\tilde{\mathrm{se}}_{200,\alpha}$ nearly equals $\hat{\mathrm{se}}_{200}$. This isn't the case for $\hat{\mathbf{v}}_2$, particularly not for the first and fifth components. Figure 7.3 shows what the trouble is. This figure indicates the empirical distribution of the 200 bootstrap replications of $\hat{v}_{ik}^*$, separately for $i = 1, 2$, $k = 1, 2, \cdots, 5$. The empirical distributions are indicated by *boxplots*. The center line of the box indicates the median of the distribution; the lower and upper ends of the box are the 25th and 75th percentiles; the whiskers extend from the lower and upper ends of the box to cover the entire range of the distribution, except for points deemed outliers according to a certain definition; these outliers are individually indicated by stars.

Table 7.3. *Bootstrap standard errors for the components of the first and second principal components,* $\hat{\mathbf{v}}_1$ *and* $\hat{\mathbf{v}}_2$; $\widehat{se}_{200}$ *is the usual bootstrap standard error estimate based on* $B = 200$ *bootstrap replications;* $\widetilde{se}_{200,.84}$ *is the standard error estimate* $\widetilde{se}_{B,\alpha}$ *of Problem 6.6, with* $B = 200, \alpha = .84$; *likewise* $\widetilde{se}_{200,.90}$ *and* $\widetilde{se}_{200,.95}$. *The values of* $\widehat{se}_{200}$ *for* $\hat{v}_{21}$ *and* $\hat{v}_{25}$ *are greatly inflated by a few outlying bootstrap replications, see Figures 7.3 and 7.4.*

	$\hat{v}_{11}$	$\hat{v}_{12}$	$\hat{v}_{13}$	$\hat{v}_{14}$	$\hat{v}_{15}$	$\hat{v}_{21}$	$\hat{v}_{22}$	$\hat{v}_{23}$	$\hat{v}_{24}$	$\hat{v}_{25}$
$\widehat{se}_{200}$	.057	.045	.029	.041	.049	.189	.138	.066	.129	.150
$\widetilde{se}_{200,.84}$	.055	.041	.028	.041	.047	.078	.122	.064	.110	.114
$\widetilde{se}_{200,.90}$	.055	.041	.027	.042	.046	.084	.129	.067	.111	.125
$\widetilde{se}_{200,.95}$	.054	.048	.029	.040	.047	.080	.130	.066	.114	.120

The large values of $\widehat{se}_{200}$ for $\hat{v}_{21}$ and $\hat{v}_{25}$ are seen to be caused by a few extreme values of $\hat{v}_{ik}^*$. The approximate confidence interval $\theta \in \hat{\theta} \pm z^{(1-\alpha)}\widehat{se}$ will be more accurate with $\widehat{se}$ equaling $\widetilde{se}_{200,\alpha}$ rather than $\widehat{se}_{200}$, at least for moderate values of α like .843. A histogram of the $\hat{v}_{21}^*$ values shows a normal-shaped central bulge with mean at $-.74$ and standard deviation .075, with a few points far away from the bulge. This indicates a small probability, perhaps 1% or 2%, that $\hat{v}_{21}$ is grossly wrong as an estimate of the true value v_{21}. If this gross error hasn't happened, then $\hat{v}_{21}$ is probably within one or two $\widetilde{se}_{200}$ units of v_{21}.

Figure 7.4 graphs the bootstrap replications $\hat{\mathbf{v}}_1^*(b)$ and $\hat{\mathbf{v}}_2^*(b)$, $b = 1, 2, \cdots, 200$, connecting the components of each vector by straight lines. This is less precise than Table 7.3 or Figure 7.3, but gives a nice visual impression of the increased variability of $\hat{\mathbf{v}}_2$. Three particular replications labeled "1", "2", "3", are seen to be outliers on several components.

A reader familiar with principal components may now see that part of the difficulty with the second eigenvector is definitional. Technically, the definition of an eigenvector applies as well to $-\mathbf{v}$ as to $\mathbf{v}$. The computer routine that calculates eigenvalues and eigenvectors makes a somewhat arbitrary choice of the signs given to $\hat{\mathbf{v}}_1, \hat{\mathbf{v}}_2, \cdots$. Replications "1" and "2" gave $\mathbf{X}^*$ matrices for which the sign convention of $\hat{\mathbf{v}}_2^*$ was reversed. This type of definitional instability is usually not important in determining the statistical properties of an estimate (though it is nice to be reminded of it by the bootstrap results). Throwing away "1" and "2", as $\widehat{se}_{200}$ does, we see that $\hat{\mathbf{v}}_2$ is still much less accurate than $\hat{\mathbf{v}}_1$.

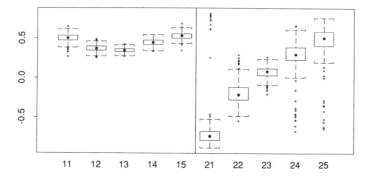

Figure 7.3. *200 bootstrap replications of the first two principal component vectors $\hat{\mathbf{v}}_1$ (left panel) and $\hat{\mathbf{v}}_2$ (right panel); for each component of the two vectors, the boxplot indicates the empirical distribution of the 200 bootstrap replications $\hat{v}_{ik}^*$. We see that $\hat{\mathbf{v}}_2$ is less accurate than $\hat{\mathbf{v}}_1$, having greater bootstrap variability for each component. A few of the bootstrap samples gave completely different results than the others for $\hat{\mathbf{v}}_2$.*

7.3 Example 2: curve fitting

In this example we will be estimating a regression function in two ways, by a standard least-squares curve and by a modern curve-fitting algorithm called "loess." We begin with a brief review of regression theory. Chapter 9 looks at the regression problem again, and gives an alternative bootstrap method for estimating regression standard errors. Figure 7.5 shows a typical data set for which regression methods are used: $n = 164$ men took part in an experiment to see if the drug cholostyramine lowered blood cholesterol levels. The men were supposed to take six packets of cholostyramine per day, but many of them actually took much less. The horizontal axis, which we will call "z", measures *Compliance*, as a percentage of the intended dose actually taken,

$$z_i = \text{percentage compliance for man } i, \quad i = 1, 2, \cdots, 164.$$

Compliance was measured by counting the number of unconsumed packets that each man returned. Men who took 0% of the dose are at the extreme left, those who took 100% are at the extreme right. The horizontal axis, labeled "y", is *Improvement*, the decrease in total blood plasma cholesterol level from the beginning

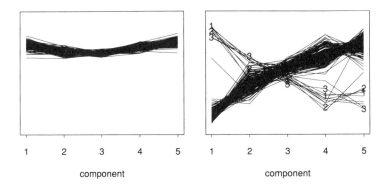

component   component

Figure 7.4. *Graphs of the 200 bootstrap replications of* $\hat{\mathbf{v}}_1$ *(left panel) and* $\hat{\mathbf{v}}_2$ *(right panel). The numbers 1, 2, 3 in the right panel follow three of the replications* $\hat{\mathbf{v}}_2^*(b)$ *that gave the most discrepant values for the first component. We see that these replications were also discrepant for other components, particularly component 5.*

to the end of the experiment,

$$y_i = \text{decrease in blood cholesterol for man } i, \quad i = 1, 2, \cdots, 164.$$

The full data set is given in Table 7.4.

The figure shows that men who took more cholostyramine tended to get bigger improvements in their cholesterol levels, just as we might hope. What we see in Figure 7.5, or at least what we think we see, is an increase in the average response y as z increases from 0 to 100%. Figure 7.6 shows the data along with two curves,

$$\hat{r}_{\text{quad}}(z) \quad \text{and} \quad \hat{r}_{\text{loess}}(z). \tag{7.16}$$

Each of these is an estimated *regression curve*. Here is a brief review of regression curves and their estimation. By definition the regression of a response variable y on an explanatory variable z is the *conditional expectation of y given z*, written

$$r(z) = \text{E}(y|z). \tag{7.17}$$

Suppose we had available the entire population $\mathcal{U}$ of men eligible for the cholostyramine experiment, and obtained the population $\mathcal{X} = (X_1, X_2, \cdots, X_N)$ of their Compliance-Improvement scores, $X_j = (Z_j, Y_j)$, $j = 1, 2, \cdots, N$. Then for each value of z, say

Table 7.4. *The cholostyramine data. 164 men were supposed to take 6 packets per day of the cholesterol-lowering drug cholostyramine. Compliance "z" is the percentage of the intended dose actually taken. Improvement "y" is the decrease in total plasma cholesterol from the beginning till the end of treatment.*

z	y	z	y	z	y	z	y
0	-5.25	27	-1.50	71	59.50	95	32.50
0	-7.25	28	23.50	71	14.75	95	70.75
0	-6.25	29	33.00	72	63.00	95	18.25
0	11.50	31	4.25	72	0.00	95	76.00
2	21.00	32	18.75	73	42.00	95	75.75
2	-23.00	32	8.50	74	41.25	95	78.75
2	5.75	33	3.25	75	36.25	95	54.75
3	3.25	33	27.75	76	66.50	95	77.00
3	8.75	34	30.75	77	61.75	96	68.00
4	8.25	34	-1.50	77	14.00	96	73.00
4	-10.25	34	1.00	78	36.00	96	28.75
7	-10.50	34	7.75	78	39.50	96	26.75
8	19.75	35	-15.75	81	1.00	96	56.00
8	-0.50	36	33.50	82	53.50	96	47.50
8	29.25	36	36.25	84	46.50	96	30.25
8	36.25	37	5.50	85	51.00	96	21.00
9	10.75	38	25.50	85	39.00	97	79.00
9	19.50	41	20.25	87	-0.25	97	69.00
9	17.25	43	33.25	87	1.00	97	80.00
10	3.50	45	56.75	87	46.75	97	86.00
10	11.25	45	4.25	87	11.50	98	54.75
11	-13.00	47	32.50	87	2.75	98	26.75
12	24.00	50	54.50	88	48.75	98	80.00
13	2.50	50	-4.25	89	56.75	98	42.25
15	3.00	51	42.75	90	29.25	98	6.00
15	5.50	51	62.75	90	72.50	98	104.75
16	21.25	52	64.25	91	41.75	98	94.25
16	29.75	53	30.25	92	48.50	98	41.25
17	7.50	54	14.75	92	61.25	98	40.25
18	-16.50	54	47.25	92	29.50	99	51.50
20	4.50	56	18.00	92	59.75	99	82.75
20	39.00	57	13.75	93	71.00	99	85.00
21	-5.75	57	48.75	93	37.75	99	70.00
21	-21.00	58	43.00	93	41.00	100	92.00
21	0.25	60	27.75	93	9.75	100	73.75
22	-10.25	62	44.50	93	53.75	100	54.00
24	-0.50	64	22.50	94	62.50	100	69.50
25	-19.00	64	-14.50	94	39.00	100	101.50
25	15.75	64	-20.75	94	3.25	100	68.00
26	6.00	67	46.25	94	60.00	100	44.75
27	10.50	68	39.50	95	113.25	100	86.75

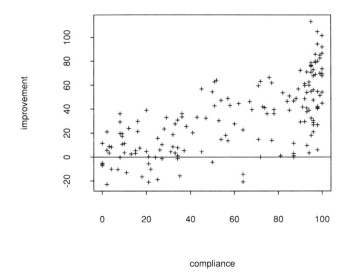

Figure 7.5. *The cholostyramine data. 164 men were supposed to take 6 packets per day of the cholesterol-lowering drug cholostyramine; horizontal axis measures Compliance, in percentage of assigned dose actually taken; vertical axis measures Improvement, in terms of blood cholesterol decrease over the course of the experiment. We see that better compliers tended to have greater improvement.*

$z = 0\%, 1\%, 2\%, \cdots, 100\%$, the regression would be the conditional expectation (7.17),

$$r(z) = \frac{\text{sum of } Y_j \text{ values for men in } \mathcal{X} \text{ with } Z_j = z}{\text{number of men in } \mathcal{X} \text{ with } Z_j = z}. \qquad (7.18)$$

In other words, $r(z)$ is the expectation of Y for the subpopulation of men having $Z = z$.

Of course we do not have available the entire population $\mathcal{X}$. We have the sample $\mathbf{x} = (\mathbf{x}_1, \mathbf{x}_2, \cdots, \mathbf{x}_{164})$, where $\mathbf{x}_i = (z_i, y_i)$, as shown in Figure 7.5 and Table 7.4. How can we estimate $r(z)$? The

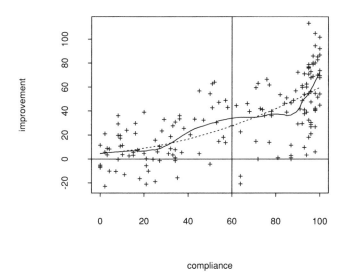

Figure 7.6. *Estimated regression curves of y = Improvement on z = Compliance. The dashed curve is $\hat{r}_{\mathrm{quad}}(z)$, the ordinary least-squares quadratic regression of y on z; the solid curve is $\hat{r}_{\mathrm{loess}}(z)$, a computer-based local linear regression. We are particularly interested in estimating the true regression r(z) at z = 60%, the average Compliance, and at z = 100%, full Compliance.*

obvious plug-in estimate is

$$\hat{r}(z) = \frac{\text{sum of } y_i \text{ values for men in } \mathbf{x} \text{ with } z_i = z}{\text{number of men in } \mathbf{x} \text{ with } z_i = z}. \qquad (7.19)$$

One can imagine drawing vertical strips of width 1% over Figure 7.5, and averaging the y_i values within each strip to get $\hat{r}(z)$. The results are shown in Figure 7.7.

This is our first example where the plug-in principle doesn't work very well. The estimated regression $\hat{r}(z)$ is much rougher than we expect the population regression $r(z)$ to be. The problem is that

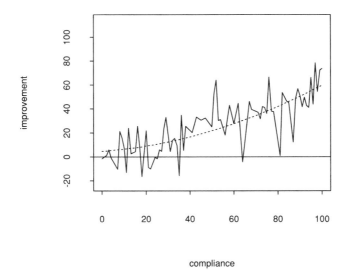

Figure 7.7. *Solid curve is plug-in estimate $\hat{r}(z)$ for the regression of improvement on compliance; averages of y_i for strips of width 1% on the z axis, as in (7.19). Some strips z are not represented because none of the 164 men had $z_i = z$. The function $\hat{r}(z)$ is much rougher than we expect the population regression curve $r(z)$ to be. The dashed curve is $\hat{r}_{\mathrm{quad}}(z)$.*

there aren't enough points in each strip of width 1% to estimate $r(z)$ very well. In some strips, like that for $z = 5\%$, there are no points at all. We could make the strip width larger, say 10% instead of 1%, but this leaves us with only a few points to plot, and, perhaps, with problems of variability still remaining. A more elegant and efficient solution is available, based on the *method of least-squares*.

The method begins by assuming that the population regression function, whatever it may be, belongs to a family $\mathcal{R}$ of smooth functions indexed by a vector parameter $\boldsymbol{\beta} = (\beta_0, \beta_1, \cdots, \beta_p)^T$. For the cholostyramine example we will consider the family of quadratic functions of z, say $\mathcal{R}_{\text{quad}}$,

$$\mathcal{R}_{\text{quad}} : \quad r_\beta(z) = \beta_0 + \beta_1 z + \beta_2 z^2, \qquad (7.20)$$

so $\boldsymbol{\beta} = (\beta_0, \beta_1, \beta_2)^T$. Later we will discuss the choice of the quadratic family $\mathcal{R}_{\text{quad}}$, but for now we will just accept it as given.

The reader can imagine choosing a trial value of $\boldsymbol{\beta}$, say $\boldsymbol{\beta} = (0, .75, .005)^T$, and plotting $r_\beta(z)$ on Figure 7.5. We would like the curve $r_\beta(z)$ to be near the data points (z_i, y_i) in some overall sense. It is particularly convenient for mathematical calculations to measure the closeness of the curve to the data points in terms of the *residual squared error*,

$$\text{RSE}(\boldsymbol{\beta}) = \sum_{i=1}^{n} [y_i - r_\beta(z_i)]^2. \qquad (7.21)$$

The residual squared error is obtained by dropping a vertical line from each point (z_i, y_i) to the curve $r_\beta(z_i)$, and summing the squared lengths of the verticals.

The *method of least-squares*, originated by Legendre and Gauss in the early 1800's, chooses among the curves in $\mathcal{R}$ by minimizing the residual squared error. The best-fitting curve in $\mathcal{R}$ is declared to be $r_{\hat{\beta}}(z)$, where $\hat{\boldsymbol{\beta}}$ minimizes $\text{RSE}(\boldsymbol{\beta})$,

$$\text{RSE}(\hat{\boldsymbol{\beta}}) = \min_{\beta} \text{RSE}(\boldsymbol{\beta}). \qquad (7.22)$$

The curve $\hat{r}_{\text{quad}}(z)$ in Figure 7.6 is $r_{\hat{\beta}}(z) = \hat{\beta}_0 + \hat{\beta}_1 z + \hat{\beta}_2 z^2$, the best-fitting quadratic curve for the cholostyramine data.

Legendre and Gauss discovered a wonderful mathematical formula for the least squares solution $\hat{\boldsymbol{\beta}}$. Let $\mathbf{C}$ be the 164×3 matrix whose ith row is

$$\mathbf{c}_i = (1, z_i, z_i^2), \qquad (7.23)$$

and let $\mathbf{y}$ be the vector of 164 y_i values. Then, in standard matrix notation,

$$\hat{\boldsymbol{\beta}} = (\mathbf{C}^T \mathbf{C})^{-1} \mathbf{C}^T y. \qquad (7.24)$$

We will examine this formula more closely in Chapter 9. For our bootstrap purposes here all we need to know is that a data set of n pairs $\mathbf{x} = (\mathbf{x}_1, \mathbf{x}_2, \cdots, \mathbf{x}_n)$ produces a quadratic least-squares curve

$r_{\hat{\beta}}(z)$ via the mapping $\mathbf{x} \rightarrow r_{\hat{\beta}}(z)$ that happens to be described by (7.23), (7.24) and (7.20).

One can think of $r_{\hat{\beta}}(z)$ as a smoothed version of the plug-in estimate $\hat{r}(z)$. Suppose that we increased the family $\mathcal{R}$ of smooth functions under consideration, say to $\mathcal{R}_{\text{cubic}}$ the class of cubic polynomials in z. Then the least-squares solution $r_{\hat{\beta}}(z)$ would come closer to the data points, but would be bumpier than the quadratic least-squares curve. As we considered higher and higher degree polynomials, $r_{\hat{\beta}}(z)$ would more and more resemble the plug-in estimate $\hat{r}(z)$. Our choice of a quadratic regression function is implicitly a choice of how smooth we believe the true regression $r(z)$ to be. Looking at Figure 7.7, we can see directly that $\hat{r}_{\text{quad}}(z)$ is much smoother than $\hat{r}(z)$, but generally follows $\hat{r}(z)$ as a function of z.

It is easy to believe that the true regression $r(z)$ is a smooth function of z. It is harder to believe that it is a quadratic function of z across the entire range of z values. The smoothing function "loess", pronounced "Low S", attempts to compromise between a *global* assumption of form, like quadraticity, and the purely *local* averaging of $\hat{r}(z)$.

A user of loess is asked to provide a number "α" that will be the proportion of the n data points used at each point of the construction. The curve $\hat{r}_{\text{loess}}(z)$ in Figure 7.6 used $\alpha = .30$. For each value of z, the value of $\hat{r}_{\text{loess}}(z)$ is obtained as follows:

(1)   The n points $\mathbf{x}_i = (z_i, y_i)$ are ranked according to $|z_i - z|$, and the $\alpha \cdot n$ nearest points, those with $|z_i - z|$ smallest, are identified. Call this neighborhood of $\alpha \cdot n$ points "$\mathcal{N}(z)$." [With $\alpha = .30$, $n = 164$, the algorithm puts 49 points into $\mathcal{N}(z)$.]

(2)   A weighted least-squares linear regression

$$\hat{r}_z(Z) = \hat{\beta}_{z,0} + \hat{\beta}_{z,1} Z \qquad (7.25)$$

is fit to the $\alpha \cdot n$ points in $\mathcal{N}(z)$. [That is, the coefficients $\hat{\beta}_{z,0}, \hat{\beta}_{z,1}$ are selected to minimize $\sum_{x_j \in \mathcal{N}(z)} w_{z,j} [y_j - (\beta_0 + \beta_1 z_j)]^2$, where the *weights* $w_{z,j}$ are positive numbers which depend on $|z_j - z|$. Letting

$$u_j = \frac{|z_j - z|}{\max_{\mathcal{N}(z)} |z_k - z|}, \qquad (7.26)$$

the weights w_j equal $(1 - u_j^3)^3$.]

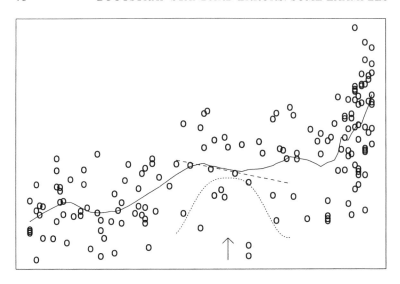

Figure 7.8. *How the Loess smoother works. The shaded region indicates the window of values around the target value (arrow). A weighted linear regression (broken line) is computed, using weights given by the "tri-cube" function (dotted curve). Repeating this process for all target values gives the solid curve.*

(3)   Finally, $\hat{r}_{\text{loess}}(z)$ is set equal to the value of $\hat{r}_z(Z)$ at $Z = z$,

$$\hat{r}_{\text{loess}}(z) = \hat{r}_z(Z = z). \qquad (7.27)$$

The components of the loess smoother are shown in Figure 7.8. Table 7.5 compares $\hat{r}_{\text{quad}}(z)$ with $\hat{r}_{\text{loess}}(z)$ at the two values of particular interest, $z = 60\%$ and $z = 100\%$. Bootstrap standard errors are given for each value. These were obtained from $B = 50$ bootstrap replications of the algorithm shown in Figure 6.1.

In this case $\hat{F}$ is the distribution putting probability $1/164$ on each of the 164 points $\mathbf{x}_i = (z_i, y_i)$. A bootstrap data set is $\mathbf{x}^* = (\mathbf{x}_1^*, \mathbf{x}_2^*, \cdots, \mathbf{x}_{164}^*)$, where each $\mathbf{x}_i^*$ equals any one of the 164 members of $\mathbf{x}$ with equal probability. Having obtained $\mathbf{x}^*$, we calculated $\hat{r}_{\text{quad}}^*(z)$ and $\hat{r}_{\text{loess}}^*(z)$, the quadratic and loess regression curves based on $\mathbf{x}^*$. Finally, we read off the values $\hat{r}_{\text{quad}}^*(60)$, $\hat{r}_{\text{loess}}^*(60)$, $\hat{r}_{\text{quad}}^*(100)$, and $\hat{r}_{\text{loess}}^*(100)$. The $B = 50$ values of $\hat{r}_{\text{quad}}^*(60)$ had sample standard error 3.03, etc., as reported in Table 7.5.

Table 7.5 shows that $\hat{r}_{\text{loess}}(z)$ is substantially less accurate than

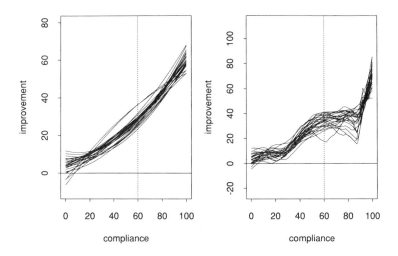

Figure 7.9. *The first 25 bootstrap replications of $\hat{r}_{quad}(z)$, left panel, and $\hat{r}_{loess}(z)$, right panel; the increased variability of $\hat{r}_{loess}(z)$ is evident.*

$\hat{r}_{quad}(z)$. This is not surprising since $\hat{r}_{loess}(z)$ is based on less data than $\hat{r}_{quad}(z)$, only α as much. See Problem 7.10. The overall greater variability of $\hat{r}_{loess}(z)$ is evident in Figure 7.9.

It is useful to plot the bootstrap curves to see if interesting features of the original curve maintain themselves under bootstrap sampling. For example, Figure 7.6 shows $\hat{r}_{loess}$ increasing much more rapidly from $z = 80\%$ to $z = 100\%$ than from $z = 60\%$ to $z = 80\%$. The difference in the average slopes is

$$\begin{aligned}
\hat{\theta} &= \frac{\hat{r}_{loess}(100) - \hat{r}_{loess}(80)}{20} - \frac{\hat{r}_{loess}(80) - \hat{r}_{loess}(60)}{20} \\
&= \frac{72.78 - 37.50}{20} - \frac{32.50 - 34.03}{20} = 1.84.
\end{aligned}$$

$$(7.28)$$

The corresponding number for $\hat{r}_{quad}$ is only 0.17. Most of the bootstrap loess curves $\hat{r}^*_{loess}(z)$ showed a similar sharp upward bend at about $z = 80\%$. None of the 50 bootstrap values $\hat{\theta}^*$ were less than 0, the minimum being .23, with most of the values > 1, see Figure 7.10.

At this point we may legitimately worry that $\hat{r}_{quad}(z)$ is *too*

Table 7.5. *Values of $\hat{r}_{\mathrm{quad}}(z)$ and $\hat{r}_{\mathrm{loess}}(z)$ at $z = 60\%$ and $z = 100\%$; also bootstrap standard errors based on $B = 50$ bootstrap replications.*

	$\hat{r}_{\mathrm{quad}}(60)$	$\hat{r}_{\mathrm{loess}}(60)$	$\hat{r}_{\mathrm{quad}}(100)$	$\hat{r}_{\mathrm{loess}}(100)$
value:	27.72	34.03	59.67	72.78
$\widehat{\mathrm{se}}_{50}$:	3.03	4.41	3.55	6.44

smooth an estimate of the true regression $r(z)$. If the value of the true slope difference

$$\theta = \frac{r(100) - r(80)}{20} - \frac{r(80) - r(60)}{20} \qquad (7.29)$$

is anywhere near $\hat{\theta} = 1.59$, then $r(z)$ will look more like $\hat{r}_{\mathrm{loess}}(z)$ than $\hat{r}_{\mathrm{quad}}(z)$ for z between 60 and 100. Estimates based on $\hat{r}_{\mathrm{loess}}(z)$ tend to be highly variable, as in Table 7.5, but they also tend to have small bias. Both of these properties come from the local nature of the loess algorithm, which estimates $r(z)$ using only data points with z_j near z.

The estimate $\hat{\theta} = 1.59$ based on $\hat{r}_{\mathrm{loess}}$ has considerable variability, $\widehat{\mathrm{se}}_{50} = .61$, but Figure 7.10 strongly suggests that the true θ, whatever it may be, is greater than the value $\hat{\theta} = .17$ based on $\hat{r}_{\mathrm{quad}}$. We will examine this type of argument more closely in Chapters 12–14 on bootstrap confidence intervals.

Table 7.5 suggests that we should also worry about the estimates $\hat{r}_{\mathrm{quad}}(60)$ and $\hat{r}_{\mathrm{quad}}(100)$, which may be substantially too low. One option is to consider higher polynomial models such as cubic, quartic, etc. Elaborate theories of model building have been put forth, in an effort to say when to go on to a bigger model and when to stop. We will consider regression models further in Chapter 9, where the cholesterol data will be looked at again. The simple bootstrap estimates of variability discussed in this chapter are often a useful step toward understanding regression models, particularly nontraditional ones like $\hat{r}_{\mathrm{loess}}(z)$.

7.4 An example of bootstrap failure

[1] Suppose we have data $X_1, X_2, \ldots X_n$ from a uniform distribution on $(0, \theta)$. The maximum likelihood estimate $\hat{\theta}$ is the largest sample value $X_{(n)}$. We generated a sample of 50 uniform numbers in the range $(0,1)$, and computed $\hat{\theta} = 0.988$. The left panel of Figure 7.11 shows a histogram of 2000 bootstrap replications of $\hat{\theta}^*$ obtained by sampling with replacement from the data. The right panel shows 2000 parametric bootstrap replications obtained by sampling from the uniform distribution on $(0, \hat{\theta})$. It is evident that the left histogram is a poor approximation to the right histogram. In particular, the left histogram has a large probability mass at $\hat{\theta}$: 62% of the values $\hat{\theta}^*$ equaled $\hat{\theta}$. In general, it is easy to show that $\mathrm{Prob}(\hat{\theta}^* = \hat{\theta}) = 1 - (1 - 1/n)^n \to 1 - e^{-1} \approx .632$ as $n \to \infty$. However, in the parametric setting of the right panel, $\mathrm{Prob}(\hat{\theta}^* = \hat{\theta}) = 0$.

What goes wrong with the nonparametric bootstrap? The difficulty occurs because the empirical distribution function $\hat{F}$ is not a good estimate of the true distribution F in the extreme tail. Either parametric knowledge of F or some smoothing of $\hat{F}$ is needed to rectify matters. Details and references on this problem may be found in Beran and Ducharme (1991, page 23). The nonparametric bootstrap can fail in other examples in which θ depends on the smoothness of F. For example, if θ is the number of atoms of F, then $\hat{\theta} = n$ is a poor estimate of θ.

7.5 Bibliographic notes

Principal components analysis is described in most books on multivariate analysis, for example Anderson (1958), Mardia, Kent and Bibby (1979), or Morrison (1976). Advanced statistical aspects of the bootstrap analysis of a covariance matrix may be found in Beran and Srivastava (1985). Curve-fitting is described in Eubank (1988), Härdle (1990), and Hastie and Tibshirani (1990). The loess method is due to Cleveland (1979), and is described in Chambers and Hastie (1991). Härdle (1990) and Hall (1992) discuss methods for bootstrapping curve estimates, and give a number of further references. Efron and Feldman (1991) discuss the cholostyramine data and the use of compliance as an explanatory

---

[1] This section contains more advanced material and may be skipped at first reading

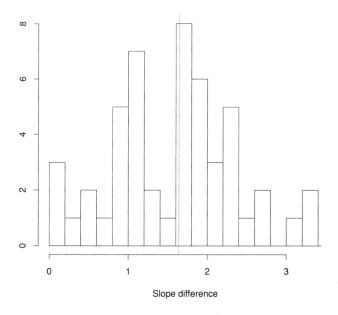

Figure 7.10. *Fifty bootstrap replications of the slope difference statistic (7.28). All of the values were positive, and most were greater than 1. The bootstrap standard error estimate is $\widehat{se}_{50}(\hat{\theta}) = .61$. The vertical line is drawn at $\hat{\theta} = 1.63$.*

variable. Leger, Politis and Romano (1992) give a number of examples illustrating the use of the bootstrap.

7.6 Problems

7.1 The sample covariance matrix of multivariate data $\mathbf{x}_1, \mathbf{x}_2, \cdots, \mathbf{x}_n$, when each $\mathbf{x}_i$ is a p-dimensional vector, is often defined to be the $p \times p$ matrix $\hat{\boldsymbol{\Sigma}}$ having j, kth element

$$\hat{\boldsymbol{\Sigma}}_{jk} = \frac{1}{n-1} \sum_{i=1}^{n} (x_{ij} - \bar{x}_j)(x_{ik} - \bar{x}_k) \qquad j, k = 1, 2, \cdots, p,$$

where $\bar{x}_j = \sum_{i=1}^{n} x_{ij}/n$ for $j = 1, 2, \cdots, p$. This differs from the empirical covariance matrix $\mathbf{G}$, (7.4), in dividing by $n-1$ rather than n.

(a) What is the first row of $\hat{\boldsymbol{\Sigma}}$ for the score data?

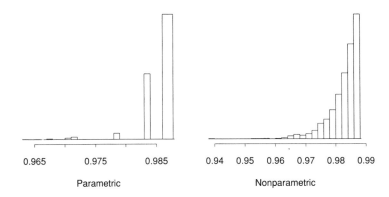

| 0.965 | 0.975 | 0.985 | | 0.94 | 0.95 | 0.96 | 0.97 | 0.98 | 0.99 |

Parametric   Nonparametric

Figure 7.11. *The left panel shows a histogram of 2000 bootstrap replications of $\hat{\theta}^* = X_{(n)}$ obtained by sampling with replacement from a sample of 50 uniform numbers. The right panel shows 2000 parametric bootstrap replications obtained by sampling from the uniform distribution on $(0, \hat{\theta})$.*

    (b)   The following fact is proved in linear algebra: the eigenvalues of matrix $c\mathbf{M}$ equal c times the eigenvalues of $\mathbf{M}$ for any constant c. (The eigenvectors of $c\mathbf{M}$ equal those of $\mathbf{M}$.) What are the eigenvalues of $\hat{\Sigma}$ for the score data? What is $\hat{\theta}$, (7.8)?

7.2 (a)   What is the sample correlation coefficient between the mechanics and vectors test scores? Between vectors and algebra?

    (b)   What is the sample correlation coefficient between the algebra test score and the sum of the mechanics and vectors test scores? (Hint: $E[(x+y)z] = E(xz) + E(yz)$ and $E[(x+y)^2] = E(x^2) + 2E(xy) + E(y^2)$.)

7.3 Calculate the probability that any particular row of the 88×5 data matrix $\mathbf{X}$ appears exactly k times in a bootstrap matrix $\mathbf{X}$, for $k = 0, 1, 2, 3$.

7.4 A random variable x is said to have the *Poisson distribution* with expectation parameter λ

$$x \sim \mathrm{Po}(\lambda), \tag{7.30}$$

if the sample space of x is the non-negative integers, and

$$\text{Prob}\{x = k\} = \frac{e^{-\lambda}\lambda^k}{k!} \quad \text{for} \quad k = 0, 1, 2, \cdots \ . \quad (7.31)$$

A useful approximation for a binomial distribution $\text{Bi}(n, p)$ is the Poisson distribution with $\lambda = np$,

$$\text{Bi}(n, p) \doteq \text{Po}(np). \quad (7.32)$$

The approximation in (7.32) becomes more accurate as n gets large and p gets small.

(a)   Suppose $\mathbf{x}^* = (\mathbf{x}_1^*, \mathbf{x}_2^*, \cdots, \mathbf{x}_n^*)$ is a bootstrap sample obtained from $\mathbf{x} = (\mathbf{x}_1, \mathbf{x}_2, \cdots, \mathbf{x}_n)$. What is the Poisson approximation for the probability that any particular member of $\mathbf{x}$ appears exactly k times in $\mathbf{x}^*$?

(b)   Give a numerical comparison with your answer to Problem 7.3.)

7.5 Notice that in the right panel of Figure 7.4, the main bundle of bootstrap curves is notably narrower half way between "1" and "2" on the horizontal axis. Suggest a reason why.

7.6 The *sample correlation matrix* corresponding to $\mathbf{G}$, (7.4), is the matrix $\mathbf{C}$ having jkth element

$$C_{jk} = G_{jk}/[G_{jj} \cdot G_{kk}]^{1/2} \quad j, k = 1, 2, \cdots, 5. \quad (7.33)$$

Principal component analyses are often done in terms of the eigenvalues and vectors of $\mathbf{C}$ rather than $\mathbf{G}$. Carry out a bootstrap analysis of the principal components based on $\mathbf{C}$, and produce the corresponding plots to Figures 7.3 and 7.4. Discuss any differences between the two analyses.

7.7 A generalized version of (7.20), called the *linear regression model*, assumes that y_i, the ith observed value of the response variable, depends on a *covariate vector* $\mathbf{c}_i = (c_{i1}, c_{i2}, \cdots, c_{ip})$ and a *parameter vector* $\boldsymbol{\beta} = (\beta_1, \beta_2, \cdots, \beta_p)^T$. The covariate $\mathbf{c}_i$ is observable, but $\boldsymbol{\beta}$ is not. The expectation of y_i is assumed to be the linear function

$$\mathbf{c}_i\boldsymbol{\beta} = \sum_{j=1}^{p} c_{ij}\beta_j. \quad (7.34)$$

[In (7.20), $c_i = (1, z_i, z_i^2)$, $\boldsymbol{\beta} = (\beta_0, \beta_1, \beta_2)^T$, and $p = 3$.] Legendre and Gauss showed that $\hat{\boldsymbol{\beta}} = (\mathbf{C}^T\mathbf{C})^{-1}\mathbf{C}^T\mathbf{y}$ minimizes $\sum_{i=1}^{n}(y_i - c_i\boldsymbol{\beta})^2$, that is $\hat{\boldsymbol{\beta}}$ as given by (7.24), is the least-squares estimate of $\boldsymbol{\beta}$. Here $\mathbf{C}$ is the $n \times p$ matrix with ith row c_i, assumed to be of full rank, and $\mathbf{y}$ is the vector of responses. Use this result to prove that $\hat{\mu} = \bar{y}$ minimizes $\sum_{i=1}^{n}(y_i - \mu)^2$ among all choices of μ.

7.8 For convenient notation, let $\mathcal{R}_2$ equal $\mathcal{R}_{quad}$, (7.20), $\mathcal{R}_3$ equal the family of cubic functions of z, $\mathcal{R}_4$ equal the set of quartic functions, etc. Define $\hat{\boldsymbol{\beta}}(j)$ as the least-squares estimate of $\boldsymbol{\beta}$ in the class $\mathcal{R}_j$, so $\hat{\boldsymbol{\beta}}(j)$ is a $j + 1$ dimensional vector, and let $\text{RSE}_j(\boldsymbol{\beta}) = \sum_{i=1}^{n}(y_i - r_{\hat{\beta}(j)}(z_i))^2$.

(a) Why is $\text{RSE}_j(\boldsymbol{\beta})$ a non-increasing function of j?

(b) Suppose that all n of the z_i values are distinct. What is the limiting value of $\text{RSE}_j(\boldsymbol{\beta})$, and for what value of j is it reached. [Hint: consider the polynomial in y, $\prod_{i=1}^{n}(y - z_i)$.]

(c) [†] Suppose the z_i are *not* distinct, as in Table 7.4. What is the limiting value of $\text{RSE}_j(\boldsymbol{\beta})$?

7.9 Problem 7.8a says that increasing the class of polynomials decreases the residual error of the fit. Give an intuitive argument why $r_{\hat{\beta}(j)}(z)$ might be a poor estimate of the true regression function $r(z)$ if we take j to be very large.

7.10 The estimate $\hat{r}_{loess}(z)$ in Table 7.5 has greater standard error than $\hat{r}_{quad}(z)$, but it only uses 30% of the available data. Suppose we randomly selected 30% of the (z_i, y_i) pairs from Table 7.4, fit a quadratic least-squares regression to this data, and called the curve $\hat{r}_{30\%}(z)$. Make a reasonable guess as to what $\hat{se}_{50}$ would be for $\hat{r}_{30\%}(z)$, $z = 60$ and 100.

† Indicates a difficult or more advanced problem.

## CHAPTER 8

More complicated data structures

8.1 Introduction

The bootstrap algorithm of Figure 6.1 is based on the simplest possible probability model for random data: the one-sample model, where a single unknown probability distribution F produces the data $\mathbf{x}$ by random sampling

$$F \to \mathbf{x} = (x_1, x_2, \cdots, x_n). \tag{8.1}$$

The individual data points x_i in (8.1) can themselves be quite complex, perhaps being numbers or vectors or maps or images or anything at all, but the probability mechanism is simple. Many data analysis problems involve more complicated data structures. These structures have names like time series, analysis of variance, regression models, multi-sample problems, censored data, stratified sampling, and so on. The bootstrap algorithm can be adapted to general data structures, as is discussed here and in Chapter 9.

8.2 One-sample problems

Figure 8.1 is a schematic diagram of the bootstrap method as it applies to one-sample problems. On the left is the real world, where an unknown distribution F has given the observed data $\mathbf{x} = (x_1, x_2, \cdots, x_n)$ by random sampling. We have calculated a statistic of interest from $\mathbf{x}$, $\hat{\theta} = s(\mathbf{x})$, and wish to know something about $\hat{\theta}$'s statistical behavior, perhaps its standard error $\mathrm{se}_F(\hat{\theta})$.

On the right side of the diagram is the bootstrap world, to use David Freedman's evocative terminology. In the bootstrap world, the empirical distribution $\hat{F}$ gives bootstrap samples $\mathbf{x}^* = (x_1^*, x_2^*, \cdots, x_n^*)$ by random sampling, from which we calcu-

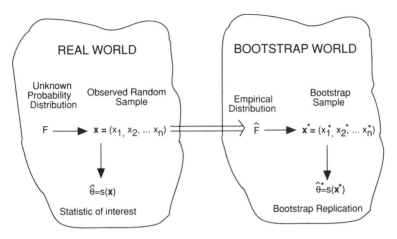

Figure 8.1. *A schematic diagram of the bootstrap as it applies to one-sample problems. In the real world, the unknown probability distribution F gives the data $\mathbf{x} = (x_1, x_2, \cdots, x_n)$ by random sampling; from $\mathbf{x}$ we calculate the statistic of interest $\hat{\theta} = s(\mathbf{x})$. In the bootstrap world, $\hat{F}$ generates $\mathbf{x}^*$ by random sampling, giving $\hat{\theta}^* = s(\mathbf{x}^*)$. There is only one observed value of $\hat{\theta}$, but we can generate as many bootstrap replications $\hat{\theta}^*$ as affordable. The crucial step in the bootstrap process is "$\Longrightarrow$", the process by which we construct from $\mathbf{x}$ an estimate $\hat{F}$ of the unknown population F.*

late bootstrap replications of the statistic of interest, $\hat{\theta}^* = s(\mathbf{x}^*)$. The big advantage of the bootstrap world is that we can calculate as many replications of $\hat{\theta}^*$ as we want, or at least as many as we can afford. This allows us to do probabilistic calculations directly, for example using the observed variability of the $\hat{\theta}^*$'s to estimate the unobservable quantity $se_F(\hat{\theta})$.

The double arrow in Figure 8.1 indicates the calculation of $\hat{F}$ from F. Conceptually, this is the crucial step in the bootstrap process, even though it is computationally simple. Every other part of the bootstrap picture is defined by analogy: F gives $\mathbf{x}$ by random sampling, so $\hat{F}$ gives $\mathbf{x}^*$ by random sampling; $\hat{\theta}$ is obtained from $\mathbf{x}$ via the function $s(\mathbf{x})$, so $\hat{\theta}^*$ is obtained from $\mathbf{x}^*$ in the same way. Bootstrap calculations for more complex probability mechanisms turn out to be straightforward, once we know how to carry out the double arrow process – estimating the entire probability mechanism from the data. Fortunately this is easy to do for all of the common data structures.

To facilitate the study of more complicated data structures, we will use the notation

$$P \rightarrow \mathbf{x} \tag{8.2}$$

to indicate that an unknown *probability model* P has yielded the observed data set $\mathbf{x}$.

8.3 The two-sample problem

To understand the notation of (8.2), consider the mouse data of Table 2.1. The probability model P can be thought of as a pair of probability distributions F and G, the first for the Treatment group and the second for the Control group,

$$P = (F, G). \tag{8.3}$$

Let $\mathbf{z} = (z_1, z_2, \cdots, z_m)$ indicate the Treatment observations, and $\mathbf{y} = (y_1, y_2, \cdots, y_n)$ indicate the Control observations with $n = 7$ and $m = 9$ (different notation than on page 10). Then the observed data comprises $\mathbf{z}$ and $\mathbf{y}$,

$$\mathbf{x} = (\mathbf{z}, \mathbf{y}). \tag{8.4}$$

We can think of $\mathbf{x}$ as a 16 dimensional vector, as long as we remember that the first seven coordinates come from F and the last nine come from G. The mapping $P \rightarrow \mathbf{x}$ is described by

$$F \rightarrow \mathbf{z} \quad \text{independently of} \quad G \rightarrow \mathbf{y}. \tag{8.5}$$

In other words, $\mathbf{z}$ is a random sample of size 7 from F, $\mathbf{y}$ is a random sample of size 9 from G, with $\mathbf{z}$ and $\mathbf{y}$ mutually independent of each other. This setup is called a *two-sample problem*.

In this case it is easy to estimate the probability mechanism P. Let $\hat{F}$ and $\hat{G}$ be the empirical distributions based on $\mathbf{z}$ and $\mathbf{y}$, respectively. Then the natural estimate of $P = (F, G)$ is

$$\hat{P} = (\hat{F}, \hat{G}). \tag{8.6}$$

Having obtained $\hat{P}$, the definition of a bootstrap sample $\mathbf{x}^*$ is obvious: the arrow in

$$\hat{P} \rightarrow \mathbf{x}^* \tag{8.7}$$

must mean the same thing as the arrow in $P \rightarrow \mathbf{x}$, (8.2). In the two-sample problem, (8.5), we have $\mathbf{x}^* = (\mathbf{z}^*, \mathbf{y}^*)$ where

$$\hat{F} \to \mathbf{z}^* \quad \text{independently of} \quad \hat{G} \to \mathbf{y}^*. \tag{8.8}$$

The sample sizes for $\mathbf{z}^*$ and $\mathbf{y}^*$ are the same as those for $\mathbf{z}$ and $\mathbf{y}$ respectively.

Figure 8.2 shows the histogram of $B = 1400$ bootstrap replications of the statistic

$$\begin{aligned}
\hat{\theta} &= \hat{\mu}_z - \hat{\mu}_y = \bar{z} - \bar{y} \\
&= 86.86 - 56.22 = 30.63, \tag{8.9}
\end{aligned}$$

the difference of the means between the Treatment and Control groups for the mouse data. This statistic estimates the parameter

$$\theta = \mu_z - \mu_y = \mathrm{E}_F(z) - \mathrm{E}_G(y). \tag{8.10}$$

If θ is really much greater than 0, as (8.9) seems to indicate, then the Treatment is a big improvement over the Control. However the bootstrap estimate of standard error for $\hat{\theta} = 30.63$ is

$$\widehat{se}_{1400} = \{\sum_{b=1}^{1400} [\hat{\theta}^*(b) - \hat{\theta}^*(\cdot)]^2 / 1399\}^{1/2} = 26.85, \tag{8.11}$$

so $\hat{\theta}$ is only 1.14 standard errors above zero, $1.14 = 30.63/26.85$. This would not usually be considered strong evidence that the true value of θ is greater than 0.

The bootstrap replications of $\hat{\theta}^*$ were obtained by using a random number generator to carry out (8.8). Each bootstrap sample $\mathbf{x}^*$ was computed as

$$\mathbf{x}^* = (\mathbf{z}^*, \mathbf{y}^*) = (z_{i_1}, z_{i_2}, \cdots, z_{i_7}, y_{j_1}, y_{j_2}, \cdots, y_{j_9}), \tag{8.12}$$

where $(i_1, i_2, \cdots, i_7)$ was a random sample of size 7 from the integers $1, 2, \cdots, 7$, and $(j_1, j_2, \cdots, j_9)$ was an independently selected random sample of size 9 from the integers $1, 2, \cdots, 9$. For instance, the first bootstrap sample had $(i_1, i_2, \cdots, i_7) = (7, 3, 1, 2, 7, 6, 3)$ and $(j_1, j_2, \cdots, j_9) = (7, 8, 2, 9, 6, 7, 8, 4, 2)$.

The standard error of $\hat{\theta}$ can be written as $se_P(\hat{\theta})$ to indicate its dependence on the unknown probability mechanism $P = (F, G)$. The bootstrap estimate of $se_P(\hat{\theta})$ is the plug-in estimate

$$se_{\hat{P}}(\hat{\theta}^*) = \{\mathrm{var}_{\hat{P}}(\bar{z}^* - \bar{y}^*)\}^{1/2}. \tag{8.13}$$

As in Chapter 6, we approximate the ideal bootstrap estimate $se_{\hat{P}}(\hat{\theta}^*)$ by $\widehat{se}_B$ of equation (6.6), in this case with $B = 1400$. The

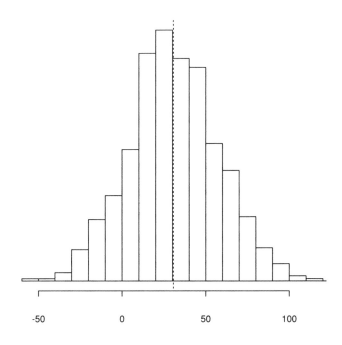

Figure 8.2. *1400 bootstrap replications of* $\hat{\theta} = \bar{z} - \bar{y}$, *the difference between the Treatment and Control means for the mouse data of Table 2.1; bootstrap estimated standard error was* $\widehat{se}_{1400} = 26.85$, *so the observed value* $\hat{\theta} = 30.63$ *(broken line) is only 1.14 standard errors above zero; 13.1% of the 1400* $\hat{\theta}^*$ *values were less than zero. This is not small enough to be considered convincing evidence that the Treatment worked better than the Control.*

fact that $\hat{\theta}^*$ is computed from two samples, $\mathbf{z}^*$ and $\mathbf{y}^*$, doesn't affect definition (6.6), namely $\widehat{se}_B = \{\sum_{b=1}^{B}[\hat{\theta}^*(b) - \hat{\theta}^*(\cdot)]^2/(B-1)\}^{1/2}$.

8.4 More general data structures

Figure 8.3 is a version of Figure 8.1 that applies to general data structures $P \to \mathbf{x}$. There is not much conceptual difference between the two figures, except for the level of generality involved. In the real world, an unknown probability mechanism P gives an observed data set $\mathbf{x}$, according to the rule of construction indicated by the

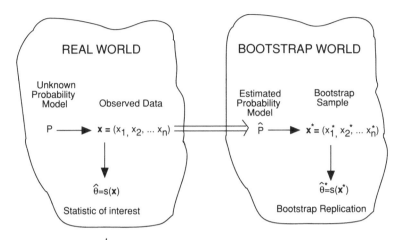

Figure 8.3. *Schematic diagram of the bootstrap applied to problems with a general data structure* $P \to \mathbf{x}$. *The crucial step "$\Longrightarrow$" produces an estimate $\hat{P}$ of the entire probability mechanism P from the observed data* $\mathbf{x}$. *The rest of the bootstrap picture is determined by the real world: "$\hat{P} \to \mathbf{x}^*$" is the same as "$P \to \mathbf{x}$"; the mapping from $\mathbf{x}^* \to \hat{\theta}^*$, $s(\mathbf{x}^*)$, is the same as the mapping from $\mathbf{x} \to \hat{\theta}$, $s(\mathbf{x})$.*

arrow "$\to$." In specific applications we need to define the arrow more carefully, as in (8.5) for the two-sample problem. The data set $\mathbf{x}$ may no longer be a single vector. It has a form dependent on the data structure, for example $\mathbf{x} = (\mathbf{z}, \mathbf{y})$ in the two-sample problem. Having observed $\mathbf{x}$, we calculate a statistic of interest $\hat{\theta}$ from $\mathbf{x}$ according to the function $s(\cdot)$.

The bootstrap side of Figure 8.3 is defined by the analogous quantities in the real world: the arrow in $\hat{P} \to \mathbf{x}^*$ is defined to mean the same thing as the arrow in $P \to \mathbf{x}$. And the function mapping $\mathbf{x}^*$ to $\hat{\theta}^*$ is the same function $s(\cdot)$ as from $\mathbf{x}$ to $\hat{\theta}$.

Two practical problems arise in actually carrying out a bootstrap analysis based on Figure 8.3:

(1) We need to estimate the entire probability mechanism P from the observed data $\mathbf{x}$. This is the step indicated by the double arrow, $\mathbf{x} \Longrightarrow \hat{P}$. It is surprisingly easy to do for most familiar data structures. No general prescription is possible, but quite natural ad hoc solutions are available in each case, for example $\hat{P} = (\hat{F}, \hat{G})$ for the two-sample problem. More examples are given in this chapter

Table 8.1. *The lutenizing hormone data.*

period	level	period	level	period	level	period	level
1	2.4	13	2.2	25	2.3	37	1.5
2	2.4	14	1.8	26	2.0	38	1.4
3	2.4	15	3.2	27	2.0	39	2.1
4	2.2	16	3.2	28	2.9	40	3.3
5	2.1	17	2.7	29	2.9	41	3.5
6	1.5	18	2.2	30	2.7	42	3.5
7	2.3	19	2.2	31	2.7	43	3.1
8	2.3	20	1.9	32	2.3	44	2.6
9	2.5	21	1.9	33	2.6	45	2.1
10	2.0	22	1.8	34	2.4	46	3.4
11	1.9	23	2.7	35	1.8	47	3.0
12	1.7	24	3.0	36	1.7	48	2.9

and the next.

(2)   We need to simulate bootstrap data from $\hat{P}$ according to the relevant data structure. This is the step $\hat{P} \to \mathbf{x}^*$ in Figure 8.3. This step is conceptually straightforward, being the same as $P \to \mathbf{x}$, but can require some care in the programming if computational efficiency is necessary. (We will see an example in the lutenizing hormone analysis below.) Usually the generation of the bootstrap data $\hat{P} \to \mathbf{x}^*$ requires less time, often *much* less time, than the calculation of $\hat{\theta}^* = s(\mathbf{x}^*)$.

8.5 Example: lutenizing hormone

Figure 8.4 shows a set of levels y_t of a lutenizing hormone for each of 48 time periods, taken from Diggle (1990); the data set is listed in Table 8.1. These are hormone levels measured on a healthy woman in 10 minute intervals over a period of 8 hours. The lutenizing hormone is one of the hormones that orchestrate the menstrual cycle and hence it is important to understand its daily variation.

It is clear that the hormone levels are not a random sample from any distribution. There is much too much structure in Figure 8.4. These data are an example of a *time series*: a data structure for which nearby values of the time parameter t indicate closely related

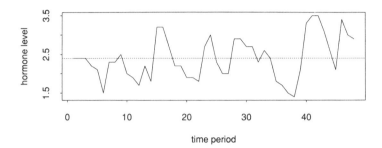

Figure 8.4. *The lutenizing hormone data. Level of lutenizing hormones y_t plotted versus time period t, for t from 1 to 48. In this plot and other plots the points are connected by lines to enhance visibility. The average value $\hat{\mu} = 2.4$ is indicated by a dashed line. Table 8.1 lists the data.*

values of the measured quantity y_t. Many interesting probabilistic models have been used to analyze time series. We will begin here with the simplest model, a *first order autoregressive scheme*.

Let μ be the expectation of y_t, assumed to be the same for all times t, and define the *centered* measurements

$$z_t = y_t - \mu. \tag{8.14}$$

All of the z_t have expectation 0. A first-order autoregressive scheme is one in which each z_t is a linear combination of the previous value z_{t-1}, and an independent disturbance term ϵ_t,

$$z_t = \beta z_{t-1} + \epsilon_t \quad \text{for} \quad t = U, U+1, U+2, \cdots, V. \tag{8.15}$$

Here β is an unknown parameter, a real number between -1 and 1.

The disturbances ϵ_t in (8.15) are assumed to be a random sample from an unknown distribution F with expectation 0,

$$F \to (\epsilon_U, \epsilon_{U+1}, \epsilon_{U+2}, \cdots, \epsilon_V) \qquad [\mathrm{E}_F(\epsilon) = 0]. \tag{8.16}$$

The dates U and V are the beginning and end of the time period under analysis. Here we have

$$U = 2 \quad \text{and} \quad V = 48. \tag{8.17}$$

Notice that the first equation in (8.15) is

$$z_U = \beta z_{U-1} + \epsilon_U \tag{8.18}$$

so we need the number z_{U-1} to get the autoregressive process started. In our case, $z_{U-1} = z_1$.

Suppose we believe that model (8.15), (8.16), the first-order autoregressive process, applies to the lutenizing hormone data. How can we estimate the value of β from the data? One answer is based on a least-squares approach. First of all, we estimate the expectation μ in (8.14) by the observed average $\bar{y}$ (this is 2.4 for the lutenizing hormone data), and set

$$z_t = y_t - \bar{y} \tag{8.19}$$

for all values of t. We will ignore the difference between definitions (8.14) and (8.19) in what follows, see Problem 8.4.

Suppose that b is any guess for the true value of β in (8.15). Define the residual squared error for this guess to be

$$\text{RSE}(b) = \sum_{t=U}^{V} (z_t - b z_{t-1})^2. \tag{8.20}$$

Using (8.15), and the fact that $E_F(\epsilon) = 0$, it is easy to show that RSE(b) has expectation $E(\text{RSE}(b)) = (b-\beta)^2 E(\sum_{t=U}^{V} z_{t-1}^2) + (V - U + 1)\text{var}_F(\epsilon)$. This is minimized when b equals the true value β. We are led to believe that RSE(b) should achieve its minimum somewhere near the true value of β.

Given the time series data, we can calculate RSE(b) as a function of b, and choose the minimizing value to be our estimate of β, [1]

$$\text{RSE}(\hat{\beta}) = \min_b \text{RSE}(b). \tag{8.21}$$

The lutenizing hormone data has least-squares estimate

$$\hat{\beta} = .586. \tag{8.22}$$

How accurate is the estimate $\hat{\beta}$? We can use the general bootstrap procedure of Figure 8.3 to answer this question. The probability mechanism P described in (8.15), (8.16) has two unknown elements, β and F, say $P = (\beta, F)$. (Here we are considering μ in

---

[1] For simplicity of exposition we use least-squares rather than normal theory maximum likelihood estimation. The difference between the two estimators is usually small.

(8.14) as known and equal to $\bar{y}$.) The data $\mathbf{x}$ consist of the observations y_t and their corresponding time periods t. We know that the rule of construction $P \to \mathbf{x}$ is described by (8.15)-(8.16). The statistic of interest $\hat{\theta}$ is $\hat{\beta}$, so the mapping $s(\cdot)$ is given implicitly by (8.21).

One step remains before we can carry out the bootstrap algorithm: the double-arrow step $\mathbf{x} \Longrightarrow \hat{P}$, in which $P = (\beta, F)$ is estimated from the data. Now β has already been estimated by $\hat{\beta}$, (8.21), so we need only estimate the distribution F of the disturbances. If we knew β, then we could calculate $\epsilon_t = z_t - \beta z_{t-1}$ for every t, and estimate F by the empirical distribution of the ϵ_t's. We don't know β, but we can use the estimated value of $\hat{\beta}$ to compute *approximate disturbances*

$$\hat{\epsilon}_t = z_t - \hat{\beta} z_{t-1} \quad \text{for} \quad t = U, U+1, U+2, \cdots, V. \qquad (8.23)$$

Let $T = V - U + 1$, the number of terms in (8.23); $T = 47$ for the choice (8.17). The obvious estimate of F is $\hat{F}$, the empirical distribution of the approximate disturbances,

$$\hat{F}: \quad \text{probability } 1/T \text{ on } \hat{\epsilon}_t \text{ for } t = U, U+1, \cdots, V. \qquad (8.24)$$

Figure 8.5 shows the histogram of the $T = 47$ approximate disturbances $\hat{\epsilon}_t = z_t - \hat{\beta} z_{t-1}$ for the first-order autoregressive scheme applied to the lutenizing data for years 2 to 48.

We see that the distribution $\hat{F}$ is not normal, having a long tail to the right. The distribution has mean 0.006 and standard deviation 0.454. It is no accident that the mean of $\hat{F}$ is near 0; see Problem 8.5. If it wasn't, we could honor the definition $E_F(\epsilon) = 0$ in (8.16) by centering $\hat{F}$; that is by changing each probability point in (8.23) from $\hat{\epsilon}_t$ to $\hat{\epsilon}_t - \bar{\epsilon}$, where $\bar{\epsilon} = \sum_U^V \hat{\epsilon}_t / T$.

Now we are ready to carry out a bootstrap accuracy analysis of the estimate $\hat{\beta} = 0.586$. A bootstrap data set $\hat{P} \to \mathbf{x}^*$ is generated by following through definitions (8.15)–(8.16), except with $\hat{P} = (\hat{\beta}, \hat{F})$ replacing $P = (\beta, F)$. We begin with the initial value $z_1 = y_1 - \bar{y}$, which is considered to be a fixed constant (like the sample size n in the one-sample problem). The bootstrap time series z_t^* is calculated recursively,

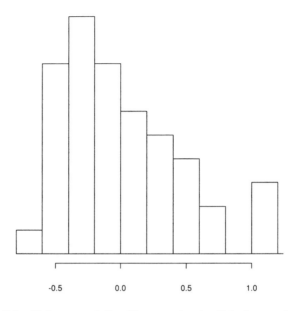

Figure 8.5. *Histogram of the 47 approximate disturbances $\hat{\epsilon}_t = z_t - \hat{\beta} z_{t-1}$, for $t = 2$ through 48; $\hat{\beta}$ equals 0.586 the least-squares estimate for the first-order autoregressive scheme. The distribution is long-tailed to the right. The disturbances averaged 0.006, with a standard deviation of 0.454, and so are nearly centered at zero.*

$$
\begin{aligned}
z_2^* &= \hat{\beta} z_1 + \epsilon_2^* \\
z_3^* &= \hat{\beta} z_2^* + \epsilon_3^* \\
z_4^* &= \hat{\beta} z_3^* + \epsilon_4^* \\
&\vdots \qquad \vdots \\
z_{48}^* &= \hat{\beta} z_{47}^* + \epsilon_{48}^*.
\end{aligned}
\tag{8.25}
$$

The bootstrap disturbance terms ϵ_t^* are a random sample from $\hat{F}$,

$$
\hat{F} \to (\epsilon_2^*, \epsilon_3^*, \cdots, \epsilon_{48}^*). \tag{8.26}
$$

In other words, each ϵ_t^* equals any one of the T approximate disturbances (8.23) with probability $1/T$.

The bootstrap process (8.25)–(8.26) was run $B = 200$ times, giving 200 bootstrap time-series. Each of these gave a bootstrap replication $\hat{\beta}^*$ for the least-squares estimate $\hat{\beta}$, (8.21). Figure 8.6 shows the histogram of the 200 $\hat{\beta}^*$ values. The bootstrap standard error estimate for $\hat{\beta}$ is $\widehat{se}_{200} = 0.116$. The histogram is fairly normal in shape.

In a first-order autoregressive scheme, each z_t depends on its predecessors only through the value of z_{t-1}. (This kind of dependence is known as a *first-order Markov process*.) A second-order autoregressive scheme extends the dependence back to z_{t-2},

$$z_t = \beta_1 z_{t-1} + \beta_2 z_{t-2} + \epsilon_t$$
$$\text{for} \quad t = U, U+1, U+2, \cdots, V. \tag{8.27}$$

Here $\boldsymbol{\beta} = (\beta_1, \beta_2)^T$ is a two-dimensional unknown parameter vector. The ϵ_t are independent random disturbances as in (8.16). Corresponding to (8.18) are initial equations

$$
\begin{aligned}
z_U &= \beta_1 z_{U-1} + \beta_2 z_{U-2} + \epsilon_U \\
z_{U+1} &= \beta_1 z_U + \beta_2 z_{U-1} + \epsilon_{U+1},
\end{aligned}
\tag{8.28}
$$

so we need the numbers z_{U-2} and z_{U-1} to get started. Now $U = 3, V = 48$, and $T = V - U + 1 = 46$.

The least-squares approach leads directly to an estimate of the vector $\boldsymbol{\beta}$. Let $\mathbf{z}$ be the T-dimensional vector $(z_U, z_{U+1}, \cdots, z_V)^T$, and let $\mathbf{Z}$ be the $T \times 2$ matrix with first column $(z_{U-1}, z_U, \cdots, z_{V-1})^T$, second column $(z_{U-2}, z_{U-1}, z_U, \cdots, z_{V-2})^T$. Then the least-squares estimate of $\boldsymbol{\beta}$ is

$$\hat{\boldsymbol{\beta}} = (\mathbf{Z}^T \mathbf{Z})^{-1} \mathbf{Z}^T \mathbf{z}. \tag{8.29}$$

For the lutenizing hormone data, the second-order autoregressive scheme had least-squares estimates

$$\hat{\boldsymbol{\beta}} = (0.771, -0.222)^T. \tag{8.30}$$

Figure 8.7 shows histograms of $B = 200$ bootstrap replications of the two components of $\hat{\boldsymbol{\beta}} = (\hat{\beta}_1, \hat{\beta}_2)^T$. The bootstrap standard errors are

$$\widehat{se}_{200}(\hat{\beta}_1) = 0.147, \quad \widehat{se}_{200}(\hat{\beta}_2) = 0.149. \tag{8.31}$$

Both histograms are roughly normal in shape. Problem 8.7 asks the reader to describe the steps leading to Figure 8.7.

A second-order autoregressive scheme with $\beta_2 = 0$ is a first-order autoregressive scheme. In doing the accuracy analysis for the second-order scheme, we check to see if $\hat{\beta}_2$ is less than 2 standard

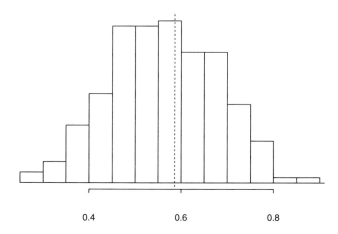

Figure 8.6. *Histogram of $B = 200$ bootstrap replications of $\hat{\beta}$, the first-order autoregressive parameter estimate for the lutenizing hormone data; from (8.25), (8.26); the bootstrap estimate of standard error is $\widehat{se}_{200} = 0.116$. The broken line is drawn at the observed value $\hat{\beta} = 0.586$.*

errors away from 0, which would usually be interpreted as $\hat{\beta}_2$ being not significantly different than zero. Here $\hat{\beta}_2$ is about 1.5 standard errors away from 0, in which case we have no strong evidence that a first-order autoregressive scheme does not give a reasonable representation of the lutenizing hormone data.

Do we know for sure that the first-order scheme gives a good representation of the lutenizing hormone series? We cannot definitively answer this question without considering still more general models such as higher-order autoregressive schemes. A rough answer can be obtained by comparison of the bootstrap time series with the actual series of Figure 8.4. Figure 8.8 shows the first four bootstrap series from the first-order scheme, left panel, and four realizations obtained by sampling with replacement from the original time series, right panel. The original data of Figure 8.4 looks quite a bit like the left panel realizations, and not at all like the right panel realizations.

Further analysis shows that the AR(1) model provides a rea-

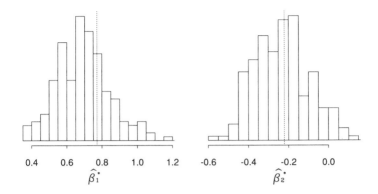

Figure 8.7. $B = 200$ *bootstrap replications of* $\hat{\beta} = (0.771, -0.222)$, *the second-order autoregressive parameter vector estimate for the lutenizing hormone data. As in the other histograms, a broken line is drawn at the parameter estimate. The histograms are roughly normal in shape.*

sonable fit to these data. However, we would need longer a time series to discriminate effectively between different models for this hormone.

In general, it pays to remember that mathematical models are conveniently simplified representations of complicated real-world phenomena, and are usually not perfectly correct. Often some compromise is necessary between the complication of the model and the scientific needs of the investigation. Bootstrap methods are particularly useful if complicated models seem necessary, since mathematical complication is no impediment to a bootstrap analysis of accuracy.

8.6 The moving blocks bootstrap

In this last section we briefly describe a different method for bootstrapping time series. Rather than fitting a model and then sampling from the residuals, this method takes an approach closer to that used for one-sample problems. The idea is illustrated in Figure 8.9. The original time series is represented by the black circles. To generate a bootstrap realization of the time series (white circles), we choose a block length ("3" in the diagram) and consider all possible contiguous blocks of this length. We sample with re-

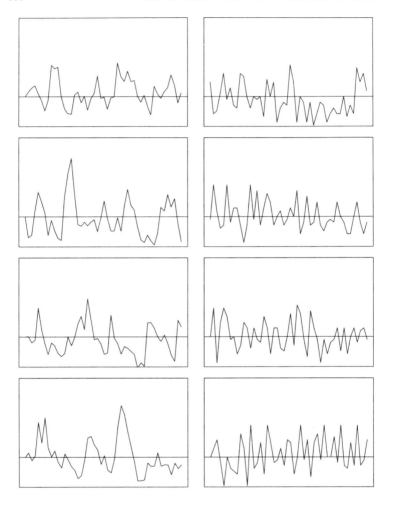

Figure 8.8. *Left panel: the first four bootstrap replications of the lut-enizing hormone data from the first-order autoregressive scheme, $y_t^* = z_t^* + 2.4$, (8.25), (8.26). Right panel: four bootstrap replications obtained by sampling with replacement from the original time series. The values from the first-order scheme look a lot more like the actual time series in Figure 8.4.*

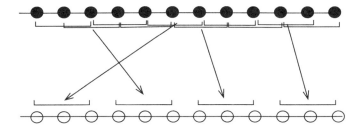

Figure 8.9. *A schematic diagram of the moving blocks bootstrap for time series. The black circles are the original time series. A bootstrap realization of the time series (white circles) is generated by choosing a block length ("3" in the diagram) and sampling with replacement from all possible contiguous blocks of this length.*

placement from these blocks and paste them together to form the bootstrap time series. Just enough blocks are sampled to obtain a series of roughly the same length as the original series. If the block length is ℓ, then we choose k blocks so that $n \approx k \cdot \ell$.

To illustrate this, we carried it out for the lutenizing hormone data. The statistic of interest was the AR(1) least-squares estimate $\hat{\beta}$. We chose a block length of 3, and used the moving blocks bootstrap to generate a bootstrap realization of the lutenizing hormone data. A typical bootstrap realization is shown in Figure 8.10, and it looks quite similar to the original time series. We then fit the AR(1) model to this bootstrap time series, and estimated the AR(1) coefficient $\hat{\beta}^*$. This entire process was repeated $B = 200$ times. (Note that the AR(1) model is being used here to estimate β, but is not being used in the generation of the bootstrap realizations of the time series.) The resulting bootstrap standard error was $\widehat{se}_{200}(\hat{\beta}) = 0.120$.[2] This is approximately the same as the value 0.116 obtained from AR(1) generated samples in the previous section. Increasing the block size to 5 caused this value to decrease to 0.103.

What is the justification for the moving blocks bootstrap? As we have seen earlier, we cannot simply resample from the individual

---

[2] If n is not exactly divisible by ℓ we need to multiply the bootstrap standard errors by $\sqrt{k\ell/n}$ to adjust for the difference in lengths of the series. This factor is 1.0 for $\ell = 3$ and 0.97 for $\ell = 5$ in our example, and hence made little difference.

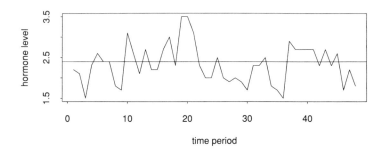

Figure 8.10. *A bootstrap realization of the lutenizing hormone data, using the moving blocks bootstrap with block length equal to 3.*

observations, as this would destroy the correlation that we're trying to capture. (Using a block size of one corresponds to sampling with replacement from the data, and gave 0.139 for the standard error estimate.) With the moving blocks bootstrap, the idea is to choose a block size ℓ large enough so that observations more than ℓ time units apart will be nearly independent. By sampling the blocks of length ℓ, we retain the correlation present in observations less than ℓ units apart.

The moving blocks bootstrap has the advantage of being less "model dependent" than the bootstrapping of residuals approach used earlier. As we have seen, the latter method is dependent on the model that is fit to the original time series (for example an AR(1) or AR(2) model). However the choice of block size ℓ can be quite important, and effective methods for making this choice have not yet been developed.

In the regression problem discussed in the next chapter, we encounter different methods for bootstrapping that are analogous to the approaches for time series that we have discussed here.

8.7 Bibliographic notes

The analysis of time series is described in many books, including Box and Jenkins (1970), Chatfield (1980) and Diggle (1990). Application of the bootstrap to time series is discussed in Efron

and Tibshirani (1986); the moving blocks method and related techniques can be found in Carlstein (1986), Künsch (1989), Liu and
Singh (1992) and Politis and Romano (1992).

8.8 Problems

8.1 If $\mathbf{z}$ and $\mathbf{y}$ are independent of each other, then $\text{var}(\bar{z} - \bar{y}) = \text{var}(\bar{z}) + \text{var}(\bar{y})$.

   (a)   How could we use the *one-sample* bootstrap algorithm
of Chapter 6 to estimate $\text{se}(\hat{\theta})$, for $\hat{\theta} = \bar{z} - \bar{y}$ as in (8.9)?

   (b)   The bootstrap data going into $\widehat{\text{se}}_{1400} = 26.85$, (8.11),
consisted of a 1400×16 matrix, each row of which was an
independent replication of (8.12). Say how your answer to
(a) would be implemented in terms of the matrix. Would
the answer still equal 26.85?

8.2 Suppose the mouse experiment was actually conducted as
follows: a large population of candidate laboratory mice were
identified, say $\mathcal{U} = (U_1, U_2, \cdots, U_N)$; a random sample of size
16 was selected, say $\mathbf{u} = (u_{k_1}, u_{k_2}, \cdots, u_{k_{16}})$; finally, a fair coin
was independently flipped sixteen times, with u_ℓ assigned to
Treatment or Control as the ℓth flip was heads or tails. Discuss how well the two-sample model (8.5) fits this situation.

8.3 Assuming model (8.14)-(8.16), show that $\text{E}(\text{RSE}(b)) = (b - \beta)^2 \text{E}(\sum_U^V z_{t-1}^2) + (V - U + 1)\text{var}_F(\epsilon)$.

8.4 The bootstrap analysis (8.25) was carried out as if $\bar{y} = 2.4$ was
the true value of $\mu = \text{E}(y_t)$. Carefully state how to calculate
$\widehat{\text{se}}(\hat{\beta})$ if we take the more honest point of view that μ is an
unknown parameter, estimated by $\bar{y}$.

8.5 Let $\bar{y}_U$ equal $\sum_{t=U}^{V} y_t / T$, $\bar{y}_{U-1}$ equal $\sum_{t=U-1}^{V-1} y_t / T$, and define $\hat{\beta}$ as in (8.19)–(8.21), except with $z_t = y_t - \bar{y}_{U-1}$.

   (a)   Show that

$$\sum_{t=U}^{V} \{(y_t - \bar{y}_U) - \hat{\beta}(y_{t-1} - \bar{y}_{U-1})\} = 0. \qquad (8.32)$$

   (b)   Why might we expect $\hat{F}$, (8.26), to have expectation
near 0?

8.6 Many statistical languages like "S" are designed for vector processing. That is, the command $\mathbf{c} = \mathbf{a} + \mathbf{b}$ to add two long vectors is carried out much more quickly than the loop

$$\text{for} \quad (i = 1 \text{ to } n)\{c_i = a_i + b_i\}. \tag{8.33}$$

This fact was used to speed the generation of the $B = 200$ bootstrap replications of the first-order autoregressive scheme for the lutenizing hormone data. How?

8.7 Give a detailed description of the bootstrap algorithm for the second-order autoregressive scheme.

CHAPTER 9

Regression models

9.1 Introduction

Regression models are among the most useful and most used of statistical methods. They allow relatively simple analyses of complicated situations, where we are trying to sort out the effects of many possible explanatory variables on a response variable. In Chapter 7 we use the one-sample bootstrap algorithm to analyze the accuracy of a regression analysis for the cholostyramine data of Table 7.4. Here we look at the regression problem more critically. The general bootstrap algorithm of Figure 8.3 is followed through, leading to a somewhat different bootstrap analysis for regression problems.

9.2 The linear regression model

We begin with the classic *linear regression model*, or *linear model*, going back to Legendre and Gauss early in the 19th century. The data set $\mathbf{x}$ for a linear regression model consists of n points $\mathbf{x}_1, \mathbf{x}_2, \cdots, \mathbf{x}_n$, where each $\mathbf{x}_i$ is itself a pair, say

$$\mathbf{x}_i = (\mathbf{c}_i, y_i). \tag{9.1}$$

Here $\mathbf{c}_i$ is a $1 \times p$ vector $\mathbf{c}_i = (c_{i1}, c_{i2}, \cdots, c_{ip})$ called the *covariate vector* or *predictor*, while y_i is a real number called the *response*.

Let μ_i indicate the conditional expectation of ith response y_i given the predictor $\mathbf{c}_i$,

$$\mu_i = \mathrm{E}(y_i | \mathbf{c}_i) \qquad (i = 1, 2, \cdots, n). \tag{9.2}$$

The key assumption in the linear model is that μ_i is a linear combination of the components of the predictor $\mathbf{c}_i$,

$$\mu_i = \mathbf{c}_i \boldsymbol{\beta} = \sum_{j=1}^{p} c_{ij} \beta_j. \tag{9.3}$$

The *parameter vector*, or *regression parameter*, $\boldsymbol{\beta} = (\beta_1, \beta_2, \cdots, \beta_p)^T$ is unknown, the usual goal of the regression analysis being to infer $\boldsymbol{\beta}$ from the observed data $\mathbf{x} = (\mathbf{x}_1, \mathbf{x}_2, \cdots, \mathbf{x}_n)$. In the quadratic regression (7.20) for the cholostyramine data, the response y_i is the improvement for the ith man, the covariate $\mathbf{c}_i$ is the vector $(1, z_i, z_i^2)$, and $\boldsymbol{\beta} = (\beta_0, \beta_1, \beta_2)^T$. Note: The "linear" in linear regression refers to the linear form of the expectation (9.3). There is no contradiction in the fact that the linear model (7.20) is a quadratic function of z.

The probability structure of the linear model is usually expressed as

$$y_i = \mathbf{c}_i \boldsymbol{\beta} + \epsilon_i \quad \text{for} \quad i = 1, 2, \cdots, n. \tag{9.4}$$

The *error terms* ϵ_i in (9.4) are assumed to be a random sample from an unknown *error distribution* F having expectation 0,

$$F \to (\epsilon_1, \epsilon_2, \cdots, \epsilon_n) = \boldsymbol{\epsilon} \quad [\mathrm{E}_F(\epsilon) = 0]. \tag{9.5}$$

Notice that (9.4), (9.5) imply

$$\begin{aligned} \mathrm{E}(y_i | \mathbf{c}_i) &= \mathrm{E}(\mathbf{c}_i \boldsymbol{\beta} + \epsilon_i | \mathbf{c}_i) = \mathrm{E}(\mathbf{c}_i \boldsymbol{\beta} | \mathbf{c}_i) + \mathrm{E}(\epsilon_i | \mathbf{c}_i) \\ &= \mathbf{c}_i \boldsymbol{\beta}, \end{aligned} \tag{9.6}$$

which is the linearity assumption (9.3). Here we have used the fact that the conditional expectation $\mathrm{E}(\epsilon_i | \mathbf{c}_i)$ is the same as the unconditional expectation $\mathrm{E}(\epsilon_i) = 0$, since the ϵ_i are selected independently of $\mathbf{c}_i$.

We want to estimate the regression parameter vector $\boldsymbol{\beta}$ from the observed data $(\mathbf{c}_1, y_1), (\mathbf{c}_2, y_2), \cdots, (\mathbf{c}_n, y_n)$. A trial value of $\boldsymbol{\beta}$, say $\mathbf{b}$, gives *residual squared error*

$$\mathrm{RSE}(\mathbf{b}) = \sum_{i=1}^{n} (y_i - \mathbf{c}_i \mathbf{b})^2, \tag{9.7}$$

as in equation (7.21). The *least-squares estimate* of $\boldsymbol{\beta}$ is the value $\hat{\boldsymbol{\beta}}$ of $\mathbf{b}$ that minimizes $\mathrm{RSE}(\mathbf{b})$,

$$\mathrm{RSE}(\hat{\boldsymbol{\beta}}) = \min_{\mathbf{b}}[\mathrm{RSE}(\mathbf{b})]. \tag{9.8}$$

Let $\mathbf{C}$ be the $n \times p$ matrix with ith row $\mathbf{c}_i$ (the *design matrix*), and let $\mathbf{y}$ be the vector $(y_1, y_2, \cdots, y_n)^T$. Then the least-squares estimate is the solution to the so-called *normal equations*

$$\mathbf{C}^T \mathbf{C} \hat{\boldsymbol{\beta}} = \mathbf{C}^T \mathbf{y} \tag{9.9}$$

Table 9.1. *The hormone data. Amount in milligrams of anti-inflammatory hormone remaining in 27 devices, after a certain number of hours of wear. The devices were sampled from 3 different manufacturing lots, called A, B, and C. Lot C looks like it had greater amounts of remaining hormone, but it also was worn the least number of hours. A regression analysis clarifies the situation.*

lot	hrs	amount	lot	hrs	amount	lot	hrs	amount
A	99	25.8	B	376	16.3	C	119	28.8
A	152	20.5	B	385	11.6	C	188	22.0
A	293	14.3	B	402	11.8	C	115	29.7
A	155	23.2	B	29	32.5	C	88	28.9
A	196	20.6	B	76	32.0	C	58	32.8
A	53	31.1	B	296	18.0	C	49	32.5
A	184	20.9	B	151	24.1	C	150	25.4
A	171	20.9	B	177	26.5	C	107	31.7
A	52	30.4	B	209	25.8	C	125	28.5
mean:	150.6	23.1		233.4	22.1		111.0	28.9

and is given by the formula [1]

$$\hat{\boldsymbol{\beta}} = (\mathbf{C}^T\mathbf{C})^{-1}\mathbf{C}^T\mathbf{y}. \tag{9.10}$$

9.3 Example: the hormone data

Table 9.1 shows a small data set which is a good candidate for regression analysis. A medical device for continuously delivering an anti-inflammatory hormone has been tested on $n = 27$ subjects. The response variable y_i is the amount of hormone remaining in the device after wearing,

$$y_i = \text{remaining amount of hormone in device } i, \quad i = 1, 2, \cdots, 27.$$

---

[1] Formula (9.10) assumes that $\mathbf{C}$ is of full rank p, as will be the case in all of our examples. We will not be using matrix-theoretic derivations in what follows. A reader unfamiliar with matrix theory can think of (9.10) simply as a function which inputs the responses $y_1, y_2, \cdots, y_n$ and the predictors $\mathbf{c}_1, \mathbf{c}_2, \cdots, \mathbf{c}_n$, and outputs the least-squares estimate $\hat{\beta}$. Similarly the bootstrap methods in Section 9.4 do not require detailed understanding of the matrix calculation in Section (9.3).

There are two predictor variables,

$$z_i = \text{number of hours the } i^{th} \text{ device was worn,}$$

and

$$L_i = \text{manufacturing lot of device } i.$$

The devices tested were randomly selected from three different manufacturing lots, called A, B, and C.

The left panel of Figure 9.1 is a scatterplot of the 27 points $(z_i, y_i) = (\text{hours}_i, \text{amount}_i)$, with the lot symbol L_i used as the plotting character. We see that longer hours of wear leads to smaller amounts of remaining hormone, as might be expected. We can quantify this observation by a regression analysis.

Consider the model where the expectation of y is a linear function of z,

$$\mu_i = \text{E}(y_i|z_i) = \beta_0 + \beta_1 z_i \qquad i = 1, 2, \cdots, 27. \qquad (9.11)$$

This model ignores the lot L_i: it is of form (9.3), with covariate vectors of dimension $p = 2$,

$$\mathbf{c}_i = (1, z_i). \qquad (9.12)$$

The unknown parameter vector $\boldsymbol{\beta}$ has been labeled (β_0, β_1) instead of (β_1, β_2) so that subscripts match powers of z as in (7.20). The normal equations (9.10) give least-squares estimate

$$\hat{\boldsymbol{\beta}} = (34.17, -.0574)^T. \qquad (9.13)$$

The estimated least-squares regression line

$$\hat{\mu}_i = \mathbf{c}_i \hat{\boldsymbol{\beta}} = \hat{\beta}_0 + \hat{\beta}_1 z_i \qquad (9.14)$$

is plotted in the right panel of Figure 9.1. Among all possible lines that could be drawn, this line minimizes the sum of 27 squared vertical distances from the points to the line.

How accurate is the estimated parameter vector $\hat{\boldsymbol{\beta}}$? An extremely useful formula, also dating back to Legendre and Gauss, provides the answer. Let $\mathbf{G}$ be the $p \times p$ *inner product matrix*,

$$\mathbf{G} = \mathbf{C}^T \mathbf{C}, \qquad (9.15)$$

the matrix with element $g_{hj} = \sum_{i=1}^{n} c_{ih} c_{ij}$ in row h, column j. Let σ_F^2 be the variance of the error terms in model (9.4),

$$\sigma_F^2 = \text{var}_F(\epsilon). \qquad (9.16)$$

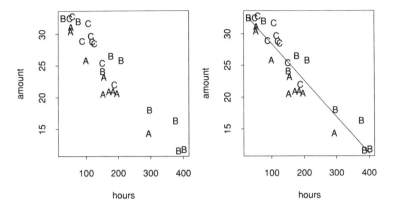

Figure 9.1. *Scatterplot of the hormone data points* (z_i, y_i) = $(hours_i, amount_i)$, *labeled by lot. It is clear that longer hours of wear result in lower amounts of remaining hormone. The right panel shows the least-squares regression of* y_i *on* z_i : $\hat{\mu}_i = \hat{\beta}_0 + \hat{\beta}_1 z_i$, *where* $\hat{\beta}$ = $(34.17, -.0574)$.

Then the standard error of the jth component of $\hat{\beta}$, the square root of its variance, is

$$\text{se}(\hat{\beta}_j) = \sigma_F \sqrt{G^{jj}} \tag{9.17}$$

when G^{jj} is the jth diagonal element of the inverse matrix $\mathbf{G}^{-1}$.

The last formula is a generalization of formula (5.4) for the standard error of a sample mean, $\text{se}_F(\bar{x}) = \sigma_F / \sqrt{n}$, see Problem 9.1. In practice, σ_F is estimated by a formula analogous to (5.11),

$$\hat{\sigma}_F = \{\sum_{i=1}^{n} (y_i - \mathbf{c}_i \hat{\beta})^2 / n\}^{1/2} = \{\text{RSE}(\hat{\beta})/n\}^{1/2} \tag{9.18}$$

or by a bias-corrected version of $\hat{\sigma}_F$,

$$\bar{\sigma}_F = \{\text{RSE}(\hat{\beta})/(n-p)\}^{1/2}. \tag{9.19}$$

The corresponding estimated standard errors for the components of $\hat{\beta}$ are

$$\widehat{\text{se}}(\hat{\beta}_j) = \hat{\sigma}_F \sqrt{G^{jj}} \quad \text{or} \quad \overline{\text{se}}(\hat{\beta}_j) = \bar{\sigma}_F \sqrt{G^{jj}}. \tag{9.20}$$

The relationship between $\widehat{\text{se}}(\hat{\beta}_j)$ and $\overline{\text{se}}(\hat{\beta}_j)$ is the same as that between formulae (5.12) and (2.2) for the mean.

Table 9.2. *Results of fitting model (9.11) to the hormone data*

	Estimate	$\widehat{\text{se}}$	$\overline{\text{se}}$
$\hat{\beta}_0$	34.17	.83	.87
$\hat{\beta}_1$	-.0574	.0043	.0045

Table 9.3. *Results of fitting model (9.21) to the hormone data.*

	Estimate	$\widehat{\text{se}}$	$\overline{\text{se}}$
$\hat{\beta}_A$	32.13	.69	.75
$\hat{\beta}_B$	36.11	.89	.97
$\hat{\beta}_C$	35.60	.60	.66
$\hat{\beta}_1$	-.0601	.0032	.0035

Most packaged linear regression programs routinely print out $\overline{\text{se}}(\hat{\beta}_j)$ along with the least-squares estimate $\hat{\beta}_j$. Applying such a program to model (9.11) for the hormone data gives the results in Table 9.2.

Looking at the right panel of Figure 9.1, most of the points for lot A lie below the fitted regression-line, while most of those for lots B and C lie above the line. This suggests a deficiency in model (9.11). If the model were accurate, we would expect about half of each lot to lie above and half below the fitted line. In the usual terminology, it looks like there is a *lot effect* in the hormone data.

It is easy to incorporate a lot effect into our linear model. We assume that the conditional expectation of y given L and z is of the form

$$\text{E}(y|L, z) = \beta_L + \beta_1 z. \qquad (9.21)$$

Here B_L equals one of three possible values, $\beta_A, \beta_B, \beta_C$, depending on which lot the device comes from. This is similar to model (9.11), except that (9.21) allows a different intercept for each lot, rather than the single intercept β_0 of (9.11). A least-squares analysis of model (9.21) gave the results in Table 9.3.

Notice that $\hat{\beta}_A$ is several standard errors less than $\hat{\beta}_B$ and $\hat{\beta}_C$, indicating that the devices in lot A contained significantly less hormone.

9.4 Application of the bootstrap

None of the calculations so far require the bootstrap. However it is useful to follow through a bootstrap analysis for the linear regression model. It will turn out that the bootstrap standard error estimates are the same as $\widehat{se}(\hat{\beta}_j)$, (9.20). Thus reassured that the bootstrap is giving reasonable answers in a case we can analyze mathematically, we can go on to apply the bootstrap to more general regression models that have no mathematical solution: where the regression function is non-linear in the parameters β, and where we use fitting methods other than least-squares.

The probability model $P \to \mathbf{x}$ for linear regression, as described by (9.4), (9.5), has two components,

$$P = (\beta, F), \tag{9.22}$$

where β is the parameter vector of regression coefficients, and F is the probability distribution of the error terms. The general bootstrap algorithm of Figure 8.3 requires us to estimate P. We already have available $\hat{\beta}$, the least-squares estimate of β. How can we estimate F? If β were known we could calculate the errors $\epsilon_i = y_i - \mathbf{c}_i\beta$ for $i = 1, 2, \cdots, n$, and estimate F by their empirical distribution. We don't know β, but we can use $\hat{\beta}$ to calculate *approximate errors*

$$\hat{\epsilon}_i = y_i - \mathbf{c}_i\hat{\beta}, \qquad \text{for} \quad i = 1, 2, \cdots, n. \tag{9.23}$$

(The $\hat{\epsilon}_i$ are also called *residuals.*) The obvious estimate of F is the empirical distribution of the $\hat{\epsilon}_i$,

$$\hat{F} : \text{ probability } 1/n \text{ on } \hat{\epsilon}_i \quad \text{for} \quad i = 1, 2, \cdots, n. \tag{9.24}$$

Usually $\hat{F}$ will have expectation 0 as required in (9.5), see Problem 9.5.

With $\hat{P} = (\hat{\beta}, \hat{F})$ in hand, we know how to calculate bootstrap data sets for the linear regression model: $\hat{P} \to \mathbf{x}^*$ must mean the same thing as $P \to \mathbf{x}$, the probability mechanism (9.4), (9.5) giving the actual data set $\mathbf{x}$. To generate $\mathbf{x}^*$, we first select a random sample of bootstrap error terms

$$\hat{F} \to (\epsilon_1^*, \epsilon_2^*, \cdots, \epsilon_n^*) = \boldsymbol{\epsilon}^*. \tag{9.25}$$

Each ϵ_i^* equals any one of the n values $\hat{\epsilon}_j$ with probability $1/n$. Then the bootstrap responses y_i^* are generated according to (9.4),

$$y_i^* = \mathbf{c}_i\hat{\beta} + \epsilon_i^* \qquad \text{for} \quad i = 1, 2, \cdots, n. \tag{9.26}$$

The reader should convince himself or herself that (9.24), (9.25), (9.26) is the same as (9.4), (9.5), except with $\hat{P} = (\hat{\beta}, \hat{F})$ replacing $P = (\beta, F)$. Notice that $\hat{\beta}$ is a fixed quantity in (9.26), having the same values for all i.

The bootstrap data set $\mathbf{x}^*$ equals $(\mathbf{x}_1^*, \mathbf{x}_2^*, \ldots, \mathbf{x}_n^*)$, where $\mathbf{x}_i^* = (\mathbf{c}_i, y_i^*)$. It may seem strange that the covariate vectors $\mathbf{c}_i$ are the same for the bootstrap data as for the actual data. This happens because we are treating the $\mathbf{c}_i$ as fixed quantities, rather than random. (The sample size n has been treated this same way in all of our examples.) This point is further discussed below.

The bootstrap least-squares estimate $\hat{\boldsymbol{\beta}}^*$ is the minimizer of the residual squared error for the bootstrap data,

$$\sum_{i=1}^{n}(y_i^* - \mathbf{c}_i\hat{\boldsymbol{\beta}}^*)^2 = \min_{\mathbf{b}} \sum_{i=1}^{n}(y_i^* - \mathbf{c}_i\mathbf{b})^2. \tag{9.27}$$

The normal equations (9.10), applied to the bootstrap data, give

$$\hat{\boldsymbol{\beta}}^* = (\mathbf{C}^T\mathbf{C})^{-1}\mathbf{C}^T\mathbf{y}^*. \tag{9.28}$$

In this case we don't need Monte Carlo simulations to figure out bootstrap standard errors for the components of $\hat{\boldsymbol{\beta}}^*$. An easy calculation gives a closed form expression for $\mathrm{se}_{\hat{F}}(\hat{\boldsymbol{\beta}}_j^*) = \widehat{\mathrm{se}}_{\infty}(\hat{\beta}_j)$, the ideal bootstrap standard error estimate:

$$
\begin{aligned}
\mathrm{var}(\hat{\boldsymbol{\beta}}^*) &= (\mathbf{C}^T\mathbf{C})^{-1}\mathbf{C}^T\mathrm{var}(\mathbf{y}^*)\mathbf{C}(\mathbf{C}^T\mathbf{C})^{-1} \\
&= \hat{\sigma}_F^2(\mathbf{C}^T\mathbf{C})^{-1},
\end{aligned} \tag{9.29}
$$

since $\mathrm{var}(\mathbf{y}^*) = \hat{\sigma}_F^2\mathbf{I}$, where $\mathbf{I}$ is the identity matrix. Therefore

$$\widehat{\mathrm{se}}_{\infty}(\hat{\beta}_j) = \hat{\sigma}_F\sqrt{G^{jj}}. \tag{9.30}$$

In other words, the bootstrap estimate of standard error for $\hat{\beta}_j$ is the same as the usual estimate [2] $\widehat{\mathrm{se}}(\hat{\beta}_j)$, (9.20).

---

[2] This implies that $\widehat{\mathrm{se}}_{\infty}(\hat{\beta}_j) = \{\frac{n-p}{n}\}^{1/2}\overline{\mathrm{se}}(\hat{\beta}_j)$, which is the same situation we encountered for the mean $\bar{x}$, cf. (5.12) and (2.2). We could adjust the bootstrap standard errors by factor $\{\frac{n}{n-p}\}^{1/2}$ to get the familiar estimates $\overline{\mathrm{se}}(\hat{\beta}_j)$, but this isn't necessarily the right thing to do in more complicated regression situations. The point gets worrisome only if p is a large fraction of n, say $p/n > .25$. In most situations the random variability in $\widehat{\mathrm{se}}_{\infty}$ is more important than the bias caused by factors like $\{\frac{n}{n-p}\}^{1/2}$.

9.5 Bootstrapping pairs vs bootstrapping residuals

The reader may have noticed an interesting fact: we now have two different ways of bootstrapping a regression model. The method discussed in Chapter 7 bootstrapped the pairs $\mathbf{x}_i = (\mathbf{c}_i, y_i)$, so that a bootstrap data set $\mathbf{x}^*$ was of the form

$$\mathbf{x}^* = \{(\mathbf{c}_{i_1}, y_{i_1}), (\mathbf{c}_{i_2}, y_{i_2}), \cdots, (\mathbf{c}_{i_n}, y_{i_n})\}, \tag{9.31}$$

for $i_1, i_2, \cdots, i_n$ a random sample of the integers 1 through n. The method discussed in this chapter, (9.24), (9.25), (9.26) can be called "bootstrapping the residuals." It produces bootstrap data sets of the form

$$\mathbf{x}^* = \{(\mathbf{c}_1, \mathbf{c}_1\hat{\boldsymbol{\beta}} + \hat{\epsilon}_{i_1}), (\mathbf{c}_2, \mathbf{c}_2\hat{\boldsymbol{\beta}} + \hat{\epsilon}_{i_2}), \cdots, (\mathbf{c}_n, \mathbf{c}_n\hat{\boldsymbol{\beta}} + \hat{\epsilon}_{i_n})\}. \tag{9.32}$$

Which bootstrap method is better? The answer depends on how far we trust the linear regression model (9.4). This model says that the error between y_i and its mean $\mu_i = \mathbf{c}_i\boldsymbol{\beta}$ doesn't depend on $\mathbf{c}_i$; it has the same distribution "F" no matter what $\mathbf{c}_i$ may be. This is a strong assumption, which can fail even if the model for the expectation $\mu_i = \mathbf{c}_i\boldsymbol{\beta}$ is correct. It *does* fail for the cholostyramine data of Figure 7.4.

Figure 9.2 shows *regression percentiles* for the cholostyramine data. For example the curve marked "75%" approximates the conditional 75th percentile of improvement y as a function of the compliance z. Near any given value of z, about 75% of the plotted points lie below the curve. Model (9.4), (9.5) predicts that these curves will be the same distance apart for all values of z. Instead the curves separate as z increases, being twice as far apart at $z = 100$ as at $z = 0$. To put it another way, the errors ϵ_i in (9.4) tend to be twice as big for $z = 100$ as for $z = 0$.

Bootstrapping pairs is less sensitive to assumptions than bootstrapping residuals. The standard error estimate obtained by bootstrapping pairs, (9.31), gives reasonable answers even if (9.4), (9.5) is completely wrong. The only assumption behind (9.31) is that the original pairs $\mathbf{x}_i = (\mathbf{c}_i, y_i)$ were randomly sampled from some distribution F, where F is a distribution on $(p + 1)$-dimensional vectors $(\mathbf{c}, y)$. Even if (9.4), (9.5) is correct, it is no disaster to bootstrap pairs as in (9.31); it can be shown that the answer given by (9.31) approaches that given by (9.32) as the number of pairs n grows large. The simple model for the hormone data (9.12) was reanalyzed bootstrapping pairs. $B = 800$ bootstrap replications

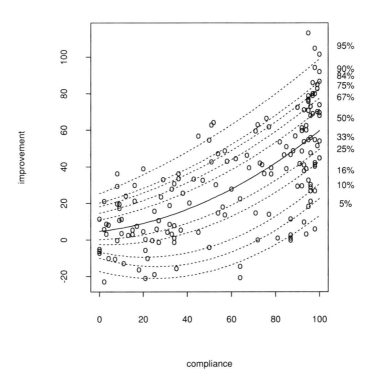

Figure 9.2. *Regression percentiles for the cholostyramine data of Figure 7.5; for example the curve labeled "75%" approximates the conditional 75th percentile of the Improvement y given the Compliance z, plotted as a function of z. The percentile curves are twice as far apart at $z = 100$ as at $z = 0$. The linear regression model (9.4), (9.5) can't be correct for this data set. (Regression percentiles calculated using asymmetric maximum likelihood, Efron, 1991.)*

gave

$$\widehat{se}_{800}(\hat{\beta}_0) = .77, \qquad \widehat{se}_{800}(\hat{\beta}_1) = .0045, \qquad (9.33)$$

not much different than Table 9.2.

The reverse argument can also be made. Model (9.4), (9.5) doesn't have to hold perfectly in order for bootstrapping residuals as in (9.32) to give reasonable results. Moreover, differences in the error distributions, as in the cholostyramine data, can be incorporated into model (9.4), (9.5), leading to a more appropriate version of bootstrapping residuals; see model (9.42). Perhaps the most important point here is that bootstrapping is not a uniquely defined concept. Figure 8.3 can be implemented in different ways for the same problem, depending on how the probability model $P \rightarrow \mathbf{x}$ is interpreted.

When we bootstrap residuals, the bootstrap data sets $\mathbf{x}^* = \{(\mathbf{c}_1, y_1^*), (\mathbf{c}_2, y_2^*), \cdots, (\mathbf{c}_n, y_n^*)\}$ have covariate vectors $\mathbf{c}_1, \mathbf{c}_2, \cdots, \mathbf{c}_n$ exactly the same as those for the actual data set $\mathbf{x}$. This seems unnatural for the hormone data, where $\mathbf{c}_i$ involves z_i, the hours worn, which is just as much a random variable as is the response variable y_i, amount remaining.

Even when covariates are generated randomly, there are reasons to do the analysis as if they are fixed. Regression coefficients have larger standard error when the covariates have smaller standard deviation. By treating the covariates as fixed constants we obtain a standard error that reflects the precision associated with the sample of covariates actually observed. However, as (9.33) shows, the difference between $\mathbf{c}_i$ fixed and $\mathbf{c}_i$ random usually doesn't affect the standard error estimate very much.

9.6 Example: the cell survival data

There are regression situations where the covariates are more naturally considered fixed rather than random. The cell survival data in Table 9.4 show such a situation. A radiologist has run an experiment involving 14 bacterial plates. The plates were exposed to various doses of radiation, and the proportion of the surviving cells measured. Greater doses lead to smaller survival proportions, as would be expected. The question mark after the response for plate 13 reflects some uncertainty in that result expressed by the investigator.

The investigator was interested in a regression analysis, with

Table 9.4. *The Cell Survival data. Fourteen cell plates were exposed to different levels of radiation. The observed response was the proportion of cells which survived the radiation exposure. The response in plate 13 was considered somewhat uncertain by the investigator.*

plate number	dose (rads/100)	survive prop.	log.surv prop.
1	1.175	0.44000	-0.821
2	1.175	0.55000	-0.598
3	2.350	0.16000	-1.833
4	2.350	0.13000	-2.040
5	4.700	0.04000	-3.219
6	4.700	0.01960	-3.219
7	4.700	0.06120	-2.794
8	7.050	0.00500	-5.298
9	7.050	0.00320	-5.745
10	9.400	0.00110	-6.812
11	9.400	0.00015	-8.805
12	9.400	0.00019	-8.568
13	14.100	0.00700?	-4.962?
14	14.100	0.00006	-9.721

predictor variable

$$\text{dose}_i = z_i \qquad i = 1, 2, \cdots, 14 \tag{9.34}$$

and response variable

$$\log(\text{survival proportion}_i) \quad = \quad y_i \qquad i = 1, 2, \cdots, 14. \tag{9.35}$$

Two different theoretical models of radiation damage were available, one of which predicted a linear regression,

$$\mu_i = \mathrm{E}(y_i|z_i) = \beta_1 z_i, \tag{9.36}$$

and the other quadratic regression,

$$\mu_i = \mathrm{E}(y_i|z_i) = \beta_1 z_i + \beta_2 z_i^2. \tag{9.37}$$

There is no intercept terms β_0 in (9.36) or (9.37) because we know that zero dose gives survival proportion 1, $y = \log(1) = 0$.

Table 9.5 shows the least-squares estimates $(\hat{\beta}_1, \hat{\beta}_2)$ and their estimated standard errors $\overline{\text{se}}(\hat{\beta}_j)$, (9.20). Two least-squares analysis are presented, one with the data from all 14 plates, the other excluding the questionable plate 13. In both analyses, the estimated quadratic regression coefficient $\hat{\beta}_2$ is positive. Is it *significantly* positive? In other words, can we reasonably conclude that $\hat{\beta}_2$ would remain positive if a great many more plates were investigated? The ratio $\hat{\beta}_2/\widehat{se}(\hat{\beta}_2)$ helps answer this question. The ratio is 2.46 for the analysis based on all 14 plates, which would usually be considered strong evidence that $\hat{\beta}_2$ is significantly greater than zero. If we believe this result, then the quadratic model (9.37) is strongly preferred to the model (9.36), which has $\beta_2 = 0$.

However removing the questionable plate 13 from the analysis reduces $\hat{\beta}_2/\widehat{se}(\hat{\beta}_j)$ to only 0.95, a non-significant result. The conclusion is not that β_2 is necessarily zero, but that it easily *could be* zero: if $\beta_2 = 0$, and if $se(\beta_2) \doteq .0091$ as on line 2 of Table 9.5, then it wouldn't be at all surprising to see a value of $\hat{\beta}_2$ as large or larger than the observed value .0086. We have no strong evidence for rejecting the linear model in favor of the quadratic model.

Statistics is the science of collecting together small pieces of information in order to get a highly informative composite result. Statisticians get nervous when they see one data point, especially a suspect one, dominating the answer to an important question. A valid criticism of least-squares regression is that one outlying point like plate 13 can have too large an effect on the fitted regression curve. This is illustrated in Figure 9.3, which plots the least-squares regression curve both with and without the data from plate 13. The powerful effect of the point "?" is evident. Even if the investigator had not questioned the validity of plate 13, we would prefer our fitted curves not to depend so much on individual data points.

9.7 Least median of squares

Least median of squares regression, abbreviated LMS, is a less sensitive fitting technique than least-squares. The only difference between least-squares and LMS is the choice of the fitting criterion. To motivate the criterion, let's divide the residual squared error (9.7) by the sample size, giving the mean squared residual

$$\frac{1}{n}\sum_{i=1}^{n}(y_i - \mathbf{c}_i\mathbf{b})^2. \tag{9.38}$$

Table 9.5. *Estimated regression coefficients and standard errors for the quadratic model (9.37) applied to the cell survival data. Least squares estimates (9.10) were obtained using all 14 plates (line 1), and also excluding plate 13 (line 2). Estimated standard errors for lines 1 and 2 are $\overline{se}(\hat{\beta}_j)$, (9.20). The estimated standard errors for the least median of squares regression (all 14 plates), line 3, were obtained from a bootstrap analysis, $B = 400$. The quadratic coefficient looks significantly nonzero in line 1, but not in lines 2 or 3. Line 4 gives the standard errors for the least median of squares estimate, based on resampling residuals from model (9.42).*

	$\hat{\beta}_1$	$(\widehat{se})$	$\hat{\beta}_2$	$(\widehat{se})$	$\hat{\beta}_2/\widehat{se}$
1. Least Squares, 14 plates	-1.05	(.159)	.0341	(.0143)	2.46
2. Least Squares, 13 plates	-0.86	(.094)	.0086	(.0091)	0.95
3. Least Median of Squares	-0.83	(.272)	.0114	(.0362)	0.32
4. (Resampling residuals)		(.141)		(.0160)	

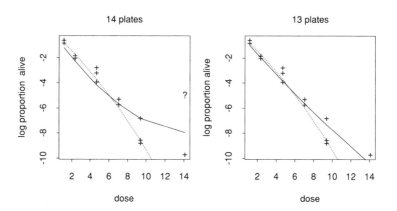

Figure 9.3. *Scatterplot of the cell survival data; solid line is the quadratic regression $\hat{\beta}_1 z + \hat{\beta}_2 z^2$ obtained by least-squares. Dashed line is quadratic regression fit by method of least median of squares (LMS). Left panel: all 14 plates; Right panel: thirteen plates, excluding the questionable result from plate 13. Plate 13, marked "?" in the left panel, has a large effect on the fitted least-squares curve. The questionable point has no effect on the LMS curve.*

Minimizing (9.38) is obviously the same as minimizing (9.7). Sample means are sensitive to influential values, but medians are not. Hence to make (9.38) less sensitive, we can replace the mean by a median, giving the *median squared residual*

$$\mathrm{MSR}(\mathbf{b}) = \mathrm{median}(y_i - \mathbf{c}_i \mathbf{b})^2. \tag{9.39}$$

The LMS estimate of $\boldsymbol{\beta}$ is the value $\hat{\boldsymbol{\beta}}$ minimizing $\mathrm{MSR}(\mathbf{b})$,

$$\mathrm{MSR}(\hat{\boldsymbol{\beta}}) = \min_{\mathbf{b}}[\mathrm{MSR}(\mathbf{b})]. \tag{9.40}$$

Notice that the difference between least-squares and LMS is not in the choice of the model, which remains (9.3), but how we measure discrepancies between the model and the observed data. $\mathrm{MSR}(\mathbf{b})$ is less sensitive than $\mathrm{RSE}(\mathbf{b})$ to outlying data points. This can be seen in Figure 9.3, where there appears to be very little difference between the quadratic LMS fit with or without point "?". In fact there is *no* difference. The estimated regression coefficients are $(\hat{\beta}_1, \hat{\beta}_2) = (-.81, .0088)$ in both cases.

It can be shown that the *breakdown* of the LMS estimator is roughly 50%. The breakdown of an estimator is the smallest proportion of the data that can have an arbitrarily large effect on its value. In other words, an estimator has breakdown α if at least $m = \alpha \cdot n$ data points must be "bad" before it breaks down. High breakdown is good, with 50% being the largest value that makes sense (if $\alpha > 50\%$, it is not clear which are the good points and which are bad). For example, the mean of a sample has breakdown $1/n$, since by changing just one data value we can force the sample mean to have any value whatsoever. The sample median has breakdown 50%, reflecting the fact that it is less sensitive to individual values. The least-squares regression estimator inherits the sensitivity of the mean, and has breakdown $1/n$, while the least median of squares estimator, like the median, has breakdown roughly 50%. The precise definition of breakdown is given in Problem 9.9.

How accurate are the LMS estimates $\hat{\beta}_1, \hat{\beta}_2$? There is no neat formula like (9.20) for LMS standard errors. (There is no neat formula for the LMS estimates themselves. They are calculated using a sampling algorithm: see Problem 9.8.) The standard errors in Table 9.5 were obtained by bootstrap methods. The standard errors in line 3 are based on resampling pairs, as in Section 7.3. A bootstrap data set was created of the form $\mathbf{x}^* = ((\mathbf{c}_1^*, y_1^*), (\mathbf{c}_2^*, y_2^*), \cdots, (\mathbf{c}_n^*, y_n^*))$, as in (9.31), where $c_i = (z_i, z_i^2)$. Having generated x^*, the bootstrap replication $\hat{\boldsymbol{\beta}}^*$ for the LMS regression vector was obtained as the minimizer of the median squared residual for the bootstrap data, that is, the minimizer over b of

$$\text{median}(y_i^* - c_i^* b)^2 \qquad (9.41)$$

$B = 400$ bootstrap replications give the estimated standard errors in line 3 of Table 9.5. Notice that $\hat{\beta}_2$ is *not* significantly greater than zero.

The covariates in the cell survival data were fixed numbers, set by the investigator: she chose the doses $1.175, 1.175, 2.35, \cdots, 14.100$ in order to have a good experiment for discriminating between the linear and quadratic radiation survival models. This makes us more interested in bootstrapping the residuals, (9.32), rather than bootstrapping pairs. Then the bootstrap data sets x^* will have the same covariate vectors $c_1, c_2, \cdots, c_{14}$ as the investigator deliberately used in the experiment.

Model (9.4), (9.5) isn't exactly right for the cell survival data. Looking at Figure 9.3, we can see that the response y_i are more dispersed for larger values of z. This is like the cholostyramine situation of Figure 9.2, except that we don't have enough points to draw good regression percentiles. As a roughly appropriate model, we will assume that the errors from the linear model increase linearly with the dose z. This amounts to replacing (9.4) with

$$y_i = c_i \boldsymbol{\beta} + z_i \epsilon_i \qquad \text{for} \quad i = 1, 2, \cdots, 14. \qquad (9.42)$$

We still assume that $(\epsilon_1, \epsilon_2, \cdots, \epsilon_n)$ is a random sample from some distribution F, (9.5). For the quadratic regression model, $c_i = (z_i, z_i^2)$.

The probability model for (9.42) is $P = (\boldsymbol{\beta}, F)$ as before; $\boldsymbol{\beta}$ was estimated by LMS, $\hat{\boldsymbol{\beta}} = (-.83, .0114)$. Then F was estimated by $\hat{F}$, the empirical distribution of the quantities $(y_i - c_i\hat{\boldsymbol{\beta}})/z_i$, $i = 1, 2, \cdots, 14$.

Line 4 of Table 9.5 reports bootstrap standard errors for the least median of squares estimates $\hat{\beta}_1$ and $\hat{\beta}_2$, obtained from $B = 200$ bootstrap replications, bootstrapping the residuals in model (9.42). The standard errors are noticeably smaller than those obtained by bootstrapping pairs. (But not small enough to make $\hat{\beta}_2$ significantly non-zero.) The standard errors in line 4 have to be regarded cautiously, since model (9.42) is only weakly suggested by the data.

The main point in presenting this model was to illustrate how boot-strapping residuals could be carried out in situations more complicated than (9.4).

9.8 Bibliographic notes

Regression is discussed in most elementary statistics texts and there are many books devoted to the topic, including Draper and Smith (1981), and Weisberg (1980). Bootstrapping of regression models is discussed at a deeper mathematical level in Freedman (1981), Shorack (1982), Bickel and Freedman (1983), Weber (1984), Wu (1986), and Shao (1988). Freedman and Peters (1984), Peters and Freedman (1984a, 1984b) examined some practical aspects. Rousseeuw (1984) introduces least median of squares estimator. Efron (1991) discusses the estimation of regression percentiles.

9.9 Problems

9.1 Show that formula (9.17) gives formula (5.4) for the standard error of the mean $\bar{x}$.

9.2 (a) Show that the least-squares estimate of β_1 in model (9.12) is

$$\hat{\beta}_1 = \sum_{i=1}^{n}(z_i - \bar{z})(y_i - \bar{y}) / \sum_{i=1}^{n}(z_i - \bar{z})^2. \qquad (9.43)$$

[For a 2×2 matrix $\mathbf{G}$, the inverse matrix is

$$\mathbf{G}^{-1} = \frac{1}{g_{22}g_{11} - g_{12}g_{21}} \begin{pmatrix} g_{22} & -g_{12} \\ g_{21} & g_{11} \end{pmatrix}.] \qquad (9.44)$$

(b) Show that $\hat{\beta}_1$ has standard error $\sigma_y / [\sum_{i=1}^{n}(z_i - \bar{z})^2]^{1/2}$.

(c) How might the allocation of doses in Table 9.4 be changed to decrease the standard error of $\hat{\beta}_1$?

9.3 Describe the matrix $\mathbf{C}$ applying to the linear model (9.21).

9.4 Often the covariate vectors $\mathbf{c}_i$ all have first component 1, as in (9.12). If this is the case, show that the empirical distribution $\hat{F}$ of the approximate errors $\hat{\epsilon}_i$, (9.24), has expectation 0.

9.5 Suppose that the empirical distribution $\hat{F}$ of the approximate errors, (9.18), has expectation 0. Derive (9.30) from (9.17).

9.6 It can be proved that expression (9.19), namely

$$\bar{\sigma}_F^2 = \sum_{i=1}^{n} \hat{\epsilon}_i^2 / (n - p), \qquad (9.45)$$

is an unbiased estimate of σ_F^2, $E(\bar{\sigma}_F^2) = \sigma_F^2$. If $\boldsymbol{\beta}$ was known, we could unbiasedly estimate σ_F^2 with the mean square average of the true errors $\epsilon_i = y_i - \mathbf{c}_i\boldsymbol{\beta}$,

$$\sum_{i=1}^{n} \epsilon_i^2 / n. \qquad (9.46)$$

(a) Comparing (9.46) with (9.45), both of which have expectations σ_F^2, shows that the approximate errors $\hat{\epsilon}_i$ tend to be collectively smaller than the true errors ϵ_i. Why do you think this is the case?

(b) The "adjusted errors"

$$\tilde{\epsilon}_i = \hat{\epsilon}_i \sqrt{\frac{n}{n - p}} \qquad (9.47)$$

are similar to the true errors in the sense that $E(\sum_{i=1}^{n} \tilde{\epsilon}_i^2) = E(\sum_{i=1}^{n} \epsilon_i^2)$. Suppose that at (9.25) we replace $\hat{F}$ by $\tilde{F}$, the empirical distribution of the adjusted errors. How would this change result (9.30)?

9.7 How would you change (9.28) to express $\hat{\boldsymbol{\beta}}$ in the situation (9.31) where we are bootstrapping pairs?

9.8 A popular method for (approximate) calculation of the least median of squares estimate is to generate a set of trial values for $\boldsymbol{\beta}$, and then choose the one that gives the smallest value of MSR(**b**) defined in (9.39). An effective way of generating the trial values is to choose p points from the data set *without replacement* and then let $\boldsymbol{\beta}$ equal the coefficients of the interpolating line or plane through the p points. Here p is the number of regressors in the model, including the intercept; there is a unique line or plane passing through p data points in p-dimensional space, as long as the points are linearly independent. Carrying out this sampling a number of times (say 100) produces a set of 100 trial values for $\hat{\boldsymbol{\beta}}$. Note that while this sampling might seem similar to the bootstrap, its purpose is quite different. It is intended to (approximately) calculate the LMS estimate itself, rather than some aspect of its distribution.

(a) Suggest why this sampling method might be an effective way of producing a set of trial values, in the sense that the minimizer among the trial values will be close to the true minimizer of MSR($\mathbf{b}$).

(b) Write a program to compute the LMS estimator for the cell survival data, fitting a linear model through the origin (recall that a quadratic model was fit in the chapter).

(c) Write a program to estimate the standard error of the LMS estimate in (b), both by resampling the data pairs and by resampling residuals. Compare your results to those in Table 9.5.

9.9 Suppose we have a data sample $\mathbf{x} = (x_1, x_2, \ldots x_n)$, and let $\mathbf{x}' = (x_1', x_2', \ldots x_n')$ be the sample obtained by replacing m data points $x_{i_1}, x_{i_2}, \ldots x_{i_m}$ by arbitrary values $y_1, y_2, \ldots y_m$. Then the breakdown of an estimator $s(\mathbf{x})$ is defined to be

$$\text{breakdown}(s(\mathbf{x})) = \frac{1}{n}\min\{m; \max_{i_1, \ldots i_m} \max|s(\mathbf{x}')| = \infty\}.$$

In other words, the breakdown is m/n, where m is the smallest number such that if we are allowed to change m data values in any way, we can force the absolute value of $s(\cdot)$ for the "perturbed" sample towards plus or minus infinity.

(a) Show that the sample mean has breakdown $1/n$, but the sample median has breakdown $(n+1)/2$ if n is odd.

(b) Consider the least-squares estimator of the slope in a simple linear regression. Show that it has breakdown $1/n$.

(c) Investigate the breakdown of the least median of squares estimator, in the simple linear regression setting, through a numerical experiment.

## CHAPTER 10

Estimates of bias

10.1 Introduction

We have concentrated on standard error as a measure of accuracy for an estimator $\hat{\theta}$. There are other useful measures of statistical accuracy (or statistical error), measuring different aspects of $\hat{\theta}$'s behavior. This chapter concerns bias, the difference between the expectation of an estimator $\hat{\theta}$ and the quantity θ being estimated. The bootstrap algorithm is easily adapted to give estimates of bias as well as of standard error. The *jackknife* estimate of bias is also introduced, though we postpone a full discussion of the jackknife until Chapter 11. One can use an estimate of bias to *bias-correct* an estimator. However this can be a dangerous practice, as discussed near the end of the chapter.

10.2 The bootstrap estimate of bias

To begin, let us assume that we are back in the nonparametric one-sample situation, as in Chapter 6. An unknown probability distribution F has given data $\mathbf{x} = (x_1, x_2, \cdots, x_n)$ by random sampling, $F \rightarrow \mathbf{x}$. We want to estimate a real-valued parameter $\theta = t(F)$. For now we will take the estimator to be any statistic $\hat{\theta} = s(\mathbf{x})$, as in Figure 6.1. Later we will be particularly interested in the plug-in estimate $\hat{\theta} = t(\hat{F})$.

The *bias* of $\hat{\theta} = s(\mathbf{x})$ as an estimate of θ is defined to be the difference between the expectation of $\hat{\theta}$ and the value of the parameter θ,

$$\text{bias}_F = \text{bias}_F(\hat{\theta}, \theta) = \text{E}_F[s(\mathbf{x})] - t(F). \tag{10.1}$$

A large bias is usually an undesirable aspect of an estimator's performance. We are resigned to the fact that $\hat{\theta}$ is a variable estimator of θ, but usually we don't want the variability to be overwhelmingly on the low side or on the high side. *Unbiased estimates*, those for which $E_F(\hat{\theta}) = \theta$, play an important role in statistical theory and practice. They promote a nice feeling of scientific objectivity in the estimation process. Plug-in estimates $\hat{\theta} = t(\hat{F})$ aren't necessarily unbiased, but they tend to have small biases compared to the magnitude of their standard errors. This is one of the good features of the plug-in principle.

We can use the bootstrap to assess the bias of any estimator $\hat{\theta} = s(\mathbf{x})$. The *bootstrap estimate of bias* is defined to be the estimate $\text{bias}_{\hat{F}}$ we obtain by substituting $\hat{F}$ for F in (10.1),

$$\text{bias}_{\hat{F}} = E_{\hat{F}}[s(\mathbf{x}^*)] - t(\hat{F}). \tag{10.2}$$

Here $t(\hat{F})$, the plug-in estimate of θ, may differ from $\hat{\theta} = s(\mathbf{x})$. In other words, $\text{bias}_{\hat{F}}$ is the plug-in estimate of bias_F, whether or not $\hat{\theta}$ is the plug-in estimate of θ. Notice that $\hat{F}$ is used twice in going from (10.1) to (10.2): it substitutes for F in $t(F)$, and it substitutes for F in $E_F[s(\mathbf{x})]$.

If $s(\mathbf{x})$ is the mean and $t(F)$ is the population mean, it is easy to show that $\text{bias}_{\hat{F}}=0$ (Problem 10.7). This makes sense because the mean is an unbiased estimate of the population mean, that is, $\text{bias}_F=0$. Typically a statistic has some bias, however, and $\text{bias}_{\hat{F}}$ provides an estimate of this bias. A simple example is the sample variance $s(\mathbf{x}) = \sum_1^n (x_i - \bar{x})^2/n$ whose bias is $(-1/n)$ times the population variance. In this case, it is easy to show that $\text{bias}_{\hat{F}} = (-1/n^2) \sum_1^n (x_i - \bar{x})^2$.

For most statistics that arise in practice, the ideal bootstrap estimate $\text{bias}_{\hat{F}}$ must be approximated by Monte Carlo simulation. We generate independent bootstrap samples $\mathbf{x}^{*1}, \mathbf{x}^{*2}, \cdots, \mathbf{x}^{*B}$ as in Figure 6.1, evaluate the bootstrap replications $\hat{\theta}^*(b) = s(\mathbf{x}^{*b})$, and approximate the bootstrap expectation $E_{\hat{F}}[s(\mathbf{x}^*)]$ by the average

$$\hat{\theta}^*(\cdot) = \sum_{b=1}^B \hat{\theta}^*(b)/B = \sum_{b=1}^B s(\mathbf{x}^{*b})/B. \tag{10.3}$$

The bootstrap estimate of bias based on the B replications $\widehat{\text{bias}}_B$, is (10.2) with $\hat{\theta}^*(\cdot)$ substituted for $E_{\hat{F}}[s(\mathbf{x}^*)]$,

$$\widehat{\text{bias}}_B = \hat{\theta}^*(\cdot) - t(\hat{F}). \tag{10.4}$$

Notice that the algorithm of Figure 6.1 applies exactly to calculation (10.4), except that at the last step we calculate $\hat{\theta}^*(\cdot) - t(\hat{F})$ rather than $\widehat{se}_B$. Of course we can calculate *both* $\widehat{se}_B$ and $\widehat{bias}_B$ from the same set of bootstrap replications.

10.3 Example: the patch data

Historically, statisticians have worried a lot about the possible biases in ratio estimators. The patch data in Table 10.1 provide a convenient example. Eight subjects wore medical patches designed to infuse a certain naturally-occurring hormone into the blood stream. Each subject had his blood levels of the hormone measured after wearing three different patches: a placebo patch, containing no hormone, an "old" patch manufactured at an older plant, and a "new" patch manufactured at a newly opened plant. The first three columns of the table show the three blood-level measurements for each subject.

The purpose of the patch experiment was to show *bioequivalence.* Patches manufactured at the old plant had already been approved for sale by the Food and Drug Administration (FDA). Patches from the new facility did not require a full new FDA investigation. They would be approved for sale if it could be shown that they were bioequivalent to those from the old facility. The FDA criterion for bioequivalence is that the expected value of the new patches match that of the old patches in the sense that

$$\frac{|\mathrm{E}(\mathrm{new}) - \mathrm{E}(\mathrm{old})|}{\mathrm{E}(\mathrm{old}) - \mathrm{E}(\mathrm{placebo})} \le .20. \tag{10.5}$$

In other words, the FDA wants the new facility to match the old facility within 20% of the amount of hormone the old drug adds to placebo blood levels.

Let θ be the parameter

$$\theta = \frac{\mathrm{E}(\mathrm{new\ patch}) - \mathrm{E}(\mathrm{old\ patch})}{\mathrm{E}(\mathrm{old\ patch}) - \mathrm{E}(\mathrm{placebo\ patch})}. \tag{10.6}$$

Chapters 12–14 consider confidence intervals for θ, an approach that leads to a full answer for the bioequivalence question "is $|\theta| \le$ .20?."[1] Here we only consider the bias and standard error of the plug-in estimate $\hat{\theta}$.

We are interested in two statistics, z_i and y_i obtained for each

[1] Chapter 25 has an extended bioequivalence analysis for this data set.

Table 10.1. *The patch data. Eight subjects wore medical patches designed to increase the blood levels of a certain natural hormone. Each subject had his blood levels of the hormone measured after wearing three different patches: a placebo patch, which had no medicine in it, an "old" patch which was from a lot manufactured at an old plant, and a "new" patch, which was from a lot manufactured at a newly opened plant. For each subject, z = oldpatch − placebo measurement, and y = newpatch − oldpatch measurement. The purpose of the experiment was to show that the new plant was producing patches equivalent to those from the old plant. Chapter 25 has an extended analysis of this data set.*

subject	placebo	oldpatch	newpatch	old-plac. z	new-old y
1	9243	17649	16449	8406	-1200
2	9671	12013	14614	2342	2601
3	11792	19979	17274	8187	-2705
4	13357	21816	23798	8459	1982
5	9055	13850	12560	4795	-1290
6	6290	9806	10157	3516	351
7	12412	17208	16570	4796	-638
8	18806	29044	26325	10238	-2719
mean:				6342	-452.3

of the eight subjects,

$$z = \text{oldpatch measurement} - \text{placebo measurement} \qquad (10.7)$$

and

$$y = \text{newpatch measurement} - \text{oldpatch measurement}. \qquad (10.8)$$

Assuming that the pairs $x_i = (z_i, y_i)$ are obtained by random sampling from an unknown bivariate distribution F, $F \rightarrow \mathbf{x} = (x_1, x_2, \cdots, x_8)$, then θ in (10.6) is the parameter

$$\theta = t(F) = \frac{\mathrm{E}_F(y)}{\mathrm{E}_F(z)}. \qquad (10.9)$$

In this case, $t(\cdot)$ is a function that inputs a probability distribution F on pairs $x = (z, y)$, and outputs the ratio of the expectations. The plug-in estimate of θ is

$$\hat{\theta} = t(\hat{F}) = \frac{\bar{y}}{\bar{z}} = \frac{\sum_{i=1}^{8} y_i/8}{\sum_{i=1}^{8} z_i/8}, \tag{10.10}$$

which we will take to be our estimator $\hat{\theta} = s(\mathbf{x})$. Notice that nothing in these definitions assumes that z and y are independent of each other. The last two columns of Table 10.1 show z_i and y_i for the eight subjects. The value of $\hat{\theta}$ is

$$\hat{\theta} = \frac{-452.3}{6342} = -.0713. \tag{10.11}$$

We see that $|\hat{\theta}|$ is considerably less than .20, so that there is some hope of satisfying the FDA's bioequivalency condition.

Figure 10.1 shows a histogram of $B = 400$ bootstrap replications of $\hat{\theta}$ obtained as in (6.1—6.2): bootstrap samples $\mathbf{x}^* = (x_1^*, x_2^*, \cdots, x_8^*) = (x_{i_1}, x_{i_2}, \cdots, x_{i_8})$ gave bootstrap replications

$$\hat{\theta}^* = \frac{\bar{y}^*}{\bar{z}^*} = \frac{\sum_{j=1}^{8} y_{i_j}/8}{\sum_{j=1}^{8} z_{i_j}/8}. \tag{10.12}$$

The 400 replications had sample standard deviation $\widehat{se}_{400} = .105$, and sample mean $\hat{\theta}^*(\cdot) = -.0670$. The bootstrap bias estimate is

$$\widehat{bias}_{400} = -.0670 - (-.0713) = .0043. \tag{10.13}$$

This is based on formula (10.4), using the fact that $\hat{\theta} = t(\hat{F})$ in this case.

The ratio of estimated bias to standard error, $\widehat{bias}_{400}/\widehat{se}_{400} = .041$ is small, indicating that in this case we don't have to worry about the bias of $\hat{\theta}$. As a rule of thumb, a bias of less than .25 standard errors can be ignored, unless we are trying to do careful confidence interval calculations. The *root mean square error* of an estimator $\hat{\theta}$ for θ, is $\sqrt{E_F[(\hat{\theta} - \theta)^2]}$, a measure of accuracy that takes into account both bias and standard error. It can be shown that the root mean square equals

$$\begin{aligned}
\sqrt{E_F[(\hat{\theta} - \theta)^2]} &= \sqrt{se_F(\hat{\theta})^2 + bias_F(\hat{\theta}, \theta)^2} \\
&= se_F(\hat{\theta}) \cdot \sqrt{1 + \left(\frac{bias_F}{se_F}\right)^2} \\
&\doteq se_F(\hat{\theta}) \cdot \left[1 + \frac{1}{2}\left(\frac{bias_F}{se_F}\right)^2\right]. \tag{10.14}
\end{aligned}$$

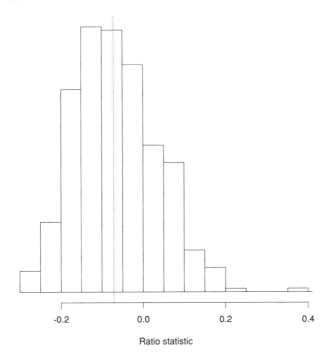

Ratio statistic

Figure 10.1. $B = 400$ *bootstrap replications of the ratio statistic (10.10),* $\hat{\theta} = \bar{y}/\bar{z}$, *for the patch data of Table 10.1. The dashed line indicates* $\hat{\theta} = -.0713$. *The 400 replications had standard deviation* $\widehat{se}_{400} = .105$ *and mean* $\hat{\theta}^*(\cdot) = -.0670$, *so the bootstrap bias estimate was* $\widehat{bias}_{400} = .0043$.

If $\text{bias}_F = 0$ then the root mean square equals its minimum value se_F. If $|\text{bias}_F/\text{se}_F| \leq .25$, then the root mean square error is no more than about 3.1% greater than se_F.

We know that $B = 400$ bootstrap replications is usually more than enough to obtain a good estimate of standard error. Is it enough to obtain a good estimate of bias? The answer in this particular case is no. Remember that $\widehat{\text{bias}}_B$, (10.4), replaces $\text{E}_{\hat{F}}(\hat{\theta}^*)$ by $\hat{\theta}^*(\cdot)$ in the definition of the ideal bootstrap bias estimate $\widehat{\text{bias}}_\infty = \text{bias}_{\hat{F}}$, (10.2). We can tell from the distribution of the bootstrap replications how well $\hat{\theta}^*(\cdot)$ estimates $\text{E}_{\hat{F}}(\hat{\theta}^*)$. An application of

(5.6) gives

$$\mathrm{Prob}_{\hat{F}}\{|\hat{\theta}^*(\cdot) - \mathrm{E}_{\hat{F}}\{\hat{\theta}^*\}| < 2\frac{\widehat{\mathrm{se}}_B}{\sqrt{B}}\}$$
$$= \mathrm{Prob}_{\hat{F}}\{|\widehat{\mathrm{bias}}_B - \widehat{\mathrm{bias}}_\infty| < 2\frac{\widehat{\mathrm{se}}_B}{\sqrt{B}}\} \doteq .95, \tag{10.15}$$

where $\widehat{\mathrm{se}}_B$ is the bootstrap standard error estimate. For the bootstrap data in Figure 10.1, with $\widehat{\mathrm{se}}_B = .105$ and $B = 400$, we obtain

$$\mathrm{Prob}_{\hat{F}}\{|\widehat{\mathrm{bias}}_{400} - \widehat{\mathrm{bias}}_\infty| < .0105\} \doteq .95 , \tag{10.16}$$

a large range of error compared to the estimated value $\widehat{\mathrm{bias}}_{400} = .0043$.

The error bound .0105 in (10.16) is small enough to show that bias isn't much of a problem here: since $\widehat{\mathrm{bias}}_{400} = .0043$, we probably have $|\widehat{\mathrm{bias}}_\infty| < .0043 + .0105 = .0148$, and so $|\widehat{\mathrm{bias}}|/\widehat{\mathrm{se}} < .0148/.106 = .14$. This is comfortably less than the rule of thumb limit .25. However we still might like to know $\widehat{\mathrm{bias}}_\infty$, or a good approximation to it, and (10.16) shows that $\widehat{\mathrm{bias}}_{400} = .0043$ can't be trusted. We could simply increase B, see Problem 10.5 , but that isn't necessary.

10.4 An improved estimate of bias

It turns out that there is a better method than (10.4) to approximate $\widehat{\mathrm{bias}}_\infty = \mathrm{bias}_{\hat{F}}$ from B bootstrap replications. The better method applies when $\hat{\theta}$ is the plug-in estimate $t(\hat{F})$ of $\theta = t(F)$. We describe the method here, and give an explanation for why it works in Chapter 23.

We need to define the notion of a resampling vector. Let P_j^* indicate the proportion of a bootstrap sample $\mathbf{x}^* = (x_1^*, x_2^*, \cdots, x_n^*)$ that equals the jth original data point,

$$P_j^* = \#\{x_i^* = x_j\}/n, \qquad j = 1, 2, \cdots, n. \tag{10.17}$$

The *resampling vector*

$$\mathbf{P}^* = (P_1^*, P_2^*, \cdots, P_n^*) \tag{10.18}$$

has non-negative components summing to one. As an example, the third bootstrap sample for the patch data was

$\mathbf{x}^* = (x_1, x_6, x_6, x_5, x_7, x_1, x_3, x_8)$, and the corresponding resampling vector is $\mathbf{P}^* = (2/8, 0, 1/8, 0, 1/8, 2/8, 1/8, 1/8)$.

A bootstrap replication $\hat{\theta}^* = s(\mathbf{x}^*)$ can be thought of as a function of the resampling vector $\mathbf{P}^*$. For example with $\hat{\theta} = \bar{y}/\bar{z}$ as in (10.10),

$$\hat{\theta}^* = \bar{y}^*/\bar{z}^* = \sum_{j=1}^{8} P_j^* y_j / \sum_{j=1}^{8} P_j^* z_j. \qquad (10.19)$$

(Notice that the original data $\mathbf{x}$ is considered fixed in this definition; the only random quantities are the P_j^*'s.) For $\hat{\theta} = t(\hat{F})$, the plug-in estimate of θ, we write

$$\hat{\theta}^* = T(\mathbf{P}^*) \qquad (10.20)$$

to indicate $\hat{\theta}^*$ as a function of the resampling vector. [2] Formula (10.19) defines $T(\cdot)$ for $\hat{\theta} = \bar{y}/\bar{z}$.

Let $\mathbf{P}^0$ indicate the vector of length n, all of whose entries are $1/n$,

$$\mathbf{P}^0 = (1/n, 1/n, \cdots, 1/n). \qquad (10.21)$$

The value of $T(\mathbf{P}^0)$ is the value of the $\hat{\theta}^*$, when each $P_j^* = 1/n$, i.e. when each original data point x_j occurs exactly once in the bootstrap sample $\mathbf{x}^*$. This means that $\mathbf{x}^* = \mathbf{x}$, except maybe for permutations of the order in which the elements $x_1, x_2, \cdots, x_n$ occur. But statistics of the form $\hat{\theta} = t(\hat{F})$ don't change when the elements of $\mathbf{x} = (x_1, x_2, \cdots, x_n)$ are reordered, because $\hat{F}$ doesn't change. In other words,

$$T(\mathbf{P}^0) = \hat{\theta} = t(\hat{F}), \qquad (10.22)$$

the observed sample value of the statistic. (This is easy to verify in (10.19).)

The B bootstrap samples $\mathbf{x}^{*1}, \mathbf{x}^{*2}, \cdots, \mathbf{x}^{*B}$ give rise to corresponding resampling vectors $\mathbf{P}^{*1}, \mathbf{P}^{*1}, \cdots, \mathbf{P}^{*B}$, each vector $\mathbf{P}^{*b}$ being of the form (10.18). Define $\bar{\mathbf{P}}^*$ to be the average of these

---

[2] We denote a plug-in statistic in two ways, $\hat{\theta} = s(\mathbf{x}) = t(\hat{F})$. Similarly, bootstrap replications are denoted $\hat{\theta}^* = s(\mathbf{x}^*) = T(\mathbf{P}^*)$. The three functions $s(\cdot), t(\cdot)$, and $T(\cdot)$ represent the same statistic, but considered as a function on three different spaces.

vectors

$$\bar{\mathbf{P}}^* = \sum_{b=1}^{B} \mathbf{P}^{*b}/B. \tag{10.23}$$

According to (10.22) we can write the bootstrap bias estimate (10.4) as

$$\widehat{\text{bias}}_B = \hat{\theta}^*(\cdot) - T(\mathbf{P}^0). \tag{10.24}$$

The *better bootstrap bias estimate*, which we will denote by $\overline{\text{bias}}_B$, is

$$\overline{\text{bias}}_B = \hat{\theta}^*(\cdot) - T(\bar{\mathbf{P}}^*). \tag{10.25}$$

The 400 resampling vectors for Figure 10.1 averaged to

$$\bar{\mathbf{P}}^* = (.1178, .1187, .1313, .1259, .1219, .1275, .1306, .1213).$$

This gives

$$T(\bar{\mathbf{P}}^*) = \sum_{j=1}^{8} \bar{P}_j^* y_j / \sum_{j=1}^{8} \bar{P}_j^* z_j = -.0750 \tag{10.26}$$

and

$$\overline{\text{bias}}_{400} = -.0670 - (-.0750) = .0080, \tag{10.27}$$

compared to $\widehat{\text{bias}}_{400} = .0043$.

Both $\widehat{\text{bias}}_B$ and $\overline{\text{bias}}_B$ converge to $\widehat{\text{bias}}_\infty = \text{bias}_{\hat{F}}$, the ideal bootstrap estimate of bias, as B goes to infinity. The convergence is much faster for $\overline{\text{bias}}_B$, which is why we have called it "better." The faster convergence is evident in Figure 10.2, which traces $\widehat{\text{bias}}_B$ and $\overline{\text{bias}}_B$ for B equaling $25, 50, 100, 200, 400, 800, 1670, 3200$. The limiting value $\widehat{\text{bias}}_\infty$ has been approximated by $\widehat{\text{bias}}_{100,000} = .0079$, shown as the dashed horizontal line. $\overline{\text{bias}}_B$ approaches the dashed line smoothly and quickly, while $\widehat{\text{bias}}_B$ is still quite variable even for $B = 3200$.

Chapter 23 discusses improved bootstrap computational methods. It will be shown there that $\overline{\text{bias}}_B$ amounts to using $\widehat{\text{bias}}_{CB}$ where C is a large constant, often 50 or greater. Problem 10.7 suggests one reason for $\overline{\text{bias}}_B$'s superiority.

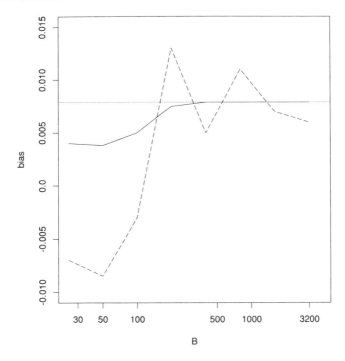

Figure 10.2. *The bootstrap bias estimate $\widehat{bias}_B$ broken line, and the better bootstrap bias estimate $\overline{bias}_B$, solid line, for $B = 25, 50, 100, \cdots, 3200;$ log scale for B; dotted line is $\widehat{bias}_{100,000} = .0079$. We see that $\overline{bias}_B$ converges much faster than $\widehat{bias}_B$ to the limiting ideal bootstrap estimate $\widehat{bias}_\infty = bias_{\hat{F}}$.*

10.5 The jackknife estimate of bias

The *jackknife* was the original computer-based method for estimating biases and standard errors. The jackknife estimate of bias, which is discussed briefly here and more completely in Chapter 11, was proposed by Maurice Quenouille in the mid 1950's. Given a data set $\mathbf{x} = (x_1, x_2, \cdots, x_n)$, the ith *jackknife sample* $\mathbf{x}_{(i)}$, is defined to be $\mathbf{x}$ with the ith data point removed,

$$\mathbf{x}_{(i)} = (x_1, x_2, \cdots, x_{i-1}, x_{i+1}, \cdots, x_n), \qquad (10.28)$$

Table 10.2. *Jackknife values for the patch data*

$\hat{\theta}_{(1)}$	$\hat{\theta}_{(2)}$	$\hat{\theta}_{(3)}$	$\hat{\theta}_{(4)}$	$\hat{\theta}_{(5)}$	$\hat{\theta}_{(6)}$	$\hat{\theta}_{(7)}$	$\hat{\theta}_{(8)}$
-.0571	-.1285	-.0215	-.1325	-.0507	-.0840	-.0649	-.0222

for $i = 1, 2, \cdots, n$. The ith *jackknife replication* $\hat{\theta}_{(i)}$ of the statistic $\hat{\theta} = s(\mathbf{x})$ is $s(\cdot)$ evaluated for $\mathbf{x}_{(i)}$, say

$$\hat{\theta}_{(i)} = s(\mathbf{x}_{(i)}) \qquad \text{for} \quad i = 1, 2, \cdots, n. \qquad (10.29)$$

For plug-in statistics $\hat{\theta} = t(\hat{F})$, $\hat{\theta}_{(i)}$ equals $t(\hat{F}_{(i)})$ where $\hat{F}_{(i)}$ is the empirical distribution of the $n - 1$ points in $\mathbf{x}_{(i)}$.

The *jackknife estimate of bias* is defined by

$$\widehat{\text{bias}}_{\text{jack}} = (n - 1)(\hat{\theta}_{(\cdot)} - \hat{\theta}) \qquad (10.30)$$

where

$$\hat{\theta}_{(\cdot)} = \sum_{i=1}^{n} \hat{\theta}_{(i)}/n. \qquad (10.31)$$

This formula applies only to plug-in statistics $\hat{\theta} = t(\hat{F})$. The formula breaks down if $t(\hat{F})$ is an unsmooth statistic like the median, but for smooth statistics like $\hat{\theta} = \bar{y}/\bar{z}$ (those for which the function $T(\mathbf{P}^*)$ in (10.20) is twice differentiable) it gives a bias estimate with only n recomputations of the function $t(\cdot)$. This compares with B recomputations for the bootstrap estimates where B needs to be at least 200 even for $\overline{\text{bias}}_B$.

For the patch data ratio statistic $\hat{\theta} = \bar{z}/\bar{y} = -.0713$, (10.10), the jackknife replications are shown in Table 10.2. These give $\hat{\theta}_{(\cdot)} = -.0702$, and

$$\widehat{\text{bias}}_{\text{jack}} = 7\{-.0702 - (-.0713)\} = .0080. \qquad (10.32)$$

It is no accident that $\widehat{\text{bias}}_{\text{jack}}$ agrees so closely with the ideal bootstrap estimate $\widehat{\text{bias}}_\infty = \text{bias}_{\hat{F}}$. Chapter 20 shows that $\widehat{\text{bias}}_{\text{jack}}$ is a quadratic Taylor series approximation to the plug-in estimate $\text{bias}_{\hat{F}}$.

The important point to remember is this: all three bias estimates, $\widehat{\text{bias}}_B, \overline{\text{bias}}_B$, and $\widehat{\text{bias}}_{\text{jack}}$, are trying to approximate the same ideal estimate, $\text{bias}_{\hat{F}}$. Chapter 20 discusses the infinitesimal jackknife, still another way to approximate $\text{bias}_{\hat{F}}$. We will also see approximations other than $\widehat{\text{se}}_B$ for the ideal standard error estimate $\text{se}_{\hat{F}}$ (though here it is harder to improve upon the straightforward Monte Carlo approximation $\widehat{\text{se}}_B$). In all of the numerical approximation methods, there is only one estimation principle at work, plugging in $\hat{F}$ for F in whatever accuracy measure we want to estimate. Executing this principle in a numerically efficient way is an important topic, but modern computers are so powerful that even inefficient ways are usually good enough to give useful answers.

The ideal estimate $\widehat{\text{bias}}_{\hat{F}}$ is not perfect. By letting $B \to \infty$, the variability in $\widehat{\text{bias}}_B$ due to Monte Carlo sampling is eliminated. There remains, however, the variability in $\widehat{\text{bias}}_\infty = \text{bias}_{\hat{F}}$ due to the randomness of $\hat{F}$ as an estimate of F. In other words, we still have the usual errors connected with estimating any parameter from a sample.

We could use the bootstrap to assess the variability in the ideal bootstrap estimate $\text{bias}_{\hat{F}}$ as in Figure 6.1, except for the practical difficulty of computing the statistic $s(\mathbf{x}) = \text{bias}_{\hat{F}}$. Instead, let us consider the simpler statistic $s(\mathbf{x}) = \widehat{\text{bias}}_{\text{jack}}$, which for $\hat{\theta} = \bar{y}/\bar{z}$ is usually close to $\text{bias}_{\hat{F}}$. The statistic $s(\mathbf{x}) = \widehat{\text{bias}}_{\text{jack}}$ is a complicated function of $\mathbf{x}$, requiring first the calculation of $\hat{\theta}$, then the $\hat{\theta}_{(i)}$, and finally (10.30), but we can still use the bootstrap to estimate the standard error of $s(\mathbf{x})$.

$B = 200$ bootstrap samples of size $n = 8$ were generated from the patch data, and for each sample the jackknife estimate of bias for the ratio statistic was calculated, say $\widehat{\text{bias}}_{\text{jack}}^{*}$. The left panel of Figure 10.3 is a histogram of the 200 $\widehat{\text{bias}}_{\text{jack}}^{*}$ values.

It is clear that the statistic $s(\mathbf{x}) = \widehat{\text{bias}}_{\text{jack}}$ is highly variable. The 200 replications $s(\mathbf{x}^*)$ had standard deviation .0081, and mean .0084, giving an estimated coefficient of variation

$$\widehat{\text{cv}}(\widehat{\text{bias}}_{\text{jack}}) = .0081/.0084 = .96. \qquad (10.33)$$

Ten percent of the $\widehat{\text{bias}}_{\text{jack}}^{*}$ values were less than zero, and 16% greater than $2 \cdot \widehat{\text{bias}}_{\text{jack}} = .0160$.

There is nothing inherently wrong with $\widehat{\text{bias}}_{\text{jack}}$, or with $\text{bias}_{\hat{F}}$, here. The trouble is that $n = 8$ data points aren't enough to accurately determine the bias of the ratio statistic in this situation.

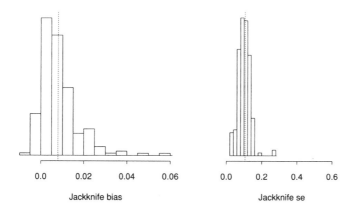

Figure 10.3. *Left panel: 200 bootstrap replications of the jackknife bias estimate (10.30) for $\hat{\theta} = \bar{y}/\bar{z}$, patch data; dashed line indicates actual estimate* $\widehat{bias}_{jack} = .0080$; *estimated coefficient of variation for* $\widehat{bias}_{jack}$ *equals .96;* $\widehat{bias}_{jack}$ *has low accuracy. Right panel: the corresponding 200 bootstrap replications of the jackknife standard error estimate for $\hat{\theta}$, (10.34); dashed line indicates actual estimate* $\widehat{se}_{jack} = .106$; *scale has been chosen so that 0 and dashed lines match left panel; estimated coefficient of variation is .33;* $\widehat{se}_{jack}$ *is about 3 times more accurate than* $\widehat{bias}_{jack}$.

Figure 10.3 makes that clear. The bias calculations weren't a complete waste of time. We are reasonably certain that the true bias of $\hat{\theta} = \bar{y}/\bar{z}$, whatever it may be, lies somewhere between -.005 and .025. The bootstrap standard error of $\hat{\theta}$ was .105, so the ratio of absolute bias to standard error is probably less than .25. Calculation (10.14) suggests that bias is not much of a worry in this case.

This calculation suggests another worry. Maybe the bootstrap estimate of standard error $\widehat{se}_{200} = .105$ is undependable too. In theory we could bootstrap $\widehat{se}_{200}$ to find out, but this is computationally difficult. However, there is a *jackknife estimate of standard error*, due to John Tukey in the late 1950's, which requires less computation than $\widehat{se}_{200}$:

$$\widehat{se}_{jack} = [\frac{n-1}{n} \sum_{i=1}^{n} (\hat{\theta}_{(i)} - \hat{\theta}_{(\cdot)})^2]^{1/2}. \qquad (10.34)$$

This formula, which applies to smoothly defined statistics like $\hat{\theta} =$

$\bar{y}/\bar{z}$, is discussed in Chapter 11. It turns out to be an alternative to $\widehat{\text{se}}_B$ for numerically approximating the ideal bootstrap estimate $\widehat{\text{se}}_\infty = \text{se}_{\hat{F}}(\hat{\theta}^*)$. For the patch data ratio statistics (10.2) gives

$$\widehat{\text{se}}_{\text{jack}} = .106, \qquad (10.35)$$

nearly the same as $\widehat{\text{se}}_{200}$. We will see that $\widehat{\text{se}}_{\text{jack}}$ is not always a good approximation to $\widehat{\text{se}}_\infty$, but for $\hat{\theta} = \bar{y}/\bar{z}$ it is quite satisfactory.

The same 200 bootstrap samples used to provide the replications of $\widehat{\text{bias}}_{\text{jack}}$ in Figure 10.3 also gave bootstrap replications of $\widehat{\text{se}}_{\text{jack}}$. The histogram of the 200 bootstrap values of $\widehat{\text{se}}_{\text{jack}}$ shown in the right panel of Figure 10.3 indicates substantial variability, but not nearly as much as for $\widehat{\text{bias}}_{\text{jack}}$. The histogram has mean .099 and standard deviation .033, giving estimated coefficient of variation

$$\widehat{\text{cv}}(\widehat{\text{se}}_{\text{jack}}) = .33, \qquad (10.36)$$

only a third of $\widehat{\text{cv}}(\widehat{\text{bias}}_{\text{jack}})$. In fact standard error is usually easier to estimate than bias, as well as being a more important determinant of the probabilistic performance of an estimator $\hat{\theta}$.

We have discussed estimating $\text{bias}_F(\hat{\theta}, \theta)$, equation (10.1). The bootstrap bias estimation procedure, which amounts to plugging in $\hat{F}$ for F in bias_F, can be generalized: 1) we can consider general probability mechanisms $P \to \mathbf{x}$, as in Figure 8.3. (Notice that here "P" means something different than the resampling vector $\mathbf{P}^*$, (10.18).) 2) We can consider general measures of bias, $\text{Bias}_P(\hat{\theta}, \theta)$, for example the median bias

$$\text{Bias}_P(\hat{\theta}, \theta) = \text{median}_P(\hat{\theta}(\mathbf{x})) - \theta(P). \qquad (10.37)$$

Figure 10.4 shows a schematic. The ideal bootstrap estimate of $\text{Bias}_P(\hat{\theta}, \theta)$ is the plug-in estimate

$$\text{Bias}_{\hat{P}}(\hat{\theta}^*, \theta(\hat{P})). \qquad (10.38)$$

Here $\hat{P} \to \mathbf{x}^*$, the bootstrap data; $\hat{\theta}^* = s(\mathbf{x}^*)$, the bootstrap replication of $\hat{\theta} = s(\mathbf{x})$; and $\theta(\hat{P})$ is the value of the parameter of interest $\theta = t(P)$ when $P = \hat{P}$, the estimated probability mechanism. (We cannot write $\theta(\hat{P}) = \hat{\theta}$ since $t(\cdot)$ might be a different function than $s(\cdot)$, see Problem 10.10.) For the median bias (10.37),

$$\text{Bias}_{\hat{P}}(\hat{\theta}^*, \theta(\hat{P})) = \text{median}_{\hat{P}}(\hat{\theta}(\mathbf{x}^*)) - \theta(\hat{P}). \qquad (10.39)$$

Usually $\text{Bias}_{\hat{P}}$ would have to be approximated by Monte Carlo methods. Improved methods like $\overline{\text{bias}}_B$ and $\widehat{\text{bias}}_{\text{jack}}$ are not usually available for general bias measures like (10.37).

10.6 Bias correction

Why would we want to estimate the bias of $\hat{\theta}$? The usual reason is to correct $\hat{\theta}$ so that it becomes less biased. If $\widehat{\text{bias}}$ is an estimate of $\text{bias}_F(\hat{\theta}, \theta)$, then the obvious *bias-corrected estimator* is

$$\bar{\theta} = \hat{\theta} - \widehat{\text{bias}}. \tag{10.40}$$

Taking $\widehat{\text{bias}}$ equal to $\widehat{\text{bias}}_B = \hat{\theta}^*(\cdot) - \hat{\theta}$ gives

$$\bar{\theta} = 2\hat{\theta} - \hat{\theta}^*(\cdot). \tag{10.41}$$

(There is a tendency, a *wrong* tendency, to think of $\hat{\theta}^*(\cdot)$ itself as the bias-corrected estimate. Notice that (10.41) says that if $\hat{\theta}^*(\cdot)$ is *greater* than $\hat{\theta}$, then the bias corrected estimate $\bar{\theta}$ should be *less* than $\hat{\theta}$.) Setting $\widehat{\text{bias}} = .0080$ for the patch data ratio statistic, equal to both $\overline{\text{bias}}_{400}$ and $\widehat{\text{bias}}_{\text{jack}}$, the bias-corrected estimate of the ratio θ is

$$\bar{\theta} = -.0713 - .0080 = -.0793. \tag{10.42}$$

Bias correction can be dangerous in practice. Even if $\bar{\theta}$ is less biased than $\hat{\theta}$, it may have substantially greater standard error. Once again, this can be checked with the bootstrap. For the patch data ratio statistic, 200 bootstrap replications of $\bar{\theta} = \hat{\theta} - \widehat{\text{bias}}_{\text{jack}}$ were compared with the corresponding replications of $\hat{\theta}$. The bootstrap standard error estimates of $\bar{\theta}$ and $\hat{\theta}$ were nearly identical, so in this case bias correction was not harmful.

To summarize, bias estimation is usually interesting and worthwhile, but the exact use of a bias estimate is often problematic. Biases are harder to estimate than standard errors, as shown in Figure 10.3. The straightforward bias correction (10.40) can be dangerous to use in practice, due to high variability in $\widehat{\text{bias}}$. Correcting the bias may cause a larger increase in the standard error, which in turn results in a larger root mean squared error (equation 10.14). If $\widehat{\text{bias}}$ is small compared to the estimated standard error $\widehat{\text{se}}$, then it is safer to use $\hat{\theta}$ than $\bar{\theta}$. If $\widehat{\text{bias}}$ is large compared to $\widehat{\text{se}}$, then it may be an indication that the statistic $\hat{\theta} = s(\mathbf{x})$ is not an appropriate estimate of the parameter θ.

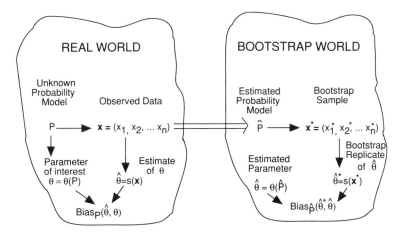

Figure 10.4. *Diagram of bootstrap bias estimation in a general frame-work, an extension of Figure 8.3. $B_{\hat{P}}(\hat{\theta}^*, \theta(\hat{P}))$ is a general bias measure. Usually $Bias_{\hat{P}}(\hat{\theta}^*, \theta(\hat{P}))$ must be approximated by Monte Carlo methods.*

Prediction error estimation is one important problem in which bias correction is useful. The bias of the obvious estimate is large relative to its standard error, and it can be effectively reduced by the addition of a correction term. Details are given in Chapter 17.

10.7 Bibliographic notes

The bootstrap estimate of bias is proposed in Efron (1979a). The improved estimate is discussed in Efron (1990). References for the jackknife are given in the bibliographic notes at the end of Chapter 11.

10.8 Problems

10.1 Suppose $F \to \mathbf{x} = (x_1, x_2, \cdots, x_8)$ where $x_i = (z_i, y_i)$ as for the patch data, but we know that z and y are independent random variables. Describe a method of bootstrapping $\hat{\theta} = \bar{y}/\bar{z}$ different than that used in Figure 10.1.

10.2 We might define the data points x_i for the patch data of Table 10.1 as

$$x_i = (p_i, o_i, n_i) \qquad i = 1, 2, \cdots, 8, \qquad (10.43)$$

where p_i = placebo, o_i = oldpatch, n_i = newpatch measurement. How would θ and $\hat{\theta}$, (10.9), and (10.10), now be defined?

10.3 Verify (10.14).

10.4 State exactly how (5.6) applies to result (10.15).

10.5 How big should B be taken in order that (10.16) becomes

$$\text{Prob}_{\hat{F}}\{|\widehat{\text{bias}}_B - \widehat{\text{bias}}_\infty| < .001\} = .95? \qquad (10.44)$$

10.6 In Figure 10.2, how accurate is $\widehat{\text{bias}}_{100,000} = .0079$ as an estimate of $\widehat{\text{bias}}_\infty$?

10.7 We know that $\hat{\theta} = \bar{x}$ is an unbiased estimate of the expectation parameter $\theta = E_F(x)$. Hence bias_F, the true bias, is zero.

(a) Show that for $\hat{\theta} = \bar{x}$, $\text{bias}_{\hat{F}} = 0$, $\overline{\text{bias}}_B = 0$ but $\widehat{\text{bias}}_B$ does not necessarily equal zero.

(b) Show that $\widehat{\text{bias}}_{\text{jack}} = 0$.

10.8 A random sample $\mathbf{x} = (x_1, x_2, \cdots, x_n)$ is observed from a probability distribution of real numbers F, and it is desired to estimate the variance $\theta = \text{var}_F(x)$. The plug-in estimate is $\hat{\theta} = \sum_{i=1}^{n}(x_i - \bar{x})^2/n$. Show that the jackknife bias-corrected estimate (10.40) is the usual unbiased estimate of variance,

$$\bar{\theta} = \hat{\theta} - \widehat{\text{bias}}_{\text{jack}} = \sum_{i=1}^{n}(x_i - \bar{x})^2/(n-1). \qquad (10.45)$$

10.9 Give a careful description of how the bootstrap replications $\widehat{\text{bias}}_{\text{jack}}^*$ and $\widehat{\text{se}}_{\text{jack}}^*$ in Figure 10.3 were generated.

10.10 Suppose we use the sample median $\text{med}(\mathbf{x})$ to estimate the population expectation $\theta = E_F(x)$. Describe $\widehat{\text{bias}}_B$.

## CHAPTER 11

The jackknife

11.1 Introduction

In Chapter 10 we mention the jackknife, a technique for estimating the bias and standard error of an estimate. The jackknife predates the bootstrap and bears close similarities to it. In this chapter we explore the jackknife method in detail. Some of the ideas presented here are pursued further in Chapters 20 and 21.

11.2 Definition of the jackknife

Suppose we have a sample $\mathbf{x} = (x_1, x_2, \ldots x_n)$ and an estimator $\hat{\theta} = s(\mathbf{x})$. We wish to estimate the bias and standard error of $\hat{\theta}$. The jackknife focuses on the samples that *leave out one observation at a time*:

$$\mathbf{x}_{(i)} = (x_1, x_2, \ldots x_{i-1}, x_{i+1}, \ldots x_n) \tag{11.1}$$

for $i = 1, 2, \ldots n$, called *jackknife samples*. The ith jackknife sample consists of the data set with the ith observation removed. Let

$$\hat{\theta}_{(i)} = s(\mathbf{x}_{(i)}) \tag{11.2}$$

be the ith *jackknife replication* of $\hat{\theta}$.

The jackknife estimate of bias is defined by

$$\widehat{\mathrm{bias}}_{\mathrm{jack}} = (n-1)(\hat{\theta}_{(\cdot)} - \hat{\theta}) \tag{11.3}$$

where

$$\hat{\theta}_{(\cdot)} = \sum_{i=1}^{n} \hat{\theta}_{(i)}/n. \tag{11.4}$$

The jackknife estimate of standard error defined by

$$\widehat{\mathrm{se}}_{\mathrm{jack}} = [\frac{n-1}{n} \sum (\hat{\theta}_{(i)} - \hat{\theta}_{(\cdot)})^2]^{1/2}. \tag{11.5}$$

Where do these formulae come from? Let's start with $\widehat{se}_{jack}$. Rather than looking at all (or some) of the data sets that can be obtained by sampling with replacement from $x_1, x_2, \ldots x_n$, the jackknife looks at the n fixed samples $\mathbf{x}_{(1)}, \ldots \mathbf{x}_{(n)}$ obtained by deleting one observation at a time. Like the bootstrap estimate of standard error, the formula for $\widehat{se}_{jack}$ looks like the sample standard deviation of these n values, except that the factor in front is $(n-1)/n$ instead of $1/(n-1)$ or $1/n$. Of course $(n-1)/n$ is much larger than $1/(n-1)$ or $1/n$. Intuitively, this "inflation factor" is needed because the jackknife deviations

$$(\hat{\theta}_{(i)} - \hat{\theta}_{(\cdot)})^2 \tag{11.6}$$

tend to be smaller than the bootstrap deviations

$$[\hat{\theta}^*(b) - \hat{\theta}^*(\cdot)]^2, \tag{11.7}$$

since the typical jackknife sample is more similar to the original data $\mathbf{x}$ than is the typical bootstrap sample.

The exact form of the factor $(n-1)/n$ is derived by considering the special case $\hat{\theta} = \bar{x}$. Then it is easy to show that

$$\widehat{se}_{jack} = \left\{ \sum_1^n (x_i - \bar{x})^2 / \{(n-1)n\} \right\}^{1/2}, \tag{11.8}$$

(Problem 11.1). That is, the factor $(n-1)/n$ is exactly what is needed to make $\widehat{se}_{jack}$ equal to the unbiased estimate of the standard error of the mean. A factor of $[(n-1)/n]^2$ would yield the plug-in estimate

$$\left\{ \sum_1^n (x_i - \bar{x})^2 / n^2 \right\}^{1/2}, \tag{11.9}$$

but this is not materially different from the unbiased estimate unless n is small. It is a somewhat arbitrary convention that $\widehat{se}_{jack}$ uses the factor $(n-1)/n$.

Similarly, the jackknife estimate of bias (11.3) is a multiple of the average of the jackknife deviations

$$\hat{\theta}_{(i)} - \hat{\theta}, \quad i = 1, 2, \ldots n. \tag{11.10}$$

The quantities (11.10) are sometimes called the *jackknife influence values*. Notice the multiplier $(n-1)$ in (11.3). This is an inflation factor similar to the one that appears in the jackknife estimate of standard error. To derive it, we cannot appeal to the special case

$\hat{\theta} = \bar{x}$, because $\bar{x}$ is unbiased and $\hat{\theta}_{(\cdot)} - \hat{\theta}$ is zero as it should be (Problem 11.7). Since this case does not tell us what the leading factor should be, we instead consider as our test case the sample variance

$$\hat{\theta} = \sum_{1}^{n}(x_i - \bar{x})^2/n. \qquad (11.11)$$

This has bias $-1/n$ times the population variance, and the factor $(n-1)$ in front of $(\hat{\theta}_{(\cdot)} - \hat{\theta})$ makes $\widehat{\text{bias}}_{\text{jack}}$ equal to $-1/n$ times $\sum(x_i - \bar{x})^2/(n-1)$, the unbiased estimate of the population variance (Problem 11.8).

11.3 Example: test score data

Let's apply the jackknife estimate of standard error to the data set on test scores for 88 students given in Table 7.1. Recall that the statistic of interest is the ratio of the largest eigenvalue of the covariance matrix over the sum of the eigenvalues as given in (7.8)

$$\hat{\theta} = \hat{\lambda}_1 / \sum_{1}^{5}\hat{\lambda}_i. \qquad (11.12)$$

To apply the jackknife, we delete each case (row) in Table 7.1 one at a time, and compute $\hat{\theta}$ for each data set of size 87. The top panel of Figure 11.1 shows a histogram of the 88 jackknife values $\hat{\theta}_{(i)}$.

We also computed 88 bootstrap values of $\hat{\theta}$. Notice how the spread of the jackknife histogram is much less than the spread of the bootstrap histogram shown in the bottom panel (we have forced the same horizontal scale to be used in all of the histograms). This exemplifies the fact that the jackknife data sets are more similar on the average to the original data set than are the bootstrap data sets. The middle panel shows a histogram of the "inflated" jackknife values

$$\sqrt{87}(\hat{\theta}_{(i)} - \hat{\theta}_{(\cdot)}) \qquad (11.13)$$

recentered at the jackknife mean $\hat{\theta}_{(\cdot)}$. With this inflation factor, the jackknife histogram looks similar to the bootstrap histogram shown in the bottom panel. The quantity $\widehat{\text{se}}_{\text{jack}}$ works out to be .049, which is just slightly larger than the value .047 for the bootstrap estimate obtained in Chapter 7.

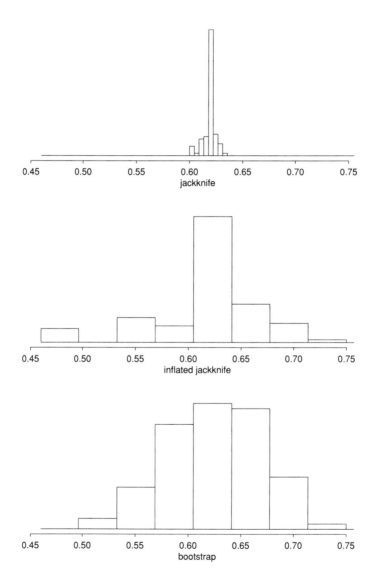

Figure 11.1. *Histogram of the 88 jackknife values for the score data of Table 7.1 (top panel); jackknife values inflated by a factor of $\sqrt{87}$ from their mean (middle panel); 88 bootstrap values for the same problem (bottom panel).*

11.4 Pseudo-values

Another way to think about the jackknife is in terms of the *pseudo-values*

$$\tilde{\theta}_i = n\hat{\theta} - (n-1)\hat{\theta}_{(i)}. \tag{11.14}$$

Notice that in the special case $\hat{\theta} = \bar{x}$, we have $\tilde{\theta}_i = x_i$, the ith data value. Furthermore, for any $\hat{\theta}$, the formula for $\widehat{\text{se}}_{\text{jack}}$ can be expressed as

$$\widehat{\text{se}}_{\text{jack}} = \left\{ \sum_1^n (\tilde{\theta}_i - \tilde{\theta})^2 / \{(n-1)n\} \right\}^{1/2}, \tag{11.15}$$

where $\tilde{\theta} = \sum \tilde{\theta}_i / n$. This looks like an estimate of the standard error of the mean for the "data" $\tilde{\theta}_i$, $i = 1, 2, \ldots n$. The idea behind (11.14) is that the pseudo-values are supposed to act as if they were n independent data values.

What happens if we try to carry this idea further and use the pseudo-values to construct a confidence interval? One reasonable approach would be to form an interval

$$\tilde{\theta} \pm t_{n-1}^{(1-\alpha)} \widehat{\text{se}}_{\text{jack}}, \tag{11.16}$$

where $t_{n-1}^{(1-\alpha)}$ is the $(1-\alpha)$th percentile to the t distribution on $n-1$ degrees of freedom. It turns out that this interval does not work very well: in particular, it is not significantly better than cruder intervals based on normal theory. More refined approaches are needed for confidence interval construction, as described in Chapters 12–14. Although pseudo-values are intriguing, it is not clear whether they are a useful way of thinking about the jackknife. We won't pursue them further here.

11.5 Relationship between the jackknife and bootstrap

Which is better, the bootstrap or jackknife? Since it requires computation of $\hat{\theta}$ only for the n jackknife data sets, the jackknife will be easier to compute if n is less than say the 100 or 200 replicates used by the bootstrap for standard error estimation. However by looking only at the n jackknife samples, the jackknife uses only limited information about the statistic $\hat{\theta}$, and thus one might guess that the jackknife is less efficient than the bootstrap. In fact it turns out that the jackknife can be viewed as an approximation to the bootstrap. This is explained in Problems 11.4 and 11.5, and in Chapter 20. Here is the essence of the idea. Consider a *linear statistic*, that is, a statistic that can be written in the form

$$\hat{\theta} = s(\mathbf{x}) = \mu + \frac{1}{n} \sum_{1}^{n} \alpha(x_i), \tag{11.17}$$

where μ is a constant and $\alpha(\cdot)$ is a function. The mean is the simple example of a linear statistic for which $\mu = 0$ and $\alpha(x_i) = x_i$. Now for such a statistic, it turns out that the jackknife and bootstrap estimate of standard errors agree, except for a minor definitional factor $\{(n-1)/n\}^{1/2}$ used by the jackknife. This is exactly what we found for $\hat{\theta} = \bar{x}$: the jackknife gives the standard error estimate $\left\{ \sum_{1}^{n}(x_i - \bar{x})^2 / \{(n-1)n\} \right\}^{1/2}$ while the bootstrap gives this value multiplied by $\{(n-1)/n\}^{1/2}$. It is not surprising that for linear statistics, there is no loss of information in using the jackknife since knowledge of a linear statistic for the n jackknife data sets $\mathbf{x}_{(i)}$ determines the value of $\hat{\theta}$ for any bootstrap data set $\mathbf{x}^*$ (Problem 11.3)

For nonlinear statistics, there is a loss of information. The jackknife makes a *linear approximation* to the bootstrap: that is, it agrees with bootstrap (except for a factor of $\{(n-1)/n\}^{1/2}$) for a certain linear statistic of the form (11.17) that approximates $\hat{\theta}$. Details of this interesting relationship are given in Problems 11.5 and 11.6, and Chapter 20. Practically speaking, these results show that accuracy of the jackknife estimate of standard error depends on how close $\hat{\theta}$ is to linearity. For highly nonlinear functions the jackknife can be inefficient, sometimes dangerously so.

Figure 11.2 shows the results of an investigation into this inefficiency in a particular example. We generated 200 samples of size 10 from a bivariate normal population with zero mean, unit variances, and correlation .7. The boxplots on the left show the bootstrap and jackknife estimates of standard error for $\hat{\theta} = \bar{x}$ while those on the right are for the correlation coefficient. The horizontal lines indicate the true standard error of $\hat{\theta}$ in each case. In both cases, the bootstrap and jackknife display little bias in estimating the standard error. The variability of the jackknife estimate is slightly larger than that of the bootstrap for the mean (a linear statistic) but is significantly larger for the correlation coefficient (a nonlinear statistic). For this reason, the bootstrap would be preferred in the

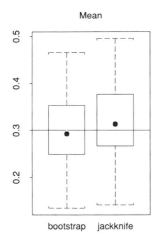
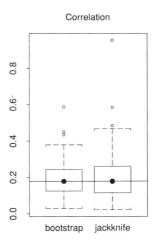

Figure 11.2. *Bootstrap and jackknife estimates of standard error for two different statistics $\hat{\theta}$, for samples of size 10 from a bivariate normal population with correlation .7. On the left $\hat{\theta} = \bar{x}$; on the right $\hat{\theta}$ is the sample correlation. Boxplots indicate the distribution of standard error estimates over 100 simulated samples.*

latter case. Problem 11.13 investigates the bootstrap and jackknife for a different nonlinear statistic.

Similarly, the jackknife estimate of bias can be shown to be an approximation to the bootstrap estimate of bias. The approximation is in terms of *quadratic* (rather than linear) statistics, which have the form

$$\hat{\theta} = s(\mathbf{x}) = \mu + \frac{1}{n} \sum_{1 \le i \le n} \alpha(x_i) + \frac{1}{n^2} \sum_{1 \le i < j \le n} \beta(x_i, x_j). \quad (11.18)$$

A simple example of a quadratic statistic is the sample variance (11.11). By expanding it out, we find that it can be expressed in the form of equation (11.18) (Problem 11.9). For such a statistic, if we know the value of $\hat{\theta}$ for $\mathbf{x}$ as well as $\mathbf{x}_{(i)}$, $i = 1, 2, \ldots n$, we can deduce the value of $\hat{\theta}$ for any bootstrap data set. As shown in Problems 11.10 — 11.11, the jackknife and bootstrap estimates of bias essentially agree for quadratic statistics.

11.6 Failure of the jackknife

To summarize so far, the jackknife often provides a simple and good approximation to the bootstrap, for estimation of standard errors and bias. However, as mentioned briefly in Chapter 10, the jackknife can fail miserably if the statistic $\hat{\theta}$ is not "smooth." Intuitively, the idea of smoothness is that small changes in the data set cause only small changes in the statistic. A simple example of a non-smooth statistic is the median. To see why the median is not smooth, consider the 9 ordered values from the control group of the mouse data (Table 2.1):

$$10, 27, 31, 40, 46, 50, 52, 104, 146. \tag{11.19}$$

The median of these values is 46. Now suppose we start increasing the value of the 4th largest value $x = 40$. The median doesn't change at all until the x becomes larger than 46, and then after that the median is equal to x, until x exceeds 50. This implies that the median is not a differentiable (or smooth) function of x.

This lack of smoothness causes the jackknife estimate of standard error to be *inconsistent* for the median. For the mouse data, the jackknife values for the median[1] are

$$48, 48, 48, 48, 45, 43, 43, 43, 43. \tag{11.20}$$

Notice that there are only 3 distinct values, a consequence of the lack of smoothness of the median and the fact that the jackknife data sets differ from the original data set by only one data point. The resulting estimate $\widehat{se}_{jack}$ is 6.68. For the mouse data, the bootstrap estimate of standard error based on $B = 100$ bootstrap samples is 9.58, considerably larger than the jackknife value of 6.68. As $n \to \infty$, it can be shown that $\widehat{se}_{jack}$ is inconsistent, that is, it fails to converge to the true standard error. The bootstrap, on the other hand, considers data sets that are less similar to the original data set than are the jackknife data sets, and consequently, is consistent for the median.

---

[1] The median of an even number of data points is the average of the middle two values.

11.7 The delete-*d* jackknife

There is a way to fix up the inconsistency of the jackknife for non-smooth statistics. Instead of leaving one observation out at a time, we leave out d observations, where $n = r \cdot d$ for some integer r. It can be shown that if $n^{1/2}/d \to 0$ and $n - d \to \infty$, then the "delete-d" jackknife is consistent for the median. Roughly speaking, one has to leave out more than $d = \sqrt{n}$, but fewer than n observations to achieve consistency for the jackknife estimate of standard error. Let $\hat{\theta}_{(s)}$ denote $\hat{\theta}$ applied to the data set with subset s removed. The formula for the delete-d jackknife estimate of standard error is

$$\left\{ \frac{r}{\binom{n}{d}} \sum (\hat{\theta}_{(s)} - \hat{\theta}_{(\cdot)})^2 \right\}^{1/2} \tag{11.21}$$

where $\hat{\theta}_{(\cdot)} = \sum \hat{\theta}_{(s)}/\binom{n}{d}$ and the sum is over all subsets s of size $n - d$ chosen without replacement from $x_1, x_2, \ldots x_n$.

In our example with $n = 9$, we can choose $d = 4 > \sqrt{9}$ and the computation of the delete-d jackknife involves finding the median for the

$$\binom{9}{4} = 126 \tag{11.22}$$

samples corresponding to leaving 4 observations out at a time. This gives an estimate of standard error of 7.16, which is somewhat closer to the bootstrap value of 9.58 than the delete-one jackknife value of 6.68.

If n is large and $\sqrt{n} < d < n$, the number of jackknife samples $\binom{n}{d}$ can be very sizable. Instead of computing $\hat{\theta}$ for all of these subsets, one can instead draw a random sample of subsets, which in turn, makes the delete-d jackknife look more like the bootstrap. Current work on the delete-d jackknife represents a revival of research on the jackknife.

An S language function for jackknifing is described in the Appendix.

11.8 Bibliographic notes

Quenouille (1949) first proposed the idea of the jackknife for estimation of bias. Tukey (1958) recognized the jackknife's potential for estimating standard errors, and gave it its name. Further development is given by Miller (1964, 1974), Gray and Schucany (1972), Hinkley (1977), Reeds (1978), Parr (1983, 1985), Hinkley and Wei (1984), Sen (1988), and Wu (1986) in the linear regression setting. Shao and Wu (1989), and Shao (1991) present general theoretical results on the delete-d jackknife.

11.9 Problems

11.1 Show that if $\hat{\theta} = \bar{x}$, the jackknife estimate of standard error is equal to the unbiased estimate (11.8).

11.2 In Problem 11.1, show that use of the factor $[(n-1)/n]^2$ in place of $(n-1)/n$ leads to the plug-in estimate of standard error.

11.3 Suppose $\hat{\theta}$ is a linear statistic of the form

$$\hat{\theta} = \mu + \frac{1}{n} \sum_{1}^{n} \alpha(x_i). \qquad (11.23)$$

Suppose we know the value of $\hat{\theta}$ for each jackknife data set $\mathbf{x}_{(i)}$, that is $s(\mathbf{x}_{(i)}) = b_i$, for $i = 1, 2, \ldots n$.

(a) Let $\alpha_i = \alpha(x_i)$ and solve the set of n linear equations

$$b_i = \mu + \sum_{j \neq i} \alpha_j / (n-1), \quad i = 1, 2, \ldots n$$

for $\alpha_1, \alpha_2, \ldots \alpha_n$.

(b) Hence deduce the value of $\hat{\theta}$ for an arbitrary bootstrap data set $x_1^*, x_2^*, \ldots x_n^*$.

11.4 *Relationship between the jackknife and bootstrap estimates of standard error.* Suppose that $\hat{\theta}$ is a linear statistic of the form (11.23). Letting $\alpha_i = \alpha(x_i)$, show that the (ideal) bootstrap estimate of standard error is

$$\left\{ \sum_{1}^{n} (\alpha_i - \bar{\alpha})^2 / n^2 \right\}^{1/2} \qquad (11.24)$$

and the jackknife estimate of standard error is

$$\left\{ \sum_{1}^{n} (\alpha_i - \bar{\alpha})^2 / \{(n-1)n\} \right\}^{1/2}. \qquad (11.25)$$

Hence these two estimates only differ by the factor $\{(n-1)/n\}^{1/2}$.

11.5 *Relationship between the jackknife and bootstrap estimates of standard error- continued.*

Suppose $\hat{\theta}$ is a nonlinear statistic, and we approximate it by the linear statistic

$$\hat{\theta}_{\text{lin}} = \mu + \frac{1}{n} \sum_1^n \alpha(x_i) \qquad (11.26)$$

that has the same value as $\hat{\theta}$ for the jackknife data sets $\mathbf{x}_{(i)}$. Find expressions for μ and $\alpha_i = \alpha(x_i)$, $i = 1, 2, \ldots n$, in terms of $\hat{\theta}_{(i)}$, $i = 1, 2, \ldots n$.

11.6 Apply the results of the previous problem to show that the jackknife estimate of standard error for $\hat{\theta}$ agrees with the (ideal) bootstrap estimate of standard error for $\hat{\theta}_{\text{lin}}$, except for a factor of $\{(n-1)/n\}^{1/2}$.

11.7 Show that for $\hat{\theta} = \bar{x}$, $\hat{\theta}_{(\cdot)} - \hat{\theta} = 0$ and hence the jackknife estimate of bias is zero.

11.8 Suppose $x_1, x_2, \ldots x_n$ are independent and identically distributed with variance σ^2.

(a) Show that the plug-in estimate of variance $\hat{\theta} = \sum_1^n (x_i - \bar{x})^2/n$ has bias equal to $-\sigma^2/n$ as an estimate of σ^2.

(b) In this case, show that $\widehat{\text{bias}}_{\text{jack}} = -s^2/n$ where $s^2 = \sum_1^n (x_i - \bar{x})^2/(n-1)$.

11.9 Show that the sample variance (11.11) is a quadratic statistic of the form (11.18), with $\mu = 0$, $\alpha(x_i) = -(n-1)x_i/n^2$ and $\beta(x_i, x_j) = -2x_i x_j/n^3$.

11.10 *Relationship between the jackknife and bootstrap estimates of bias.*

Suppose that $\hat{\theta}$ is a quadratic statistic of the form (11.18). Derive the (ideal) bootstrap and jackknife estimates of bias and show that they only differ by the factor $(n-1)/n$.

11.11 *Relationship between the jackknife and bootstrap estimates of bias– continued.*

Suppose that $\hat{\theta}$ is not a quadratic statistic, and we approximate it by the quadratic statistic $\hat{\theta}_{\text{quad}}$ of the form (11.18), having the same value as $\hat{\theta}$ for the jackknife data sets $\mathbf{x}_{(i)}$, as well as for the original data set $\mathbf{x}$.

(a) Find expressions for $\alpha_i = \alpha(x_i)$, $i = 1, 2, \ldots n$ and $\beta_{ij} = \beta(x_i, x_j)$ in terms of $\hat{\theta}$ and $\hat{\theta}_{(i)}$, $i = 1, 2, \ldots n$.

(b) Apply the results of the previous problem to show that the jackknife and (ideal) bootstrap estimates of bias for $\hat{\theta}_{\text{quad}}$ agree, except for a factor of $(n-1)/n$.

11.12 Calculate the jackknife estimates of standard error and bias for the correlation coefficient of the law school data. Compare these to the bootstrap estimates of the same quantities.

11.13 Generate 100 samples $X_1, X_2, \ldots X_{20}$ from a normal population $N(\theta, 1)$ with $\theta = 1$.

(a) For each sample compute the bootstrap and jackknife estimate of variance for $\hat{\theta} = \bar{X}$ and compute the mean and standard deviation of these variance estimates over the 100 samples.

(b) Repeat (a) for the statistic $\hat{\theta} = \bar{X}^2$, and compare the results. Give an explanation for your findings.

## CHAPTER 12

Confidence intervals based on bootstrap "tables"

12.1 Introduction

Most of our work so far has concerned the computation of bootstrap standard errors. Standard errors are often used to assign approximate confidence intervals to a parameter θ of interest. Given an estimate $\hat{\theta}$ and an estimated standard error $\widehat{se}$, the usual 90% confidence interval for θ is

$$\hat{\theta} \pm 1.645 \cdot \widehat{se}. \qquad (12.1)$$

The number 1.645 comes from a standard normal table, as will be reviewed briefly below. Statement (12.1) is called an *interval estimate* or *confidence interval* for θ. An interval estimate is often more useful than just a *point estimate* $\hat{\theta}$. Taken together, the point estimate and the interval estimate say what is the best guess for θ, and how far in error that guess might reasonably be.

In this chapter and the next two chapters we describe different techniques for constructing confidence intervals using the bootstrap. This area has been a major focus of theoretical work on the bootstrap; an overview of this work is given later in the book (Chapter 22).

Suppose that we are in the one-sample situation where the data are obtained by random sampling from an unknown distribution F, $F \to \mathbf{x} = (x_1, x_2, \cdots, x_n)$, as in Chapter 6. Let $\hat{\theta} = t(\hat{F})$ be the plug-in estimate of a parameter of interest $\theta = t(F)$, and let $\widehat{se}$ be some reasonable estimate of standard error for $\hat{\theta}$, based perhaps on bootstrap or jackknife computations. Under most circumstances it turns out that as the sample size n grows large, the distribution of $\hat{\theta}$ becomes more and more normal, with mean near θ and variance near $\widehat{se}^2$, written $\hat{\theta} \overset{.}{\sim} N(\theta, \widehat{se}^2)$ or equivalently

$$\frac{\hat{\theta} - \theta}{\widehat{se}} \overset{.}{\sim} N(0, 1). \tag{12.2}$$

The *large-sample*, or *asymptotic*, result (12.2) usually holds true for general probability models $P \to \mathbf{x}$ as the amount of data gets large, and for statistics other than the plug-in estimate, but we shall stay with the one-sample plug-in situation for most of this chapter.

Let $z^{(\alpha)}$ indicate the $100 \cdot \alpha$th percentile point of a $N(0, 1)$ distribution, as given in a standard normal table, $z^{(.025)} = -1.960$, $z^{(.05)} = -1.645$, $z^{(.95)} = 1.645$, $z^{(.975)} = 1.960$, etc.

If we take approximation (12.2) to be exact, then

$$\text{Prob}_F\{z^{(\alpha)} \leq \frac{\hat{\theta} - \theta}{\widehat{se}} \leq z^{(1-\alpha)}\} = 1 - 2\alpha, \tag{12.3}$$

which can be written as

$$\text{Prob}_F\{\theta \in [\hat{\theta} - z^{(1-\alpha)} \cdot \widehat{se}, \ \hat{\theta} - z^{(\alpha)} \cdot \widehat{se}]\} = 1 - 2\alpha. \tag{12.4}$$

Interval (12.1) is obtained from (12.4), with $\alpha = .05$, $1 - 2\alpha = .90$. In general

$$[\hat{\theta} - z^{(1-\alpha)} \cdot \widehat{se}, \ \hat{\theta} - z^{(\alpha)} \cdot \widehat{se}] \tag{12.5}$$

is called the *standard confidence interval* with *coverage probability*[1] equal $1 - 2\alpha$, or *confidence level* $100 \cdot (1 - 2\alpha)\%$. Or, more simply, it is called a $1 - 2\alpha$ confidence interval for θ. Since $z^{(\alpha)} = -z^{(1-\alpha)}$ we can write (12.5) in the more familiar form

$$\hat{\theta} \pm z^{(1-\alpha)} \cdot \widehat{se}. \tag{12.6}$$

As an example, consider the $n = 9$ Control group mice of Table 2.1. Suppose we want a confidence interval for the expectation θ of the Control group distribution. The plug-in estimate is the mean $\hat{\theta} = 56.44$, with estimated standard error $\widehat{se} = 13.33$ as in (5.12). The 90% standard confidence interval for θ, (12.1), is

$$56.22 \pm 1.645 \cdot 13.33 = [34.29, 78.15]. \tag{12.7}$$

---

[1] It would be more precise to call (12.5) an <u>approximate</u> confidence interval since the coverage probability will usually <u>not exactly</u> equal the desired value $100(1 - 2\alpha)$. The bootstrap intervals discussed in this chapter are also approximate but in general are better approximations than the standard intervals.

The coverage property of this interval implies that 90% of the time, a random interval constructed in this way will contain the true value θ. Of course (12.2) is only an approximation in most problems, and the standard interval is only an approximate confidence interval, though a very useful one in an enormous variety of situations. We will use the bootstrap to calculate better approximate confidence intervals. As $n \to \infty$, the bootstrap and standard intervals converge to each other, but in any given situation like that of the mouse data the bootstrap may make substantial corrections. These corrections can significantly improve the inferential accuracy of the interval estimate.

12.2 Some background on confidence intervals

Before beginning the bootstrap exposition, we review the logic of confidence intervals, and what it means for a confidence interval to be "accurate." Suppose that we are in the situation where an estimator $\hat{\theta}$ is normally distributed with unknown expectation θ,

$$\hat{\theta} \sim N(\theta, \text{se}^2), \tag{12.8}$$

with the standard error "se" known. (There is no dot over the "$\sim$" sign because we are assuming that (12.8) holds exactly.) Then an exact version of (12.2) is true: the random quantity equaling $(\hat{\theta} - \theta)/\text{se}$ has a standard normal distribution,

$$Z = \frac{\hat{\theta} - \theta}{\text{se}} \sim N(0,1). \tag{12.9}$$

The equality $\text{Prob}\{|Z| \leq z^{(1-\alpha)}\} = 1 - 2\alpha$ is algebraically equivalent to

$$\text{Prob}_\theta\{\theta \in [\hat{\theta} - z^{(1-\alpha)} \cdot \text{se}, \hat{\theta} - z^{(\alpha)} \cdot \text{se}]\} = 1 - 2\alpha. \tag{12.10}$$

The notation "$\text{Prob}_\theta\{\ \}$" emphasizes that probability calculation (12.10) is done with the true mean equaling θ, so $\hat{\theta} \sim N(\theta, \text{se}^2)$.

For convenience we will denote confidence intervals by $[\hat{\theta}_{\text{lo}}, \hat{\theta}_{\text{up}}]$, so $\hat{\theta}_{\text{lo}} = \hat{\theta} - z^{(1-\alpha)} \cdot \text{se}$ and $\hat{\theta}_{\text{up}} = \hat{\theta} - z^{(\alpha)} \cdot \text{se}$ for the interval in (12.10). In this case we can see that the interval $[\hat{\theta} - z^{(1-\alpha)} \cdot \text{se}, \hat{\theta} - z^{(\alpha)} \cdot \text{se}]$ has probability exactly $1 - 2\alpha$ of containing the true value of θ. More precisely, the probability that θ lies below the lower limit is exactly α, as is the probability that θ exceeds the upper limit,

$$\text{Prob}_\theta\{\theta < \hat{\theta}_{\text{lo}}\} = \alpha, \qquad \text{Prob}_\theta\{\theta > \hat{\theta}_{\text{up}}\} = \alpha. \tag{12.11}$$

The fact that (12.11) holds for every possible value of θ is what we mean when we say that a $(1 - 2\alpha)$ confidence interval $(\hat{\theta}_{\mathrm{lo}}, \hat{\theta}_{\mathrm{up}})$ is accurate. It is important to remember that θ is a constant in probability statements (12.11), the random variables being $\hat{\theta}_{\mathrm{lo}}$ and $\hat{\theta}_{\mathrm{up}}$.

A $1 - 2\alpha$ confidence interval $(\hat{\theta}_{\mathrm{lo}}, \hat{\theta}_{\mathrm{up}})$ with property (12.11) is called *equal-tailed*. This refers to the fact that the coverage error 2α is divided up evenly between the lower and upper ends of the interval. Confidence intervals are almost always constructed to be equal-tailed and we will restrict attention to equal-tailed intervals in our discussion. Notice also that property (12.11) implies property (12.10), but not vice-versa. That is, (12.11) requires that the one-sided miscoverage of the interval be α on each side, rather that just an overall coverage of $1 - 2\alpha$. This forces the interval to be the right shape, that is, to extend the correct distance above and below $\hat{\theta}$. We shall aim for correct one-sided coverage in our construction of approximate confidence intervals.

12.3 Relation between confidence intervals and hypothesis tests

There is another way to interpret the statement that $(\hat{\theta}_{\mathrm{lo}}, \hat{\theta}_{\mathrm{up}})$ is a $1 - 2\alpha$ confidence interval for θ. Suppose that the true θ were equal to $\hat{\theta}_{\mathrm{lo}}$, say

$$\hat{\theta}^* \sim N(\hat{\theta}_{\mathrm{lo}}, \mathrm{se}^2). \tag{12.12}$$

Here we have used $\hat{\theta}^*$ to denote the random variable, to avoid confusion with the observed estimate $\hat{\theta}$. The quantity $\hat{\theta}_{\mathrm{lo}}$ is considered to be fixed in (12.12), only $\hat{\theta}^*$ being random. It is easy to see that the probability that $\hat{\theta}^*$ exceeds the actual estimate $\hat{\theta}$ is α,

$$\mathrm{Prob}_{\hat{\theta}_{\mathrm{lo}}}\{\hat{\theta}^* \geq \hat{\theta}\} = \alpha. \tag{12.13}$$

Then for any value of θ less than $\hat{\theta}_{\mathrm{lo}}$ we have

$$\mathrm{Prob}_{\theta}\{\hat{\theta}^* \geq \hat{\theta}\} < \alpha \qquad [\text{for any } \theta < \hat{\theta}_{\mathrm{lo}}]. \tag{12.14}$$

The probability calculation in (12.14) has $\hat{\theta}$ fixed at its observed value, and $\hat{\theta}^*$ random, $\hat{\theta}^* \sim N(\theta, \mathrm{se}^2)$, see Problem 12.2. Likewise,

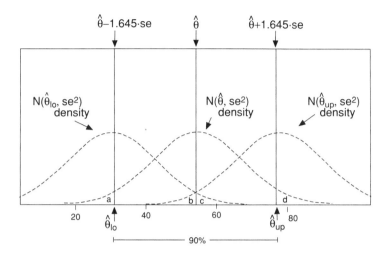

Figure 12.1. *90% confidence interval for the expectation of a normal distribution. We observe $\hat\theta \sim N(\theta, se^2)$ and want a confidence interval for the unknown parameter θ; the standard error se is assumed known. The confidence interval is given by $(\hat\theta_{\text{lo}}, \hat\theta_{\text{up}}) = \hat\theta \pm 1.645se$. Notice that $\hat\theta$ is the 95th percentile of the distribution $N(\hat\theta_{lo}, se^2)$, so region c has probability .05 for $N(\hat\theta_{lo}, se^2)$. Likewise $\hat\theta$ is the 5th percentile of $N(\hat\theta_{\text{up}}, se^2)$, so region b has probability .05 for $N(\hat\theta_{\text{up}}, se^2)$. In this figure, $\hat\theta = 56.22, se = 13.33$ as in (12.7).*

for any value of θ greater than the upper limit $\hat\theta_{\text{up}}$,

$$\text{Prob}_\theta\{\hat\theta^* \le \hat\theta\} < \alpha \qquad [\text{for any } \theta > \hat\theta_{\text{up}}]. \tag{12.15}$$

The logic of the confidence interval $(\hat\theta_{\text{lo}}, \hat\theta_{\text{up}})$ can be stated in terms of (12.14)—(12.15). We choose a small probability α which is our "threshold of plausibility." We decide that values of the parameter θ less than $\hat\theta_{\text{lo}}$ are implausible, because they give probability less than α of observing an estimate as large as the one actually seen, (12.14). We decide that values of θ greater than $\hat\theta_{\text{up}}$ are implausible because they give probability less than α of observing an estimate as small as the one actually seen, (12.15). To summarize:

The $1 - 2\alpha$ confidence interval $(\hat\theta_{\text{lo}}, \hat\theta_{\text{up}})$ is the set of plausible values of θ having observed $\hat\theta$, those values not ruled out by either of the plausibility tests (12.14) or (12.15).

The situation is illustrated in Figure 12.1. We assume that $\hat{\theta} \sim N(\theta, se^2)$ as in (12.8), and take $\alpha = .05$, $1 - 2\alpha = .90$. Having observed $\hat{\theta}$, the 90% confidence interval (12.10) has endpoints

$$\hat{\theta}_{\text{lo}} = \hat{\theta} - 1.645 \cdot se, \qquad \hat{\theta}_{\text{up}} = \hat{\theta} + 1.645 \cdot se. \qquad (12.16)$$

The dashed curve having its highest point at $\hat{\theta}_{\text{lo}}$ indicates part of the probability density of the normal distribution $N(\hat{\theta}_{\text{lo}}, se^2)$. The 95th percentile of the distribution $N(\hat{\theta}_{\text{lo}}, se^2)$ occurs at $\hat{\theta}$. Another way to say this is that the region under the $N(\hat{\theta}_{\text{lo}}, se^2)$ density curve to the right of $\hat{\theta}$, labeled "c", has area .05. Likewise the dashed curve that has its highest point at $\hat{\theta}_{\text{up}}$ indicates the probability density of $N(\hat{\theta}_{\text{up}}, se^2)$; $\hat{\theta}$ is the 5th percentile of the distribution; and region "b" has area .05.

The plausibility tests (12.14) and (12.15) are also the significance levels for the related *hypothesis test*. The value in (12.14) is the significance level for the one-sided alternative hypothesis that the true parameter is greater than θ, and (12.15) is the significance level for the one-sided alternative hypothesis that the true parameter is less than θ. In many situations a hypothesis test can be carried out by constructing a confidence interval and then checking whether the null value is in the interval. Hypothesis testing is the subject of Chapters 15 and 16.

12.4 Student's t interval

With this background, let's see how we can improve upon the standard confidence interval $[\hat{\theta} - z^{(1-\alpha)} \cdot \widehat{se}, \hat{\theta} - z^{(\alpha)} \cdot \widehat{se}]$. As we have seen, this interval is derived from the assumption that

$$Z = \frac{\hat{\theta} - \theta}{\widehat{se}} \ \dot{\sim} \ N(0, 1). \qquad (12.17)$$

This is valid as $n \to \infty$, but is only an approximation for finite samples. Back in 1908, for the case $\hat{\theta} = \bar{x}$, Gosset derived the better approximation

$$Z = \frac{\hat{\theta} - \theta}{\widehat{se}} \ \dot{\sim} \ t_{n-1}, \qquad (12.18)$$

Table 12.1. *Percentiles of the t distribution with 5, 8, 20, 50 and 100 degrees of freedom, the $N(0,1)$ distribution and the bootstrap distribution of $Z^*(b)$ (for the control group of the mouse data).*

Percentile	5%	10%	16%	50%	84%	90%	95%
t_5	-2.01	-1.48	-1.73	0.00	1.73	1.48	2.01
t_8	-1.86	-1.40	-1.10	0.00	1.10	1.40	1.86
t_{20}	-1.73	-1.33	-1.06	0.00	1.06	1.33	1.73
t_{50}	-1.68	-1.30	-1.02	0.00	1.02	1.30	1.68
t_{100}	-1.66	-1.29	-1.00	0.00	1.00	1.29	1.66
Normal	-1.65	-1.28	-0.99	0.00	0.99	1.28	1.65
Bootstrap-t	-4.53	-2.01	-1.32	-.025	0.86	1.19	1.53

where t_{n-1} represents the Student's t distribution on $n-1$ degrees of freedom. Using this approximation, our interval is

$$[\hat{\theta} - t_{n-1}^{(1-\alpha)} \cdot \widehat{se}, \hat{\theta} - t_{n-1}^{(\alpha)} \cdot \widehat{se}], \tag{12.19}$$

with $t_{n-1}^{(\alpha)}$ denoting the αth percentile of the t distribution on $n-1$ degrees of freedom. That is to say, we look up the appropriate percentile in a t_{n-1} table rather than a normal table.

Table 12.1 shows the percentiles of the t_{n-1} and $N(0,1)$ distribution for various degrees of freedom. (The values in the last line of the table are the "bootstrap-t" percentiles" discussed below.) When $\hat{\theta} = \bar{x}$, this approximation is exact if the observations are normally distributed, and has the effect of widening the interval to adjust for the fact that the standard error is unknown. But notice that if $n \geq 20$, the percentiles of t_n distribution don't differ much from those of $N(0,1)$. In our example with $n = 9$, use of the 5% and 95% percentiles from the t table with 8 degrees of freedom leads to the interval

$$56.22 \pm 1.86 \cdot 13.33 = (31.22, 81.01),$$

which is a little wider than the normal interval $(34.29, 78.15)$.

The use of the t distribution doesn't adjust the confidence interval to account for skewness in the underlying population or other errors that can result when $\hat{\theta}$ is not the sample mean. The next section describes the bootstrap-t interval, a procedure which does adjust for these errors.

12.5 The bootstrap-t interval

Through the use of the bootstrap we can obtain accurate intervals without having to make normal theory assumptions like (12.17). In this section we describe one way to get such intervals, namely the "bootstrap-t" approach. This procedure estimates the distribution of Z directly from the data; in essence, it builds a table like Table 12.1 *that is appropriate for the data set at hand.* [1] This table is then used to construct a confidence interval in exactly the same way that the normal and t tables are used in (12.17) and (12.18). The bootstrap table is built by generating B bootstrap samples, and then computing the bootstrap version of Z for each. The bootstrap table consists of the percentiles of these B values.

Here is the bootstrap-t method in more detail. Using the notation of Figure 8.3 we generate B bootstrap samples $\mathbf{x}^{*1}, \mathbf{x}^{*2}, \cdots, \mathbf{x}^{*B}$ and for each we compute

$$Z^*(b) = \frac{\hat{\theta}^*(b) - \hat{\theta}}{\widehat{se}^*(b)}, \qquad (12.20)$$

where $\hat{\theta}^*(b) = s(\mathbf{x}^{*b})$ is the value of $\hat{\theta}$ for the bootstrap sample $\mathbf{x}^{*b}$ and $\widehat{se}^*(b)$ is the estimated standard error of $\hat{\theta}^*$ for the bootstrap sample $\mathbf{x}^{*b}$. The αth percentile of $Z^*(b)$ is estimated by the value $\hat{t}^{(\alpha)}$ such that

$$\#\{Z^*(b) \le \hat{t}^{(\alpha)}\}/B = \alpha. \qquad (12.21)$$

For example, if $B = 1000$, the estimate of the 5% point is the 50th largest value of the $Z^*(b)$s and the estimate of the 95% point is the 950th largest value of the $Z^*(b)$s. Finally, the "bootstrap-t" confidence interval is

$$(\hat{\theta} - \hat{t}^{(1-\alpha)} \cdot \widehat{se}, \hat{\theta} - \hat{t}^{(\alpha)} \cdot \widehat{se}). \qquad (12.22)$$

This is suggested by the same logic that gave (12.19) from (12.18).

If $B \cdot \alpha$ is not an integer, the following procedure can be used. Assuming $\alpha \le .5$, let $k = [(B+1)\alpha]$, the largest integer $\le (B+1)\alpha$. Then we define the empirical α and $1 - \alpha$ quantiles by the kth

---

[1] The idea behind the bootstrap-t method is easier to describe than the percentile-based bootstrap intervals of the next two chapters, which is why we discuss the bootstrap-t procedure first. In practice, however, the bootstrap-t can give somewhat erratic results, and can be heavily influenced by a few outlying data points. The percentile based methods of the next two chapters are more reliable.

largest and $(B + 1 - k)$th largest values of $Z^*(b)$, respectively.

The last line of Table 12.1 shows the percentiles of $Z^*(b)$ for $\hat{\theta}$ equal to the mean of the control group of the mouse data, computed using 1000 bootstrap samples. It is important to note that $B = 100$ or 200 is not adequate for confidence interval construction, see Chapter 19. Notice that the bootstrap-t points greatly differ from the normal and t percentiles! The resulting 90% bootstrap-t confidence interval for the mean is

$$[56.22 - 1.53 \cdot 13.33, 56.22 + 4.53 \cdot 13.33] = [35.82, 116.74]$$

The lower endpoint is close to the standard interval, but the upper endpoint is much greater. This reflects the two very large data values 104 and 146.

The quantity $Z = (\hat{\theta} - \theta)/\widehat{se}$ is called an *approximate pivot*: this means that its distribution is approximately the same for each value of θ. In fact, this property is what allows us to construct the interval (12.22) from the bootstrap distribution of $Z^*(b)$, using the same argument that gave (12.5) from (12.3).

Some elaborate theory (Chapter 22) shows that in large samples the coverage of the bootstrap-t interval tends to be closer to the desired level (here 90%) than the coverage of the standard interval or the interval based on the t table. It is interesting that like the t approximation, the gain in accuracy is at the price of generality. The standard normal table applies to all samples, and all sample sizes; the t table applies all samples of a fixed size n; the bootstrap-t table applies *only to the given sample*. However with the availability of fast computers, it is not impractical to derive a "bootstrap table"' for each new problem that we encounter.

Notice also that the normal and t percentage points in Table 12.1 are symmetric about zero, and as a consequence the resulting intervals are symmetric about the point estimate $\hat{\theta}$. In contrast, the bootstrap-t percentiles can by asymmetric about 0, leading to intervals which are longer on the left or right. This asymmetry represents an important part of the improvement in coverage that it enjoys.

The bootstrap-t procedure is a useful and interesting generalization of the usual Student's t method. It is particularly applicable to *location statistics* like the sample mean. A location statistic is one for which increasing each data value x_i by a constant c increases the statistic itself by c. Other location statistics are the median, the trimmed mean, or a sample percentile.

The bootstrap-t method, at least in its simple form, cannot be trusted for more general problems, like setting a confidence interval for a correlation coefficient. We will present more dependable bootstrap confidence interval methods in the next two chapters. In the next section we describe the use of transformations to improve the bootstrap-t approach.

12.6 Transformations and the bootstrap-t

[1] There are both computational and interpretive problems with the bootstrap-t confidence procedure. In the denominator of the statistic $Z^*(b)$ we require $\widehat{se}^*(b)$, the standard deviation of $\hat{\theta}^*$ for the bootstrap sample $\mathbf{x}^{*b}$. For the mouse data example, where $\hat{\theta}$ is the mean, we used the plug-in estimate

$$\widehat{se}^*(b) = \left\{ \sum_1^n (x_i^{*b} - \bar{x}^{*b})^2 / n^2 \right\}^{1/2}, \qquad (12.23)$$

$x_1^{*b}, x_1^{*b} \ldots x_n^{*b}$ being a bootstrap sample.

The difficulty arises when $\hat{\theta}$ is a more complicated statistic, for which there is no simple standard error formula. As we have seen in Chapter 5, standard error formulae exist for very few statistics, and thus we would need to compute a bootstrap estimate of standard error *for each bootstrap sample*. This implies two nested levels of bootstrap sampling. Now for the estimation of standard error, $B = 25$ might be sufficient, while $B = 1000$ is needed for the computation of percentiles. Hence the overall number of bootstrap samples needed is perhaps $25 \cdot 1000 = 25,000$, a formidable number if $\hat{\theta}$ is costly to compute.

A second difficulty with the bootstrap-t interval is that it may perform erratically in small-sample, nonparametric settings. This trouble can be alleviated. Consider for example the law school data of Table 3.1, for which $\hat{\theta}$ is the sample correlation coefficient. In constructing a bootstrap-t interval, we used for $\widehat{se}^*(b)$ the bootstrap estimate of standard error with $B = 25$ bootstrap samples. As mentioned above, the overall procedure involves two nested levels of bootstrap sampling. A total of 1000 values of $\hat{\theta}^*$ were generated, so that a total $25,000$ bootstrap samples were used. The resulting 90% bootstrap-t confidence was $[-.026, .90]$. For the correlation

---

[1] This section contains more advanced material and may be skipped at first reading.

coefficient, it is well known (cf. page 54) that if we construct a confidence interval for the transformed parameter

$$\phi = .5 \log\left(\frac{1+\theta}{1-\theta}\right) \tag{12.24}$$

and then transform the endpoints back with the inverse transformation $(e^{2\phi} - 1)/(e^{2\phi} + 1)$, we obtain a better interval. For the law school data, if we compute a 90% bootstrap-*t* confidence interval for ϕ and then transform it back, we obtain the interval $[.45, .93]$ for θ, which is much shorter than the interval obtained without transformation. In addition, if we look at more extreme confidence points, for example a 98% interval, the endpoints are $[-.66, 1.03]$ for the interval that doesn't use a transformation and $[.17, .95]$ for the one that does. Notice that the first interval falls outside of the allowable range for a correlation coefficient! In general, use of the (untransformed) bootstrap-*t* procedure for this and other problems can lead to intervals which are often too wide and fall outside of the allowable range for a parameter.

To put it another way, the bootstrap-*t* interval is not *transformation-respecting*. It makes a difference which scale is used to construct the interval, and some scales are better than others. In the correlation coefficient example, the transformation (12.24) is known to be the appropriate one if the data are bivariate normal, and works well in general for this problem. For most problems, however, we don't know what transformation to apply, and this is a major stumbling block to the general use of the bootstrap-*t* for confidence interval construction.

One way out of this dilemma is to use the bootstrap to estimate the appropriate transformation from the data itself, and then use this transformation for the construction of a bootstrap t interval. Let's see how this can be done. With θ equal to the correlation coefficient, define $\hat{\phi} = .5 \cdot \log[(1 + \hat{\theta})/(1 - \hat{\theta})]$, $\phi = .5 \cdot \log[(1 + \theta)/(1 - \theta)]$. Then

$$\hat{\phi} - \phi \,\dot{\sim}\, N(0, \frac{1}{n-3}). \tag{12.25}$$

This transformation approximately *normalizes* and *variance stabilizes* the estimate $\hat{\theta}$. We would like to have an automatic method for finding such transformations. It turns out, however, that it is not usually possible to both normalize and variance stabilize an estimate. It seems that for bootstrap-*t* intervals, it is the second property that is important: *bootstrap-t intervals work better for variance stabilized parameters.* Now if X is a random variable with mean θ and standard deviation $s(\theta)$ that varies as a function of θ, then a Taylor series argument (Problem 12.4) shows that the transformation $g(x)$ with derivative

$$g'(x) = \frac{1}{s(x)} \qquad (12.26)$$

has the property that the variance of $g(X)$ is approximately constant. Equivalently,

$$g(x) = \int^x \frac{1}{s(u)} du. \qquad (12.27)$$

In the present problem, X is $\hat{\theta}$ and for each u, we need to know $s(u)$, the standard error of $\hat{\theta}$ when $\theta = u$, in order to apply (12.27). We will write $s(u) = \mathrm{se}(\hat{\theta}|\theta = u)$. Of course, $\mathrm{se}(\hat{\theta}|\theta = u)$ is usually unknown; however we can use the bootstrap to estimate it. We then compute a bootstrap-t interval for the parameter $\phi = g(\theta)$, and transform it back via the mapping g^{-1} to obtain the interval for θ. The details of this process are shown in Algorithm 12.1. Further details of the implementation may be found in Tibshirani (1988).

The left panel of Figure 12.2 shows an example for the law school data. $B_1 = 100$ bootstrap samples were generated, and for each one the correlation coefficient and its bootstrap estimate of standard error were computed using $B_2 = 25$ second-level bootstrap samples; this entails a nested bootstrap with a total of $100 \cdot 25 = 2500$ bootstrap samples (empirical evidence suggests that 100 first level samples are adequate). Notice the strong dependence of $\mathrm{se}(\hat{\theta}^*)$ on $\hat{\theta}^*$. We drew a smooth curve through this plot to obtain an estimate of $s(u) = \mathrm{se}(\hat{\theta}|\theta = u)$, and applied formula (12.27) to obtain the estimated transformation $g(\hat{\theta})$ indicated by the solid curve in the middle panel. The broken curve in the middle panel is the transformation (12.24). The curves are roughly similar but different; we would expect them to coincide if the bootstrap sampling was carried out from a bivariate normal population. The right panel is the same as the left panel, for $\hat{\phi}^* = g(\hat{\theta}^*)$ instead of $\hat{\theta}^*$. Notice how the dependence has been reduced.

Using $B_3 = 1000$ bootstrap samples, the resulting 90% and 98% confidence intervals for the correlation coefficient turn out to be $[.33, .92]$ and $[.07, .95]$. Both intervals are shorter than those ob-

Algorithm 12.1

Computation of the variance-stabilized bootstrap-t interval

1. Generate B_1 bootstrap samples, and for each sample $\mathbf{x}^{*b}$ compute the bootstrap replication $\hat{\theta}^*(b)$. Take B_2 bootstrap samples from $\mathbf{x}^{*b}$ and estimate the standard error $\widehat{\text{se}}(\hat{\theta}^*(b))$.

2. Fit a curve to the points $[\hat{\theta}^*(b), \widehat{\text{se}}(\hat{\theta}^*(b))]$ to produce a smooth estimate of the function $s(u) = \text{se}(\hat{\theta}|\theta = u)$.

3. Estimate the variance stabilizing transformation $g(\hat{\theta})$ from formula (12.27), using some sort of numerical integration.

4. Using B_3 new bootstrap samples, compute a bootstrap-t interval for $\phi = g(\theta)$. Since the standard error of $g(\hat{\theta})$ is roughly constant as a function of θ, we don't need to estimate the denominator in the quantity $(g(\hat{\theta}^*) - g(\hat{\theta}))/\widehat{\text{se}}^*$ and can set it equal to one.

5. Map the endpoints of the interval back to the θ scale via the transformation g^{-1}.

---

tained without transformation, and lie within the set of permissible values $[-1, 1]$ for a correlation coefficient. The total number of bootstrap samples was $2500 + 1000 = 3500$, far less than the $25,000$ figure for the usual bootstrap-t procedure.

An important by-product of the transformation $\hat{\phi} = g(\hat{\theta})$ is that it allows us to ignore the denominator of the t statistic in step 4. This is because the standard error of $\hat{\phi}$ is approximately constant, and thus can be assumed to be 1. As a consequence, once the transformation $\hat{\phi} = g(\hat{\theta})$ has been obtained, the construction of the bootstrap-t interval based on $\hat{\phi}$ does not require nested bootstrap sampling.

The other approach to remedying the problems with the bootstrap-t interval is quite different. Instead of focusing on a statistic of the form $Z = (\hat{\theta} - \theta)/\widehat{\text{se}}$, we work directly with the bootstrap distribution of $\hat{\theta}$ and derive a transformation-respecting confidence procedure from them. This approach is described in the

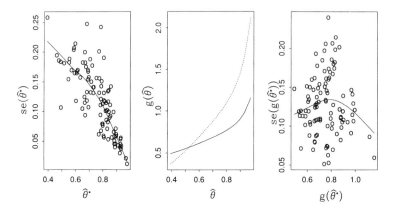

Figure 12.2. *Law school data: left panel shows a plot of se($\hat{\theta}^*$) versus $\hat{\theta}^*$, and a smooth curve se($\hat{\theta}^*$) drawn through it. The middle panel shows the estimated variance stabilizing transformation g($\hat{\theta}$) (solid curve) derived from se($\hat{\theta}^*$) and formula (12.27). The broken curve is the (standardized) transformation (12.24) that would be appropriate if the data came from a bivariate normal distribution. The right panel is the same as the left panel, with g($\hat{\theta}^*$) taking the place of $\hat{\theta}^*$. Notice how the transformation g($\cdot$) has stabilized the standard deviation.*

next two chapters, culminating in the "BC$_a$" procedure of Chapter 14. Like the bootstrap-t method, the BC$_a$ interval produces more accurate intervals than the standard normal or t intervals.

An S language function for computing bootstrap-t confidence intervals is described in the Appendix. It includes an option for automatic variance stabilization.

12.7 Bibliographic notes

Background references on bootstrap confidence intervals are given in the bibliographic notes at the end of Chapter 22.

12.8 Problems

12.1 Derive the second relation in (12.3) from the first, and then prove (12.4).

12.2 Let Z indicate a $N(0,1)$ random variable. It is then true that

$$a + bZ \sim N(a, b^2) \tag{12.28}$$

for any constants a and b.

(a) Derive (12.13) from (12.12).

(b) Derive (12.14) and (12.15).

12.3 Derive (12.22) from (12.20) and (12.21).

12.4 Suppose X is a random variable with mean θ and standard deviation $s(\theta)$, and we consider applying a transformation $g(x)$ to X.

(a) Expand $g(X)$ in a Taylor series to show that

$$\mathrm{var}(g(X)) \approx g'(\theta)^2 \mathrm{var}(X). \tag{12.29}$$

(b) Hence show that the transformation given in (12.27) has the property $\mathrm{var}(g(X)) \approx \mathit{constant}$.

(c) If X_i/θ are independently and identically distributed as χ_1^2 for $i = 1, 2, \ldots n$ and $\hat{\theta} = \bar{X}$, show that the approximate variance stabilizing transformation for $\hat{\theta}$ is

$$g(\hat{\theta}) = (n/2)^{1/2} \log \hat{\theta}. \tag{12.30}$$

12.5 Suppose X_i/θ are independently and identically distributed as χ_1^2 for $i = 1, 2, \ldots 20$. Carry out a small simulation study to compare the following intervals for θ based on $\bar{X}$, assuming that the true value of θ is one:

(a) the exact interval based on $20 \cdot \hat{\theta}/\theta \sim \chi_{20}^2$

(b) the standard interval based on $(\hat{\theta} - \theta)/\widehat{se} \,\dot\sim\, N(0,1)$ where $\widehat{se}$ is the plug-in estimate of the standard error of the mean

(c) the bootstrap-t interval based on $(\hat{\theta} - \theta)/\widehat{se}$.

(d) the bootstrap-t interval based the asymptotic variance stabilizing transformation $\hat{\phi} = \log \hat{\theta}$ (from part (c) of the previous problem).

Use at least 1000 samples in your simulation, and for each interval compute the miscoverage in each tail and the overall miscoverage, as well as the mean and standard deviation of the interval length. Discuss the results. Relate this problem to that of inference for the variance of a normal distribution.

## CHAPTER 13

Confidence intervals based on bootstrap percentiles

13.1 Introduction

In this chapter and the next, we describe another approach to boot-strap confidence intervals based on percentiles of the bootstrap distribution of a statistic. For motivation we take a somewhat different view of the standard normal-theory interval, and this leads to a generalization based on the bootstrap, the "percentile" interval. This interval is improved upon in Chapter 14, and the result is a bootstrap confidence interval with good theoretical coverage properties as well as reasonable stability in practice.

13.2 Standard normal intervals

Let $\hat{\theta}$ be the usual plug-in estimate of a parameter θ and $\widehat{se}$ be its estimated standard error. Consider the standard normal confidence interval $[\hat{\theta} - z^{(1-\alpha)} \cdot \widehat{se}, \hat{\theta} - z^{(\alpha)} \cdot \widehat{se}]$. The endpoints of this interval can be described in a way that is particularly convenient for bootstrap calculations. Let $\hat{\theta}^*$ indicate a random variable drawn from the distribution $N(\hat{\theta}, \widehat{se}^2)$,

$$\hat{\theta}^* \sim N(\hat{\theta}, \widehat{se}^2). \qquad (13.1)$$

Then $\hat{\theta}_{lo} = \hat{\theta} - z^{(1-\alpha)} \cdot \widehat{se}$ and $\hat{\theta}_{up} = \hat{\theta} - z^{(\alpha)} \cdot \widehat{se}$ are the 100αth and $100(1-\alpha)$th percentiles of $\hat{\theta}^*$. In other words,

$$\begin{aligned} \hat{\theta}_{lo} &= \hat{\theta}^{*(\alpha)} = 100 \cdot \alpha^{th} \text{ percentile of } \hat{\theta}^*\text{'s distribution} \\ \hat{\theta}_{up} &= \hat{\theta}^{*(1-\alpha)} = 100 \cdot (1-\alpha)^{th} \text{ percentile of } \hat{\theta}^*\text{'s distribution.} \end{aligned}$$
$$(13.2)$$

Consider for example the treated mice of Table 2.1 and let $\hat{\theta} = 86.85$, the mean of the 7 treated mice. The bootstrap standard

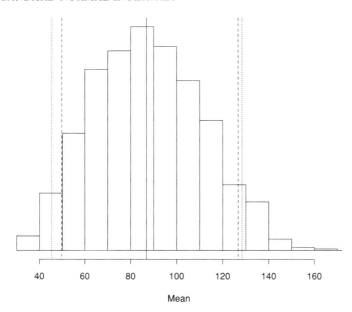

Mean

Figure 13.1. *Histogram of 1000 bootstrap replications of* $\hat{\theta}$, *the mean of the 7 treated mice in Table 2.1. The solid line is drawn at* $\hat{\theta}$. *The dotted vertical lines show standard normal 90% interval* [86.85 − 1.645 · 25.23, 86.85 + 1.645 · 25.23] = [45.3, 128.4]. *The dashed vertical lines are drawn at 49.7 and 126.7, the 5% and 95% percentiles of the histogram. Since the histogram is roughly normal-shaped, the broken and dotten lines almost coincide, in accordance with equation (13.2)*

error of $\hat{\theta}$ is 25.23, so if we choose say $\alpha = .05$, then the standard 90% normal confidence interval for the true mean θ is [86.85 − 1.645 · 25.23, 86.85 + 1.645 · 25.23] = [45.3, 128.4].

Figure 13.1 shows a histogram of 1000 bootstrap replications $\hat{\theta}^*$. This histogram looks roughly normal in shape, so according to equation (13.2) above, the 5% and 95% percentiles of this histogram should be roughly 45.3 and 128.4, respectively. This isn't a bad approximation: as shown in Table 13.1, the 5% and 95% percentiles are actually 49.7 and 126.7.

Table 13.1. *Percentiles of $\hat{\theta}^*$ based on 1000 bootstrap replications, where $\hat{\theta}$ equals the mean of the treated mice of Table 2.1.*

2.5%	5%	10%	16%	50%	84%	90%	95%	97.5%
45.9	49.7	56.4	62.7	86.9	112.3	118.7	126.7	135.4

13.3 The percentile interval

[1] The previous discussion suggests how we might use the percentiles of the bootstrap histogram to define confidence limits. This is exactly how the *percentile interval* works. Suppose we are in the general situation of Figure 8.3. A bootstrap data set $\mathbf{x}^*$ is generated according to $\hat{P} \rightarrow \mathbf{x}^*$, and bootstrap replications $\hat{\theta}^* = s(\mathbf{x}^*)$ are computed. Let $\hat{G}$ be the cumulative distribution function of $\hat{\theta}^*$. The $1 - 2\alpha$ *percentile interval* is defined by the α and $1 - \alpha$ percentiles of $\hat{G}$:

$$[\hat{\theta}_{\%,\mathrm{lo}}, \hat{\theta}_{\%,\mathrm{up}}] = [\hat{G}^{-1}(\alpha), \hat{G}^{-1}(1 - \alpha)]. \tag{13.3}$$

Since by definition $\hat{G}^{-1}(\alpha) = \hat{\theta}^{*(\alpha)}$, the $100 \cdot \alpha$th percentile of the bootstrap distribution, we can also write the percentile interval as

$$[\hat{\theta}_{\%,\mathrm{lo}}, \hat{\theta}_{\%,\mathrm{up}}] = [\hat{\theta}^{*(\alpha)}, \hat{\theta}^{*(1-\alpha)}]. \tag{13.4}$$

Expressions (13.3) and (13.4) refer to the ideal bootstrap situation in which the number of bootstrap replications is infinite. In practice we must use some finite number B of replications. To proceed, we generate B independent bootstrap data sets $\mathbf{x}^{*1}, \mathbf{x}^{*2}, \cdots, \mathbf{x}^{*B}$ and compute the bootstrap replications $\hat{\theta}^*(b) = s(\mathbf{x}^{*b})$, $b = 1, 2, \ldots B$. Let $\hat{\theta}_B^{*(\alpha)}$ be the $100 \cdot \alpha$th empirical percentile of the $\hat{\theta}^*(b)$ values, that is, the $B \cdot \alpha$th value in the ordered list of the B replications of $\hat{\theta}^*$. So if $B = 2000$ and $\alpha = .05$, $\hat{\theta}_B^{*(\alpha)}$ is the 100th ordered value of the replications. (If $B \cdot \alpha$ is not an integer, we may use the convention given after equation (12.22) of Chapter 12.) Likewise let $\hat{\theta}_B^{*(1-\alpha)}$ be the $100 \cdot (1 - \alpha)$th empirical percentile.

---

[1] The BC_a interval of Chapter 14 is more difficult to explain than the percentile interval, but not much more difficult to calculate. It gives more accurate confidence limits than the percentile method and is preferable in practice.

Table 13.2. *Percentiles of $\hat{\theta}^*$ based on 1000 bootstrap replications, where $\hat{\theta}$ equals* $\exp(\bar{x})$ *for a normal sample of size 10.*

2.5%	5%	10%	16%	50%	84%	90%	95%	97.5%
0.75	0.82	0.90	0.98	1.25	1.61	1.75	1.93	2.07

The approximate $1 - 2\alpha$ percentile interval is

$$[\hat{\theta}_{\%,\mathrm{lo}}, \hat{\theta}_{\%,\mathrm{up}}] \approx [\hat{\theta}_B^{*(\alpha)}, \hat{\theta}_B^{*(1-\alpha)}]. \tag{13.5}$$

If the bootstrap distribution of $\hat{\theta}^*$ is roughly normal, then the standard normal and percentile intervals will nearly agree (as in Figure 13.1). The central limit theorem tells us that as $n \to \infty$, the bootstrap histogram will become normal shaped, but for small samples it may look very non-normal. Then the standard normal and percentile intervals will differ. Which one should we use?

Let's examine this question in an artificial example where we know what the correct confidence interval should be. We generated a sample $X_1, X_2, \ldots X_{10}$ from a standard normal distribution. The parameter of interest θ was chosen to be e^μ, where μ is the population mean. The true value of θ was $e^0 = 1$, while the sample value $\hat{\theta} = e^{\bar{x}}$ equaled 1.25. The left panel of Figure 13.2 shows the bootstrap histogram of $\hat{\theta}^*$ based on 1000 replications (Although the population is Gaussian in this example, we didn't presuppose knowledge of this and therefore used nonparametric bootstrap sampling.)

The distribution is quite asymmetric, having a long tail to the left. Empirical percentiles of the 1000 $\hat{\theta}^*$ replications are shown in Table 13.2.

The .95 percentile interval for θ is

$$[\hat{\theta}_{\%,\mathrm{lo}}, \hat{\theta}_{\%,\mathrm{up}}] = [0.75, 2.07]. \tag{13.6}$$

This should be compared with the .95 standard interval based on $\widehat{\mathrm{se}}_{1000} = 0.34$,

$$1.25 \pm 1.96 \cdot 0.34 = [0.59, 1.92]. \tag{13.7}$$

Notice the large discrepancy between the standard normal and percentile intervals. There is a good reason to prefer the percentile interval (13.6) to the standard interval (13.7). First note that there is an obvious objection to (13.7). The left panel of Figure 13.2

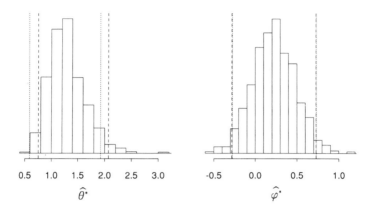

Figure 13.2. *Left panel: $B = 1000$ bootstrap replications of $\hat{\theta} = \exp(\bar{x})$, from a standard normal sample of size 10. The vertical dotted lines show the standard normal interval $1.25 \pm 1.96 \cdot 0.34 = [.59, 1.92]$, while the dashed lines are drawn at the 2.5% and 97.5% percentiles 0.75 and 2.07. These percentiles give the .95 percentile confidence interval, namely $[0.75, 2.07]$. Right panel: Same as left panel, except that $\phi = \log \theta$ and $\hat{\phi} = \log \hat{\theta}$ replace θ and $\hat{\theta}$ respectively.*

shows that the normal approximation $\hat{\theta} \dot\sim N(\theta, \widehat{se}^2)$ which underlies the standard intervals just isn't very accurate in this case. Clearly the logarithmic transformation makes the distribution of $\hat{\theta}$ normal. The right panel of Figure 13.2 shows the bootstrap histogram of 1000 values of $\hat{\phi}^* = \log(\hat{\theta}^*)$, along with the standard normal and percentile intervals for ϕ. Notice that the histogram is much more normal in shape than that for $\hat{\theta}^*$. This isn't surprising since $\hat{\phi}^* = \bar{x}^*$. The standard normal interval for $\phi = \log(\theta)$ is $[-0.28, 0.73]$ while the percentile interval is $[-0.29, 0.73]$. Because of the normal shape of the histogram, these intervals agree more closely than they do in the left panel. Since the histogram in the right panel of Figure 13.2 appears much more normal than that in the left panel, it seems reasonable to base the standard interval on $\hat{\phi}$, and then map the endpoints back to the θ scale, rather than to base them directly on $\hat{\theta}$.

The inverse mapping of the logarithm is the exponential function. Using the exponential function to map the standard interval back to the θ scale gives $[0.76, 2.08]$. This interval is closer to the percentile interval $[0.75, 2.07]$ than is the standard interval

$[0.59, 1.92]$ constructed using $\hat{\theta}$ directly.

We see that the percentile interval for θ agrees well with a standard normal interval constructed on an appropriate transformation of θ and then mapped to the θ scale. The difficulty in improving the standard method in this way is that we need to know a different transformation like the logarithm for each parameter θ of interest. The percentile method can be thought of as an algorithm for automatically incorporating such transformations.

The following result formalizes the fact that the percentile method always "knows" the correct transformation:

Percentile interval lemma. Suppose the transformation $\hat{\phi} = m(\hat{\theta})$ perfectly normalizes the distribution of $\hat{\theta}$:

$$\hat{\phi} \sim N(\phi, c^2) \qquad (13.8)$$

for some standard deviation c. Then the percentile interval based on $\hat{\theta}$ equals $[m^{-1}(\hat{\phi} - z^{(1-\alpha)}c), m^{-1}(\hat{\phi} - z^{(\alpha)}c)]$.

In the setup of Figure 8.3 in Chapter 8, where the probability mechanism P associated with parameter θ gives the data x. we are assuming that $\hat{\phi} = m(\hat{\theta})$ and $\phi = m(\theta)$ satisfy (13.8) for every choice of P. Under this assumption, the lemma is little more than a statement that the percentile method transforms endpoints correctly. See Problems 13.1 and 13.2.

The reader can think of the percentile method as a computational algorithm for extending the range of effectiveness of the standard intervals. In situations like that of Figure 13.1, $\hat{\theta} \sim N(\theta, \widehat{se}^2)$, where the standard intervals are nearly correct, the percentile intervals agree with them. In situations like that of the left panel of Figure 13.2, where the standard intervals would be correct if we transformed parameters from θ to ϕ, the percentile method automatically makes this transformation. The advantage of the percentile method is that we don't need to know the correct transformation. All we assume is that such a transformation exists.

In the early 1920's Sir Ronald Fisher developed maximum likelihood theory, which automatically gives efficient estimates $\hat{\theta}$ and standard errors $\widehat{se}$ in a wide variety of situations. (Chapter 21 discusses the close connection between maximum likelihood theory and the bootstrap.) Fisher's theory greatly increased the use of the standard intervals, by making them easier to calculate and better justified. Since then, statisticians have developed many tricks for improving the practical performance of the standard intervals. Among these is a catalogue of transformations that make certain types of problems better fit the ideal situation $\hat{\theta} \sim N(\theta, \widehat{se}^2)$. The percentile interval extends the usefulness of the standard normal interval without requiring explicit knowledge of this catalogue of transformations.

13.4 Is the percentile interval backwards?

The percentile interval uses $\hat{G}^{-1}(\alpha)$ as the left endpoint of the confidence interval for θ and $\hat{G}^{-1}(1 - \alpha)$ as the right endpoint. The bootstrap-t approach of the previous chapter uses the bootstrap to estimate the distribution of a studentized (approximate) pivot, and then inverts the pivot to obtain a confidence interval. To compare this with the percentile interval, consider what happens if we simplify the bootstrap-t and base the interval on $\hat{\theta} - \theta$. That is, we set the denominator of the pivot equal to 1. It is easy to show (Problem 13.5) that the resulting interval is

$$[2\hat{\theta} - \hat{G}^{-1}(1 - \alpha), 2\hat{\theta} - \hat{G}^{-1}(\alpha)]. \tag{13.9}$$

Notice that if $\hat{G}$ has a long *right* tail then this interval is long on the *left*, opposite in behavior to the percentile interval.

Which is correct? Neither of these intervals works well in general: in the latter case we should start with $(\hat{\theta} - \theta)/\widehat{se}$ rather than $\hat{\theta} - \theta$ (see Section 22.3), while the percentile interval may need further refinements as described in the next chapter. However in some simple examples we can see that the percentile interval is more appropriate. For the correlation coefficient discussed in Chapter 12 (in the normal model), the quantity $\hat{\phi} - \phi$, where ϕ is Fisher's transform (12.24), is well approximated by a normal distribution and hence the percentile interval is accurate. In contrast, the quantity $\hat{\theta} - \theta$ is far from pivotal so that the interval (13.9) is not very accurate. Another example concerns inference for the median. The percentile interval matches closely the order statistic-based interval, while (13.9) is backwards. Details are in Efron (1979a).

13.5 Coverage performance

The arguments in favor of the percentile interval should translate into better coverage performance. Table 13.3 investigates this in

Table 13.3. *Results of 300 confidence interval realizations for* $\theta = \exp(\mu)$ *from a standard normal sample of size 10. The table shows the percentage of trials that the indicated interval missed the true value 1.0 on the left or right side. For example, "Miss left" means that the left endpoint was* > 1.0. *The desired coverage is 95%, so the ideal values of Miss left and Miss right are both 2.5%.*

Method	% Miss left	% Miss right
Standard normal $\hat{\theta} \pm 1.96\widehat{\text{se}}$	1.2	8.8
Percentile (Nonparametric)	4.8	5.2

the context of the normal example of Figure 13.2.

It shows the percentage of times that the standard and percentile intervals missed the true value on the left and right sides, in 500 simulated samples. The target miscoverage is 2.5% on each side. The standard interval overcovers on the left and undercovers on the right. The percentile interval achieves better balance in the left and right sides, but like the standard interval it still undercovers overall. This is a consequence of non-parametric inference: the percentile interval has no knowledge of the underlying normal distribution and uses the empirical distribution in its place. In this case, it underestimates the tails of the distribution of $\hat{\theta}^*$. More advanced bootstrap intervals like those discussed in Chapters 14 and 22 can partially correct this undercoverage.

13.6 The transformation-respecting property

Let's look back again at the right panel of Figure 13.2. The 95% percentile interval for ϕ turns out to be $[-0.29, 0.73]$. What would we get if we transformed this back to the θ scale via the inverse transformation $(\exp \phi)$? The transformed interval is $[-0.75, 2.07]$, which is exactly the percentile interval for θ. In other words, the percentile interval is *transformation-respecting*: the percentile interval for any (monotone) parameter transformation $\phi = m(\theta)$ is simply the percentile interval for θ mapped by $m(\theta)$:

$$[\hat{\phi}_{\%,\text{lo}}, \hat{\phi}_{\%,\text{up}}] = [m(\hat{\theta}_{\%,\text{lo}}), m(\hat{\theta}_{\%,\text{up}})]. \qquad (13.10)$$

The same property holds for the empirical percentiles based on B bootstrap samples (Problem 13.3).

As we have seen in the correlation coefficient example above, the standard normal interval is not transformation-respecting. This property is an important practical advantage of the percentile method.

13.7 The range-preserving property

For some parameters, there is a restriction on the values that the parameter can take. For example, the values of the correlation coefficient lie in the interval $[-1, 1]$. Clearly it would be desirable if a confidence procedure always produced intervals that fall within the allowable range: such an interval is called *range-preserving*. The percentile interval is range-preserving, since a) the plug-in estimate $\hat{\theta}$ obeys the same range restriction as θ, and b) its endpoints are values of the bootstrap statistic $\hat{\theta}^*$, which again obey the same range restriction as θ. In contrast, the standard interval need not be range-preserving. Confidence procedures that are range-preserving tend to be more accurate and reliable.

13.8 Discussion

The percentile method is not the last word in bootstrap confidence intervals. There are other ways the standard intervals can fail, besides non-normality. For example $\hat{\theta}$ might be a *biased* normal estimate,

$$\hat{\theta} \sim N(\theta + \text{bias}, \widehat{\text{se}}^2), \tag{13.11}$$

in which case no transformation $\phi = m(\theta)$ can fix things up. Chapter 14 discusses an extension of the percentile method that automatically handles both bias and transformations. A further extension allows the standard error in (13.11) to vary with θ, rather than being forced to stay constant. This final extension will turn out to have an important theoretical advantage.

13.9 Bibliographic notes

Background references on bootstrap confidence intervals are given in the bibliographic notes at the end of Chapter 22.

13.10 Problems

13.1 Prove the transformation-respecting property of the percentile interval (13.10). Use this to verify the percentile interval lemma.

13.2 (a) Suppose we are in the one-sample nonparametric setting of Chapter 6, where $F \to \mathbf{x}$, $\hat{\theta} = t(\hat{F})$. Why can relation (13.8) not hold exactly in this case?

(b) Give an example of a parametric situation $\hat{P} \to \mathbf{x}$ in which (13.8) holds exactly.

13.3 Prove that the <u>approximate</u> percentile interval (13.4) is transformation-respecting, as defined in (13.10).

13.4 Carry out a simulation study like that in Table 13.3 for the following problem: $x_1, x_2, \ldots x_{20}$ are each independent with an exponential distribution having mean θ. (An exponential variate with mean θ may be defined as $-\theta \log U$ where U is a standard uniform variate on $[0, 1]$.) The parameter of interest is $\theta = 1$. Compute the coverage of the standard and percentile intervals, and give an explanation for your results.

13.5 Suppose that we estimate the distribution of $\hat{\theta} - \theta$ by the bootstrap distribution of $\hat{\theta}^* - \hat{\theta}$. Denote the α-percentile of $\hat{\theta}^* - \hat{\theta}$ by $\hat{H}^{-1}(\alpha)$. Show that the interval for θ that results from inverting the relation

$$\hat{H}^{-1}(\alpha) \leq \hat{\theta} - \theta \leq \hat{H}^{-1}(1 - \alpha) \qquad (13.12)$$

is given by expression (13.9).

## CHAPTER 14

Better bootstrap confidence intervals

14.1 Introduction

One of the principal goals of bootstrap theory is to produce good confidence intervals automatically. "Good" means that the bootstrap intervals should closely match exact confidence intervals in those special situations where statistical theory yields an exact answer, and should give dependably accurate coverage probabilities in all situations. Neither the bootstrap-t method of Chapter 12 nor the percentile method of Chapter 13 passes these criteria. The bootstrap-t intervals have good theoretical coverage probabilities, but tend to be erratic in actual practice. The percentile intervals are less erratic, but have less satisfactory coverage properties.

This chapter discusses an improved version of the percentile method called BC_a, the abbreviation standing for *bias-corrected and accelerated*. The BC_a intervals are a substantial improvement over the percentile method in both theory and practice. They come close to the criteria of goodness given above, though their coverage accuracy can still be erratic for small sample sizes. (Improvements are possible, as shown in Chapter 25.) A simple computer algorithm called **bcanon**, listed in the Appendix, produces the BC_a intervals on a routine basis, with little more effort required than for the percentile intervals. We also discuss a method called ABC, standing for *approximate bootstrap confidence* intervals, which reduces by a large factor the amount of computation required for the BC_a intervals. The chapter ends with an application of these methods to a real data analysis problem.

14.2 Example: the spatial test data

Our next example, the *spatial test data*, demonstrates the need for improvements on the percentile and bootstrap-t methods. Twenty-six neurologically impaired children have each taken two tests of spatial perception, called "A" and "B." The data are listed in Table 14.1 and displayed in Figure 14.1. Suppose that we wish to find a 90% central confidence interval for $\theta = \text{var}(A)$, the variance of a random A score.

The plug-in estimate of θ based on the $n = 26$ data pairs $x_i = (A_i, B_i)$ in Table 14.1 is

$$\widehat{\theta} = \sum_{i=1}^{n} (A_i - \bar{A})^2/n = 171.5 , \qquad (\bar{A} = \sum_{1}^{n} A_i/n). \quad (14.1)$$

Notice that this is slightly smaller than the usual unbiased estimate of θ,

$$\bar{\theta} = \sum_{i=1}^{n} (A_i - \bar{A})^2/(n-1) = 178.4. \quad (14.2)$$

The plug-in estimate $\widehat{\theta}$ is biased downward. The BC_a method automatically corrects for bias in the plug-in estimate, which is one of its advantages over the percentile method. [1]

A histogram of 2000 bootstrap replications $\widehat{\theta}^*$ appears in the left panel of Figure 14.2. The replications are obtained as in Figure 6.1: if $\mathbf{x} = (x_1, x_2, \cdots, x_{26})$ represents the original data set of Table 14.1, where $x_i = (A_i, B_i)$ for $i = 1, 2, \cdots, 26$, then a bootstrap data set $\mathbf{x}^* = (x_1^*, x_2^*, \cdots, x_{26}^*)$ is a random sample of size 26 drawn with replacement from $\{x_1, x_2, \cdots, x_{26}\}$; the bootstrap replication $\widehat{\theta}^*$ is the variance of the A components of $\mathbf{x}^*$, with $x_i^* = (A_i^*, B_i^*)$,

$$\widehat{\theta}^* = \sum_{i=1}^{n} (A_i^* - \bar{A}^*)^2/n \qquad (\bar{A}^* = \sum_{i=1}^{n} A_i^*/n). \quad (14.3)$$

$B = 2000$ bootstrap samples $\mathbf{x}^*$ gave the 2000 bootstrap replications $\widehat{\theta}^*$ in Figure 14.2. [2] These are *nonparametric* bootstrap

---

[1] The discussion in this chapter, and the algorithms **bcanon** and **abcnon** in the Appendix, assume that the statistic is of the plug-in form $\widehat{\theta} = t(\widehat{F})$.

[2] We don't need the second components of the x_i^* for this particular calculation, see Problem 14.2.

Table 14.1. *Spatial Test Data; n = 26 children have each taken two tests of spatial ability, called A and B.*

	1	2	3	4	5	6	7	8	9	10	11	12	13
A	48	36	20	29	42	42	20	42	22	41	45	14	6
B	42	33	16	39	38	36	15	33	20	43	34	22	7

	14	15	16	17	18	19	20	21	22	23	24	25	26
A	0	33	28	34	4	32	24	47	41	24	26	30	41
B	15	34	29	41	13	38	25	27	41	28	14	28	40

replications, the kind we have discussed in the previous chapters. Later we also discuss *parametric* bootstrap replications, referring in this chapter to a Normal, or Gaussian model for the data. In the notation of Chapter 6, a nonparametric bootstrap sample is generated by random sampling from $\widehat{F}$,

$$\widehat{F} \to \mathbf{x}^* = (x_1^*, x_2^*, \cdots, x_n^*), \tag{14.4}$$

where $\widehat{F}$ is the empirical distribution, putting probability $1/n$ on each x_i.

The top panel of Table 14.2 shows five different approximate 90% nonparametric confidence intervals for θ: the standard interval $\widehat{\theta} \pm 1.645\widehat{\sigma}$, where $\widehat{\sigma} = 41.0$, the bootstrap estimate of standard error; the percentile interval $(\widehat{\theta}^{*(.05)}, \widehat{\theta}^{*(.95)})$ based on the left histogram in Figure 14.2; the BC_a and ABC intervals, discussed in the next two sections; and the bootstrap-t intervals of Chapter 12. Each interval $(\widehat{\theta}_{\text{lo}}, \widehat{\theta}_{\text{up}})$ is described by its length and shape,

$$\text{length} = \widehat{\theta}_{\text{up}} - \widehat{\theta}_{\text{lo}}, \qquad \text{shape} = \frac{\widehat{\theta}_{\text{up}} - \widehat{\theta}}{\widehat{\theta} - \widehat{\theta}_{\text{lo}}}. \tag{14.5}$$

"Shape" measures the asymmetry of the interval about the point estimate $\widehat{\theta}$. Shape > 1.00 indicates greater distance from $\widehat{\theta}_{\text{up}}$ to $\widehat{\theta}$ than from $\widehat{\theta}$ to $\widehat{\theta}_{\text{lo}}$. The standard intervals are symmetrical about $\widehat{\theta}$, having shape = 1.00 by definition. Exact intervals, when they exist, are often quite asymmetrical. The most serious errors made by standard intervals are due to their enforced symmetry.

In the spatial test problem the standard and percentile intervals are almost identical. They are both quite different than the BC_a

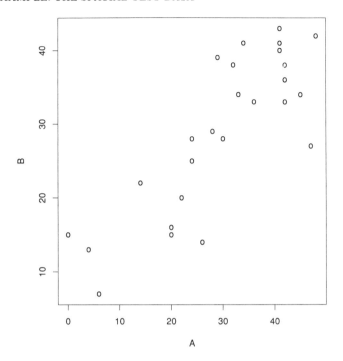

Figure 14.1. *The spatial test data of Table 14.1.*

and ABC intervals, which are longer and asymmetric to the right of $\hat{\theta}$. A general result quoted in Section 13.2 strongly suggests the superiority of the BC_a and ABC intervals, but there is no gold standard by which we can make a definitive comparison.

We can obtain a gold standard by considering the problem of estimating $\text{var}(A)$ in a normal, [3] or Gaussian, parametric framework. To do so, we assume that the data points $x_i = (A_i, B_i)$ are a random sample from a two-dimensional normal distribution F_{norm},

$$F_{\text{norm}} \to \mathbf{x} = (x_1, x_2, \cdots, x_n). \tag{14.6}$$

In the normal-theory framework we can construct an exact con-

---

[3] In fact the normal distribution gives a poor fit to the spatial test data. This does not affect the comparisons below, which compare how well the various methods would approximate the exact interval *if* the normal assumption were valid. However if we compare the normal and nonparametric intervals, the latter are preferable for this data set.

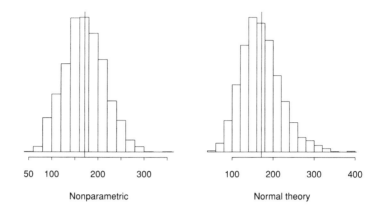

50   100   200   300   100   200   300   400

Nonparametric   Normal theory

Figure 14.2. *Left panel: 2000 nonparametric bootstrap replications of the variance* $\widehat{\theta}$*, (14.2); Right panel: 2000 normal-theory parametric bootstrap replications of* $\widehat{\theta}$*. A solid vertical line is drawn at* $\hat{\theta}$ *in each histogram. The parametric bootstrap histogram is long-tailed to the right. These histograms are used to form the percentile and* BC_a *intervals in Table 14.2.*

fidence interval for $\theta = \mathrm{var}(A)$. See Problem 14.4. This interval, called "exact" in Table 14.2, is a gold standard for judging the various approximate intervals, in the parametric setting.

Normal-theory parametric bootstrap samples are obtained by sampling from the bivariate normal distribution $\widehat{F}_{\mathrm{norm}}$ that best fits the data $\mathbf{x}$, instead of from the empirical distribution $\widehat{F}$,

$$\widehat{F}_{\mathrm{norm}} \to \mathbf{x}^* = (x_1^*, x_2^*, \cdots, x_n^*). \qquad (14.7)$$

See Problem 14.3. Having obtained $\mathbf{x}^*$, the bootstrap replication $\widehat{\theta}^*$ equals $\sum_1^n (A_i^* - \bar{A}^*)^2/n$ as in (14.3). The right panel of Figure 14.2 is the histogram of 2000 normal-theory bootstrap replications. Compared to the nonparametric case, this histogram is longer-tailed to the right, and wider, having $\hat{\sigma} = 47.1$ compared to the nonparametric standard error of 41.0.

Looking at the bottom of Table 14.2, we see that the BC_a and ABC intervals [4] do a much better job than the standard or percentile methods of matching the exact gold standard. This is not an accident or a special case. As a matter of fact bootstrap theory, described briefly in Section 14.3, says that we should expect

---

[4] Parametric BC_a and ABC methods are discussed in Chapter 22, with algorithms given in the Appendix.

Table 14.2. _Top:_ *five different approximate 90% nonparametric confidence intervals for* $\theta = var(A)$; *in this case the standard and percentile intervals are nearly the same; the* BC_a *and ABC intervals are longer, and asymmetric around the point estimate* $\hat{\theta} = 171.5$. _Bottom:_ *parametric normal-theory intervals. In the normal case there is an exact confidence interval for* θ. *Notice how much better the exact interval is approximated by the* BC_a *and ABC intervals.* _Bottom line:_ *the bootstrap-t intervals are nearly exact in the parametric case, but give too large an upper limit nonparametrically.*

Nonparametric

method	0.05	0.95	length	shape
standard	98.8	233.6	134.8	1.00
percentile	100.8	233.9	133.1	0.88
BC_a	115.8	259.6	143.8	1.58
ABC	116.7	260.9	144.2	1.63
bootstrap-t	112.3	314.8	202.5	2.42

Parametric (Normal-Theory)

method	0.05	0.95	length	shape
standard	91.9	251.2	159.3	1.00
percentile	95.0	248.6	153.6	1.01
BC_a	114.6	294.7	180.1	2.17
ABC	119.3	303.4	184.1	2.52
exact	118.4	305.2	186.8	2.52
bootstrap-t	119.4	303.6	184.2	2.54

superior performance from the BC_a/ABC intervals.

Bootstrap-t intervals for θ appear in the bottom lines of Table 14.2. These were based on 1000 bootstrap replications of the t-like statistic $(\hat{\theta} - \theta)/\hat{se}$, with a denominator suggested by standard statistical theory,

$$\hat{se} = [\frac{U_4 - U_2^2}{26}]^{1/2} \qquad (U_h = \sum_{i=1}^{26}(A_i - \bar{A})^h/26). \qquad (14.8)$$

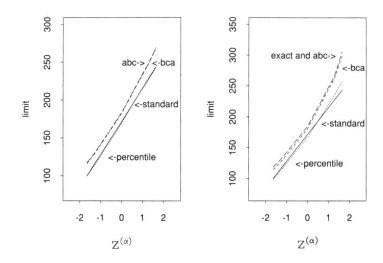

Figure 14.3. *A comparison of various approximate confidence intervals for $\theta = var(A)$, spatial test data; interval endpoint $\widehat{\theta}[\alpha]$ is plotted versus $\Phi^{-1}(\alpha) = z^{(\alpha)}$. Left panel: nonparametric intervals. Right panel: normal-theory parametric intervals. In the parametric case we can see that the* BC_a *and* ABC *endpoints are close to the exact answer.*

The resulting intervals, (12.19), are almost exactly right in the normal-theory situations. However the upper limit of the nonparametric interval appears to be much too large, though it is difficult to be certain in the absence of a nonparametric gold standard. At the present level of development the bootstrap-t cannot be recommended for general nonparametric problems.

14.3 The BC_a method

This section describes the construction of the BC_a intervals. These are more complicated to define than the percentile intervals, but almost as easy to use. The algorithm **bcanon** given in the Appendix produces the nonparametric BC_a intervals on a routine automatic basis.

Let $\widehat{\theta}^{*(\alpha)}$ indicate the $100 \cdot \alpha$th percentile of B bootstrap replications $\widehat{\theta}^*(1), \widehat{\theta}^*(2), \cdots, \widehat{\theta}^*(B)$, as in (13.5). The percentile interval $(\widehat{\theta}_{\mathrm{lo}}, \widehat{\theta}_{\mathrm{up}})$ of intended coverage $1 - 2\alpha$, is obtained directly from these percentiles,

$$\text{percentile method:} \quad (\widehat{\theta}_{\mathrm{lo}}, \widehat{\theta}_{\mathrm{up}}) = (\widehat{\theta}^{*(\alpha)}, \widehat{\theta}^{*(1-\alpha)}).$$

For example, if $B = 2000$ and $\alpha = .05$, then the percentile interval $(\widehat{\theta}^{*(.05)}, \widehat{\theta}^{*(.95)})$ is the interval extending the 100th to the 1900th ordered values of the 2000 numbers $\widehat{\theta}^*(b)$.

The BC$_a$ interval endpoints are also given by percentiles of the bootstrap distribution, but not necessarily the same ones as in (14.8). The percentiles used depend on two numbers $\hat{a}$ and $\hat{z}_0$, called the *acceleration* and *bias-correction*. (BC$_a$ stands for *bias-corrected and accelerated*.) Later we will describe how $\hat{a}$ and $\hat{z}_0$ are obtained, but first we give the definition of the BC$_a$ interval endpoints.

The BC$_a$ *interval of intended coverage* $1 - 2\alpha$, *is given by*

$$\text{BC}_a \; : \; (\widehat{\theta}_{\mathrm{lo}}, \widehat{\theta}_{\mathrm{up}}) = (\widehat{\theta}^{*(\alpha_1)}, \widehat{\theta}^{*(\alpha_2)}), \tag{14.9}$$

where

$$
\begin{aligned}
\alpha_1 &= \Phi\left(\hat{z}_0 + \frac{\hat{z}_0 + z^{(\alpha)}}{1 - \hat{a}(\hat{z}_0 + z^{(\alpha)})}\right) \\
\alpha_2 &= \Phi(\hat{z}_0 + \frac{\hat{z}_0 + z^{(1-\alpha)}}{1 - \hat{a}(\hat{z}_0 + z^{(1-\alpha)})})
\end{aligned} \tag{14.10}
$$

Here $\Phi(\cdot)$ is the standard normal cumulative distribution function and $z^{(\alpha)}$ is the 100αth percentile point of a standard normal distribution. For example $z^{(.95)} = 1.645$ and $\Phi(1.645) = .95$.

Formula (14.10) looks complicated, but it is easy to compute. Notice that if $\hat{a}$ and $\hat{z}_0$ equal zero, then

$$\alpha_1 = \Phi(z^{(\alpha)}) = \alpha \quad \text{and} \quad \alpha_2 = \Phi(z^{(1-\alpha)}) = 1 - \alpha, \tag{14.11}$$

so that the BC$_a$ interval (14.9) is the same as the percentile interval (13.4). Non-zero values of $\hat{a}$ or $\hat{z}_0$ change the percentiles used for the BC$_a$ endpoints. These changes correct certain deficiencies of the standard and percentile methods, as explained in Chapter 22. The nonparametric BC$_a$ intervals in Table 14.2 are based on the values

$$(\hat{a}, \hat{z}_0) = (.061, .146), \tag{14.12}$$

giving

$$(\alpha_1, \alpha_2) = (.110, .985) \qquad (14.13)$$

according to (14.10). In this case the 90% BC_a interval is $(\widehat{\theta}^{*(.110)}, \widehat{\theta}^{*(.985)})$, the interval extending from the 220th to the 1970th ordered value of the 2000 numbers $\widehat{\theta}^*(b)$.

How are $\hat{a}$ and $\hat{z}_0$ computed? The value of the bias-correction $\hat{z}_0$ is obtained directly from the proportion of bootstrap replications less than the original estimate $\widehat{\theta}$,

$$\hat{z}_0 = \Phi^{-1}\left(\frac{\#\{\widehat{\theta}^*(b) < \widehat{\theta}\}}{B}\right), \qquad (14.14)$$

$\Phi^{-1}(\cdot)$ indicating the inverse function of a standard normal cumulative distribution function, e.g., $\Phi^{-1}(.95) = 1.645$. The left histogram of Figure 14.2 has 1116 of the 2000 $\widehat{\theta}^*$ values less than $\widehat{\theta} = 171.5$, so $\hat{z}_0 = \Phi^{-1}(.558) = .146$. Roughly speaking, $\hat{z}_0$ measures the median bias of $\widehat{\theta}^*$, that is, the discrepancy between the median of $\widehat{\theta}^*$ and $\widehat{\theta}$, in normal units. We obtain $\hat{z}_0 = 0$ if exactly half of the $\widehat{\theta}^*(b)$ values are less than or equal to $\widehat{\theta}$.

There are various ways to compute the acceleration $\hat{a}$. The easiest to explain is given in terms of the *jackknife values* of a statistic $\widehat{\theta} = s(\mathbf{x})$. Let $\mathbf{x}_{(i)}$ be the original sample with the ith point x_i deleted, let $\widehat{\theta}_{(i)} = s(\mathbf{x}_{(i)})$, and define $\widehat{\theta}_{(\cdot)} = \sum_{i=1}^n \widehat{\theta}_{(i)}/n$, as discussed at the beginning of Chapter 11. A simple expression for the acceleration is

$$\hat{a} = \frac{\sum_{i=1}^n (\widehat{\theta}_{(\cdot)} - \widehat{\theta}_{(i)})^3}{6\{\sum_{i=1}^n (\widehat{\theta}_{(\cdot)} - \widehat{\theta}_{(i)})^2\}^{3/2}}. \qquad (14.15)$$

The statistic $s(\mathbf{x}) = \sum_{i=1}^n (A_i - \bar{A})^2/n$, (14.2), has $\hat{a} = .061$ for the spatial test data. Both $\hat{a}$ and $\hat{z}_0$ are computed automatically by the nonparametric BC_a algorithm bcanon. The quantity $\hat{a}$ is called the *acceleration* because it refers to the rate of change of the standard error of $\widehat{\theta}$ with respect to the true parameter value θ. The standard normal approximation $\widehat{\theta} \sim N(\theta, \text{se}^2)$ assumes that the standard error of $\widehat{\theta}$ is the same for all θ. However, this is often unrealistic and the acceleration constant $\hat{a}$ corrects for this. For instance, in the present example where $\widehat{\theta}$ is the variance, it is clear in the normal theory case that $\text{se}(\widehat{\theta}) \sim \theta$ (Problem 14.4). In actual fact, $\hat{a}$ refers to the rate of change of the standard error of $\widehat{\theta}$ with respect to the true parameter value θ, *measured on a normalized scale*. It is not all obvious why the formula (14.15) should provide an estimate of the acceleration of the standard error: some discussion of this may be found in Efron (1987).

The BC$_a$ method can be shown to have two important theoretical advantages. First of all, it is *transformation respecting*, [5] as in (13.10). This means that the BC$_a$ endpoints transform correctly if we change the parameter of interest from θ to some function of θ. For example, the BC$_a$ confidence intervals for $\sqrt{\operatorname{var}(A)} = \sqrt{\theta}$ are obtained by taking the square roots of the BC$_a$ endpoints in Table 14.2. The transformation-respecting property saves us from concerns like those in Section 12.6, where we worried about the proper choice of scale for the bootstrap-t intervals. A transformation-respecting method like BC$_a$ in effect automatically chooses its own best scale.

The second advantage of the BC$_a$ method concerns its accuracy. A central $1 - 2\alpha$ confidence interval $(\widehat{\theta}_{\text{lo}}, \widehat{\theta}_{\text{up}})$ is supposed to have probability α of not covering the true value of θ from above or below,

$$\operatorname{Prob}\{\theta < \widehat{\theta}_{\text{lo}}\} = \alpha \quad \text{and} \quad \operatorname{Prob}\{\theta > \widehat{\theta}_{\text{up}}\} = \alpha. \qquad (14.16)$$

Approximate confidence intervals can be graded on how accurately they match (14.16). The BC$_a$ intervals can be shown to be *second-order accurate*. This means that its errors in matching (14.16) go to zero at rate $1/n$ in terms of the sample size n,

$$\operatorname{Prob}\{\theta < \widehat{\theta}_{\text{lo}}\} \doteq \alpha + \frac{c_{\text{lo}}}{n} \quad \text{and} \quad \operatorname{Prob}\{\theta > \widehat{\theta}_{\text{up}}\} \doteq \alpha + \frac{c_{\text{up}}}{n} \qquad (14.17)$$

for two constants c_{lo} and c_{up}. The standard and percentile methods are only *first-order accurate*, meaning that the errors in matching (14.16) are an order of magnitude larger,

$$\operatorname{Prob}\{\theta < \widehat{\theta}_{\text{lo}}\} \doteq \alpha + \frac{c_{\text{lo}}}{\sqrt{n}} \quad \text{and} \quad \operatorname{Prob}\{\theta > \widehat{\theta}_{\text{up}}\} \doteq \alpha + \frac{c_{\text{up}}}{\sqrt{n}}, \qquad (14.18)$$

---

[5] This statement is strictly true if we modify definition (14.15) of $\hat{a}$ to use derivatives instead of finite differences, as in Chapter 22. In practice, this modification makes little difference.

the constants c_{lo} and c_{up} being possibly different from those above. The difference between first and second order accuracy is not just a theoretical nicety. It leads to much better approximations of exact endpoints when exact endpoints exist, as seen on the right side of Table 14.2.

The bootstrap-t method is second-order accurate, but not transformation respecting. The percentile method is transformation respecting but not second-order accurate. The standard method is neither, while the BC_a method is both. At the present level of development, the BC_a intervals are recommended for general use, especially for nonparametric problems. This is not to say that they are perfect or cannot be made better: in the case study of Chapter 25, Section 25.6 uses a second layer of bootstrap computations to improve upon the BC_a/ABC intervals. Problem 14.13 describes a difficulty that can occur with the BC_a interval in extreme situations.

A typical call to the S language function bcanon has the form

$$bcanon(x, nboot, theta), \qquad (14.19)$$

where x is the data, nboot is the number of bootstrap replications, and theta computes the statistic of interest $\widehat{\theta}$. More details may be found in the Appendix.

14.4 The ABC method

The main disadvantage of the BC_a method is the large number of bootstrap replications required. The discussion in Chapter 19 shows that at least $B = 1000$ replications are needed in order to sufficiently reduce the Monte Carlo sampling error. ABC, standing for *approximate bootstrap confidence* intervals, is a method of approximating the BC_a interval endpoints analytically, without using any Monte Carlo replications at all. The approximation is usually quite good, as seen in Table 14.2. (The differences between the BC_a and ABC endpoints in Table 14.2 are due largely to Monte Carlo fluctuations in the BC_a endpoints. Increasing B to 10,000 parametric replications gave a BC_a interval (118.4, 303.8), nearly identical to the ABC interval.)

The ABC method is explained in Chapter 22. It works by approximating the bootstrap random sampling results by Taylor series expansions. These require that the statistic $\widehat{\theta} = s(\mathbf{x})$ be defined smoothly in $\mathbf{x}$. An example of an unsmooth statistic is the sample median. For most commonly occurring statistics the ABC approximation is quite satisfactory. (A counterexample appears in Section 14.5.) The ABC endpoints are both transformation respecting and second-order accurate, like their BC_a counterparts. In Table 14.2, the ABC endpoints required only 3% of the computational effort for the BC_a intervals.

The nonparametric ABC endpoints in Table 14.2 were obtained from the algorithm abcnon given in the Appendix. In order to use this algorithm, the statistic $\widehat{\theta} = s(\mathbf{x})$ must be represented in *resampling form*. The resampling form plays a key role in advanced explanations of the bootstrap, as seen in Chapter 20. We defined the resampling form in Section 10.4. With the original sample $\mathbf{x} = (x_1, x_2, \cdots, x_n)$ considered to be fixed, we write the bootstrap value $\widehat{\theta}^* = s(\mathbf{x}^*)$ as a function of the resampling vector $\mathbf{P}^*$, say

$$\widehat{\theta}^* = T(\mathbf{P}^*). \tag{14.20}$$

The vector $P^* = (P_1^*, P_2^*, \cdots, P_n^*)$ consists of the proportions

$$P_i^* = N_i^*/n = \frac{\#\{x_j^* > x_i\}}{n} \qquad (i = 1, 2, \cdots, n). \tag{14.21}$$

The statistic $\widehat{\theta}^* = \sum_{i=1}^n (A_i^* - \bar{A}^*)^2/n$, (14.3), can be expressed in form (14.20) as

$$\widehat{\theta}^* = \sum_{i=1}^n P_i^* (A_i - \bar{A}^*)^2 \quad \text{where} \quad \bar{A}^* = \sum_{i=1}^n P_i^* A_i. \tag{14.22}$$

The function $T(\mathbf{P}^*)$ in (14.20) is the resampling form of the statistic used in the ABC algorithm abcnon. Recall that the special resampling vector

$$\mathbf{P}^0 = (1/n, 1/n, \cdots, 1/n) \tag{14.23}$$

has $T(\mathbf{P}^0) = \widehat{\theta}$, the original value of the statistic, since $\mathbf{P}^0$ corresponds to choosing each x_i once in the bootstrap sample: $\mathbf{x}^* = \mathbf{x}$. The algorithm abcnon requires $T(\mathbf{P}^*)$ to be smoothly defined for $\mathbf{P}^*$ near $\mathbf{P}^0$. This happens naturally, as in (14.22), for plug-in statistics $\widehat{\theta} = t(\widehat{F})$.

A typical call to the S language function abcnon has the form

$$\texttt{abcnon}(\mathbf{x}, \mathtt{tt}) \tag{14.24}$$

where x is the data and tt is the resampling version of the statistic of interest $\widehat{\theta}^*$. More details may be found in the Appendix.

To summarize this section, the ABC intervals are transformation respecting, second-order accurate, and good approximations to the BC_a intervals for most reasonably smooth statistics $\widehat{\theta}^* = s(\mathbf{x}^*)$. The nonparametric ABC algorithm abconon requires that the statistic be expressed in the resampling form $\widehat{\theta}^* = T(\mathbf{P}^*)$, but aside from this it is as easy and automatic to use as the BC_a algorithm bcanon, and requires only a few percent as much computation.

14.5 Example: the tooth data

[6] We conclude this chapter with a more complicated example that shows both the power and limitations of nonparametric BC_a/ABC confidence intervals.

Table 14.3 displays the *tooth data*. Thirteen accident victims each lost from 1 to 4 healthy teeth. The strength of these teeth was measured by a destructive testing method that could not be used under ordinary circumstances. "Strength", the last column of Table 14.3, records the average measured strength (on a logarithmic scale) for each patient's teeth.

The investigators wanted to predict tooth strength using variables that could be obtained on a routine basis. Four such variables are shown in Table 14.3, labeled D_1, D_2, E_1, E_2. The pair of variables (D_1, D_2) are difficult and expensive to obtain, while the pair (E_1, E_2) are easy and cheap. The investigators wished to answer the following question: how well do the Easy variables (E_1, E_2) predict strength, compared to the Difficult variables (D_1, D_2)?

We can phrase this question in a crisp way by using linear models, as in Chapters 7 and 9. Each row x_i of the data matrix in Table 14.3 consists of five numbers, the two D measurements, the two E measurements, and the strength measurement, say

$$x_i = (d_{i1}, d_{i2}, e_{i1}, e_{i2}, y_i) \qquad (i = 1, 2, \cdots, 13). \quad (14.25)$$

Let $\mathbf{D}$ be the matrix we would use for ordinary linear regression of y_i on just the D variables, including an intercept term, so $\mathbf{D}$ is the 13×3 matrix with ith row

$$(1, d_{i1}, d_{i2}). \qquad (14.26)$$

---

[6] Some of the material in this section is more advanced. It may be skipped at first reading.

Table 14.3. *The tooth data*. Thirteen accident victims have had the strength of their teeth measured, right column. It is desired to predict tooth strength from measurements not requiring destructive testing. Four such variables have been measured for each subject: the pair labeled (D_1, D_2), are difficult to obtain, the pair labeled (E_1, E_2) are easy to obtain. Do the Easy variables predict strength as well as the Difficult ones?

patient	D_1	D_2	E_1	E_2	strength
1	-5.288	10.091	12.30	13.08	36.05
2	-5.944	10.001	11.41	12.98	35.51
3	-5.607	10.184	11.76	13.19	35.35
4	-5.413	10.131	12.09	12.75	35.95
5	-5.198	8.835	10.72	11.73	34.64
6	-5.598	9.837	11.74	12.80	33.99
7	-6.120	10.052	11.10	12.87	34.60
8	-5.572	9.900	11.85	12.72	34.62
9	-6.056	9.966	11.78	13.06	35.05
10	-5.010	10.449	12.91	13.15	35.85
11	-6.090	10.294	11.63	12.97	35.53
12	-5.900	10.252	11.91	13.15	34.86
13	-5.620	9.316	10.89	12.25	34.75

The least-squares predictor of y_i in terms of the D variables is

$$\widehat{y}_i(D) = \widehat{\beta}_0(D) + \widehat{\beta}_1(D)d_{i1} + \widehat{\beta}_2(D)d_2 \qquad (14.27)$$

where $\widehat{\beta}(D) = (\widehat{\beta}_0(D), \widehat{\beta}_1(D), \widehat{\beta}_2(D))$ is the least-squares solution (9.28),

$$\widehat{\beta}(D) = (\mathbf{D}^T\mathbf{D})^{-1}\mathbf{D}^T\mathbf{y}, \qquad (14.28)$$

$\mathbf{y}$ being the vector $(y_1, y_2, \cdots, y_{13})$. The residual squared error RSE(D) is the total squared difference between the predictions $\widehat{y}_i(D)$ and the observations y_i for the $n = 13$ patients,

$$\mathrm{RSE}(D) = \sum_{n=1}^{n}(y_i - \widehat{y}_i(D))^2. \qquad (14.29)$$

Small values of RSE(D) indicate good prediction, the best possible value RSE(D) = 0 corresponding to a perfect prediction for every patient.

In a similar way we can predict the y_i from just the E measurements, and compute

$$\text{RSE}(E) = \sum_{i=1}^{n}(y_i - \widehat{y}_i(E))^2. \tag{14.30}$$

The investigator's question, how do the D and E variables compare as predictors of strength, can be phrased as a comparison between RSE(D) and RSE(E). A handy comparison statistic is

$$\widehat{\theta} = \frac{1}{n}[\text{RSE}(E) - \text{RSE}(D)]. \tag{14.31}$$

A positive value of $\widehat{\theta}$ would indicate that the E variables are not as good as the D variables for predicting strength. (If the number of E and D measures were not the same, $\hat{\theta}$ should be modified: see Problem 14.12).

The actual RSE values were RSE(D) = 2.761 and RSE(E) = 3.130, giving

$$\widehat{\theta} = .0285. \tag{14.32}$$

This suggests that the D variables are better predictors, since $\widehat{\theta}$ is greater than 0, but we can't decide if this is really true until we understand the statistical variability of $\widehat{\theta}$. We will use the BC$_a$ and ABC methods for this purpose. Figure 14.4 suggests that it will be a close call, since the predicted values $\widehat{y}_i(D)$ and $\widehat{y}_i(E)$ are quite similar on a case-by-case basis. Notice also that the difference between RSE(E) and RSE(D) is only about 10% as big as the RSE values themselves, so even if the difference is statistically significant, it may not be of great practical importance. Confidence intervals are a good way to answer both the significance and importance questions.

The left panel of Figure 14.5 is a histogram of 2000 nonparametric bootstrap replications of the RSE difference statistics $\widehat{\theta}$, (14.31). Let $\mathbf{x} = (x_1, x_2, \cdots, x_{13})$ indicate the tooth data matrix in Table 14.3, x_i being the ith row of the matrix, (14.25). A nonparametric bootstrap sample $\mathbf{x}^* = (x_1^*, x_2^*, \cdots, x_{13}^*)$ has each row x_i^* randomly drawn with replacement from $\{x_1, x_2, \cdots, x_{13}\}$. Equiva-

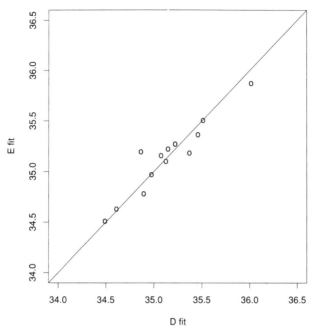

Figure 14.4. *The least-squares predictions $\widehat{y}(D)$, horizontal axis, versus $\widehat{y}_i(E)$, vertical axis, for the 13 patients in Table 14.3. The 45° line is shown for reference. The two sets of predictions appear quite similar.*

lently,

$$\widehat{F} \to \mathbf{x}^* = (x_1^*, x_2^*, \cdots, x_{13}^*), \tag{14.33}$$

where $\widehat{F}$ is the empirical distribution, putting probability $1/13$ on each $\mathbf{x}_i$.

By following definitions (14.25)-(14.30) the bootstrap matrix $\mathbf{x}^*$ gives $\mathbf{y}^*, \mathbf{D}^*, \widehat{\beta}(\mathbf{D})^*, \widehat{y}_i(D)^*$ and then

$$\text{RSE}(D)^* = \sum_{i=1}^{13} (y_i^* - \widehat{y}_i(D)^*)^2, \tag{14.34}$$

and likewise $\text{RSE}(E)^* = \sum_{i=1}^{13} (y_i^* - \widehat{y}_i(E)^*)^2$. The bootstrap replication of $\widehat{\theta}$ is

$$\widehat{\theta}^* = \frac{1}{13} [\text{RSE}(E)^* - \text{RSE}(D)^*]. \tag{14.35}$$

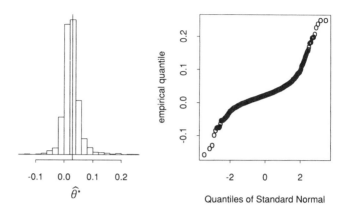

Figure 14.5. *Left panel: 2000 nonparametric bootstrap replications of the RSE difference statistic $\widehat{\theta}$, (14.31); bootstrap standard error estimate is $\widehat{se}_{2000} = .0311$; 1237 of the 2000 $\widehat{\theta}^*$ values are less than $\widehat{\theta} = .0285$, so $\hat{z}_0 = .302$. Right panel: quantile-quantile plot of the $\hat{\theta}^*$ values. Their distribution has much heavier tails than a normal distribution.*

As always, $\widehat{\theta}^*$ is computed by the same program that gives the original estimator $\widehat{\theta}$. All that changes is the data matrix, from $\mathbf{x}$ to $\mathbf{x}^*$.

The bootstrap histogram contains the information we need to answer questions about the significance and importance of $\widehat{\theta}$. Before going on to construct confidence intervals, we can say quite a bit just by inspection. The bootstrap standard error estimate (6.6) is

$$\widehat{se}_{2000} = .0311. \qquad (14.36)$$

This means that $\widehat{\theta} = .0285$ is less than one standard error above zero, so we shouldn't expect a conclusive significance level against the hypothesis that the true value of θ equals 0. On the other hand, the estimate is biased downward, 62% of the $\widehat{\theta}^*$ values being less than $\widehat{\theta}$. This implies that the significance level will be more conclusive than the value $.18 = 1 - \Phi(.0285/.0311)$ suggested by the normal approximation $\widehat{\theta} \sim N(\theta, .0311^2)$.

The bootstrap histogram makes it seem likely that θ is no greater than .0.10. How important is this difference? We need to say exactly what the parameter θ measures in order to answer this question. If F indicates the true five-dimensional distribution of the vector (d_1, d_2, e_1, e_2, y), then

$$\begin{aligned}
\theta_D &= \min_{\beta_D} \mathrm{E}_F[y - (\beta_{D_0} + \beta_{D_1} d_1 + \beta_{D_2} d_2)]^2 \quad \text{and} \\
\theta_E &= \min_{\beta_E} \mathrm{E}_F[y - (\beta_{E_0} + \beta_{E_1} e_1 + \beta_{E_2} e_2)]^2 \quad\quad (14.37)
\end{aligned}$$

are the true squared prediction errors using the D or E variables, respectively. The parameter θ corresponding to the plug-in estimate $\widehat{\theta}$, (14.31), is

$$\theta = \theta_E - \theta_D. \quad\quad (14.38)$$

The plug-in estimate of θ_D is $\widehat{\theta}_D = \mathrm{RSE}(D)/13 = .212$. Our belief that $\theta \leq .10$ gives

$$\frac{\theta_E - \theta_D}{\theta_D} \doteq \frac{\theta_E - \theta_D}{\widehat{\theta}_D} < \frac{.10}{.212} = .47. \quad\quad (14.39)$$

To summarize, the E variables are probably no better than the D variables for the prediction of strength, and are probably no more than roughly 50% worse.

The first column of Table 14.4 shows the BC_a confidence limits for θ based on the 2000 nonparametric bootstrap replications.

Confidence limits $\widehat{\theta}[\alpha]$ are given for eight values of the significance level α, $\alpha = .025, .05, \cdots, .975$. Confidence intervals are obtained using pairs of these limits, for example $(\widehat{\theta}[.05], \widehat{\theta}[.95])$ for a 90% interval. (So .05 corresponds to α and .95 corresponds to $1 - \alpha$ in (14.10).) Formulas (14.14) and (14.15) give a small acceleration and a large bias-correction in this case, $\hat{a} = .040$ and $\hat{z}_0 = .302$.

Notice that the .05 nonparametric limit is positive, $\widehat{\theta}[.05] = .004$. As mentioned earlier, this has a lot to do with the large bias-correction. If the BC_a method were exact, we could claim that the null hypothesis $\theta = 0$ was rejected at the .05 level, one-sided. The method is not exact, and it pays to be cautious about such claims. Nonparametric BC_a intervals are often a little too short, especially when the sample size is small, as it is here. If the hypothesis test were of crucial importance it would pay to improve the BC_a significance level with calibration, as in Section 25.6.

As a check on the nonparametric intervals, another 2000 bootstrap samples were drawn, this time according to a multivariate normal model: assume that the rows x_i of the tooth data matrix were obtained by sampling from a five-dimensional normal distribution F_{norm}; fit the best such distribution $\widehat{F}_{\mathrm{norm}}$ to the data (see

Table 14.4. *Bootstrap confidence limits for θ, (14.31); limits $\widehat{\theta}[\alpha]$ given for significance levels $\alpha = .025, .05, \cdots, .975$, so central 90% interval is $(\widehat{\theta}[.05], \widehat{\theta}[.95])$. Left panel: nonparametric bootstrap (14.33); Center panel: normal theory bootstrap (14.7); Right panel: linear model bootstrap, (9.25), (9.26). BC_a limits based on 2000 bootstrap replications for each of the three models; ABC limits obtained from the programs* abcnon *and* abcpar *in the Appendix (assuming normal errors for the linear model case); values of $\hat{a}$ and $\hat{z}_0$ vary depending on the details of the program used. The ABC limits are much too short in the nonparametric case because of the very heavy tails of the bootstrap distribution shown in Figure 14.5. Notice that in the nonparametric case the bootstrap estimate of standard error is nearly twice as big as the estimate used in the ABC calculations.*

| α | nonparametric | | normal theory | | linear model | |
	BC_a	ABC	BC_a	ABC	BC_a	ABC
0.025	-0.002	0.004	-0.010	-0.010	-.031	-0.019
0.05	0.004	0.008	-0.004	-0.004	-.020	-0.012
0.1	0.010	0.012	0.004	0.003	-.008	-0.004
0.16	0.015	0.016	0.010	0.010	.000	0.003
0.84	0.073	0.053	0.099	0.092	.070	0.067
0.9	0.095	0.061	0.113	0.111	.083	0.079
0.95	0.155	0.072	0.145	0.139	.098	0.094
0.975	0.199	0.085	0.192	0.167	.118	0.108
$\widehat{se}$	.0311	.0170	.0349	.0336	.0366	.0316
$\hat{a}$	.040	.056	.062	.062	0	0
$\hat{z}_0$	.302	.203	.353	.372	.059	.011

Problem 14.3); and sample $\mathbf{x}^*$ from $\widehat{F}_{\text{norm}}$ as in (14.7). Then $\widehat{\theta}^*$ is obtained as before, (14.34). The histogram of the 2000 normal-theory $\widehat{\theta}^*$'s, left panel of Figure 14.6, looks much like the histogram in Figure 14.5, except that the tails are less heavy.

The BC_a intervals are computed as before, using (14.9), (14.10). The bias-correction formula (14.14) is also unchanged. The acceleration constant $\hat{a}$ is calculated from a parametric version of (14.15) appearing in the parametric ABC program abcpar. In this case the normal-theory BC_a limits, center panel of Table 14.4, are not

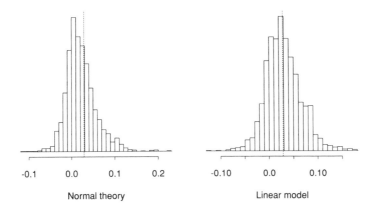

Figure 14.6. *Bootstrap replications giving the normal theory and linear model BC$_a$ confidence limits in Table 14.4. Left panel: normal theory; Right panel: linear model. A broken line is drawn at the parameter estimate.*

much different than the nonparametric BC$_a$ limits. The difference is large enough, though, so that the hypothesis $\theta = 0$ is no longer rejected at the .05 one-sided level.

There is nothing particularly normal-looking about the tooth data. The main reason for computing the normal-theory bootstraps is the small sample size, $n = 13$. In very small samples, even a badly fitting parametric analysis may outperform a nonparametric analysis, by providing less variable results at the expense of a tolerable amount of bias. That isn't the case here, where the two analyses agree.

Chapter 9 discusses linear regression models. We can use the linear regression model to develop a different bootstrap analysis of the RSE difference statistic $\widehat{\theta}$. Using the notation in (14.25), let c_i be the vector

$$\mathbf{c}_i = (1, d_{1i}, d_{2i}, c_{1i}, c_{2i}), \tag{14.40}$$

and consider the linear model (9.4), (9.5),

$$y_i = \mathbf{c}_i \boldsymbol{\beta} + \epsilon_i \qquad (i = 1, 2, \cdots, 13). \tag{14.41}$$

Bootstrap samples $\mathbf{y}^* = (y_1^*, y_2^*, \cdots, y_{13}^*)$ are constructed by resampling residuals as in (9.25, 9.26). The bootstrap replication $\widehat{\theta}^*$ is still given by (14.35). Notice though that the calculation of $\widehat{y}_i(D)^*$

and $\widehat{y}_i(E)^*$ is somewhat different.

The right panel of Figure 14.6 shows the bootstrap distribution based on 2000 replications of $\widehat{\theta}^*$. The tails of the histogram are much lighter than those in Figure 14.5. This is reflected in narrower bootstrap confidence intervals, as shown in the right panel of Table 14.4. Even though the intervals are narrower, the hypothesis $\theta = 0$ is rejected less strongly than before, at only the $\alpha = .16$ level. This happens because now $\widehat{\theta}$ does not appear to be biased strongly downward, $\widehat{z}_0$ equaling only .059 compared to .302 for the nonparametric case.

Confidence intervals and hypothesis tests are delicate tools of statistical inference. As such, they are more affected by model choice than are simple standard errors. This is particularly true in small samples. Exploring the relationship of five variables based on 13 observations is definitely a small sample problem. Even if the BC_a intervals were perfectly accurate, which they aren't, different model choices would still lead to different confidence intervals, as seen in Table 14.4.

Table 14.4 shows ABC limits for all three model choices. These were obtained using the programs abcnon and abcpar in the Appendix. The nonparametric ABC limits are much too short in this case. This happens because of the unusually heavy tails on the nonparametric bootstrap distribution. In traditional statistical language the ABC method can correct for skewness in the bootstrap distribution, but not for kurtosis. This is all it needs to do to achieve second order accuracy, (14.17). However the asymptotic accuracy of the ABC intervals doesn't guarantee good small-sample behavior.

Standard errors for $\widehat{\theta}$ are given for each of the six columns in Table 14.4. The BC_a entries are the usual bootstrap standard errors. ABC standard errors are given by the delta method, Chapter 21, a close relative of the jackknife standard error, (11.5). The BC_a standard error is nearly double that for the ABC in the nonparametric case, strongly suggesting that the ABC intervals will be too short. (The greater BC_a standard error became obvious after the first 100 bootstrap replications.) Usually the ABC approximations work fine as in Table 14.2, but it is reassuring to check the standard errors with 100 or so bootstrap replications.

14.6 Bibliographic notes

Background references on bootstrap confidence intervals are given in the bibliographic notes at the end of Chapter 22.

14.7 Problems

14.1 Verify that (14.1) is the plug-in estimate for $\theta = \text{var}(A)$.

14.2 The estimate $\widehat{\theta}$, (14.1), only involves the A_i components of the x_i pairs. In this case we might throw away the B_i components, and consider the data to be $\mathbf{A} = (A_1, A_2, \cdots, A_n)$.

   (a) Describe how this would change the nonparametric bootstrap sample (14.4).

   (b) Show that the nonparametric bootstrap intervals for θ would stay the same.

14.3 $\widehat{F}_{\text{norm}}$ in (14.7) is the bivariate normal distribution with mean vector $(\bar{A}, \bar{B})$ and covariance matrix

$$\frac{1}{26} \begin{pmatrix} \sum (A_i - \bar{A})^2 & \sum (A_i - \bar{A})(B_i - \bar{B}) \\ \sum (A_i - \bar{A})(B_i - \bar{B}) & \sum (B_i - \bar{B})^2 \end{pmatrix}$$

What would $\widehat{F}_{\text{norm}}$ be if we reduced the data to $\mathbf{A}$ as in problem 14.2?

14.4[†] In the normal-theory case it can be shown that $\widehat{\theta}$ is distributed according to a constant multiple of the chi-square distribution with $n - 1$ degrees of freedom,

$$\widehat{\theta} \sim \theta \frac{\chi^2_{n-1}}{n}. \tag{14.42}$$

   (a) Show that $[\text{var}(\widehat{\theta})]^{1/2} \propto \theta$.

   (b) Use (14.42) to calculate the exact interval endpoints in Table 14.2.

14.5 For the normal-theory bootstrap of the spatial test data, $(\hat{a}, \hat{z}_0) = (.092, .189)$. What were the values of α_1 and α_2 in (14.10)?

14.6 Explain why it makes sense that having 1118 out of 2000 $\widehat{\theta}^*$ values less than $\widehat{\theta}$ leads to a positive bias-correction at (14.14).

14.7 A plug-in statistic $\widehat{\theta} = t(\widehat{F})$ does not depend on the order of the points x_i in $\mathbf{x} = (x_1, x_2, \cdots, x_n)$. Rearranging the order of the points does not change the value $\widehat{\theta}$. Why is this important for the resampling representation (14.20)?

14.8 Suppose we take $\widehat{\theta}$ equal to the sample correlation coefficient for the spatial data,

$$\widehat{\theta} = [\sum_{1}^{n}(A_i - \bar{A})(B_i - \bar{B})]/[\sum_{1}^{n}(A_i - \bar{A})^2 \sum_{1}^{n}(B_i - \bar{B})^2]^{1/2}.$$

What is the resampling form (14.20) in this case?

14.9 Explain why θ as given by (14.38) is the parameter corresponding to $\widehat{\theta}$, (14.31). Why is the factor $1/n$ included in definition (14.31)?

14.10 We substituted $\widehat{\theta}_D$ for θ_D in the denominator of (14.39). What is a better way to get an approximate upper limit for $(\theta_E - \theta_D)/\theta_E$?

14.11 Explain how $\widehat{y}_i(D)^*$ is calculated in (14.34), as opposed to its calculation in finding $\widehat{\theta}^*$ in model (14.41).

14.12 Suppose there were p_E E measures and p_D D measures. Show that an appropriate definition for $\hat{\theta}$ is

$$\hat{\theta} = \frac{\text{RSE}(E)}{n - p_E - 1} - \frac{\text{RSE}(D)}{n - p_D - 1}. \tag{14.43}$$

14.13 [†] *Non-monotonicity of the* BC_a *point.*

Consider the BC_a confidence point for a parameter θ, as defined in equation (14.9). Define

$$z[\alpha] = \hat{z}_0 + \frac{\hat{z}_0 + z^{(\alpha)}}{1 - \hat{a}(\hat{z}_0 + z^{(\alpha)})} \tag{14.44}$$

For simplicity assume $\hat{\theta} = 0$ and $\hat{z}_0 = 0$.

(a) Set the acceleration constant $\hat{a} = 0$ and plot $z[\alpha]$ against α for 100 equally spaced α values between .001 and .999. Observe that $z[\alpha]$ is monotone increasing in α, so that the BC_a confidence point is also monotone increasing in α.

(b) Repeat part (a) for $\hat{a} = \pm 0.1, \pm 0.2, \ldots \pm 0.5$. For what values of $\hat{a}$ and α does $z[\alpha]$ fail to be monotone?

(c) To get some idea of how large a value of $\hat{a}$ one might expect in practice, generate a standard normal sample $x_1, x_2, \ldots x_{20}$. Compute the acceleration $\hat{a}$ for $\hat{\theta} = \bar{x}$. Create a more skewed sample by defining $y_i = \exp(x_i)$, and compute the acceleration $\hat{a}$ for $\hat{\theta} = \bar{y}$. Repeat this for $z_i = \exp(y_i)$. How large a value of $\hat{a}$ seems likely to occur in practice?

14.14 For the tooth data, compute the percentile and BC_a confidence intervals for the parameter $\theta = E(D_1 - D_2)$.

14.15 For the spatial test data, compute BC_a confidence intervals for $\theta_1 = \log E(A/B)$ and $\theta_2 = E\log(A/B)$. Are intervals the same? Explain.

† Indicates a difficult or more advanced problem.

CHAPTER 15

Permutation tests

15.1 Introduction

Permutation tests are a computer-intensive statistical technique that predates computers. The idea was introduced by R.A. Fisher in the 1930's, more as a theoretical argument supporting Student's t-test than as a useful statistical method in its own right. Modern computational power makes permutation tests practical to use on a routine basis. The basic idea is attractively simple and free of mathematical assumptions. There is a close connection with the bootstrap, which is discussed later in the chapter.

15.2 The two-sample problem

The main application of permutation tests, and the only one that we discuss here, is to the two-sample problem (8.3)-(8.5): We observe two independent random samples $\mathbf{z} = (z_1, z_2, \cdots, z_n)$ and $\mathbf{y} = (y_1, y_2, \cdots, y_m)$ drawn from possibly different probability distributions F and G,

$$
\begin{aligned}
F \rightarrow \quad & \mathbf{z} = (z_1, z_2, \cdots, z_n) \quad \text{independently of} \\
G \rightarrow \quad & \mathbf{y} = (y_1, y_2, \cdots, y_m).
\end{aligned}
$$
(15.1)

Having observed $\mathbf{z}$ and $\mathbf{y}$, we wish to test the null hypothesis H_0 of no difference between F and G,

$$
H_0 : \ F = G. \tag{15.2}
$$

The equality $F = G$ means that F and G assign equal probabilities to all sets, $\text{Prob}_F\{A\} = \text{Prob}_G\{A\}$ for A any subset of the common sample space of the z's and y's. If H_0 is true, then there is no difference between the probabilistic behavior of a random z or a random y.

Hypothesis testing is a useful tool for situations like that of the mouse data, Table 2.1. We have observed a small amount of data, $n = 7$ Treatment measurements and $m = 9$ Controls. The difference of the means,

$$\hat{\theta} = \bar{z} - \bar{y} = 30.63, \qquad (15.3)$$

encourages us to believe that the Treatment distribution F gives longer survival times than does the Control distribution G. As a matter of fact the experiment was designed to demonstrate exactly this result.

In this situation the *null hypothesis* (15.2), that $F = G$, plays the role of a devil's advocate. If we cannot decisively reject the possibility that H_0 is true (as will turn out to be the case for the mouse data), then we have not successfully demonstrated the superiority of Treatment over Control. An *hypothesis test*, of which a permutation test is an example, is a formal way of deciding whether or not the data decisively reject H_0.

An hypothesis test begins with a *test statistic* $\hat{\theta}$ such as the mean difference (15.3). For convenience we will assume here that if the null hypothesis H_0 is not true, we expect to observe larger values of $\hat{\theta}$ than if H_0 is true. If the Treatment works better than the Control in the mouse experiment, as intended, then we expect $\hat{\theta} = \bar{z} - \bar{y}$ to be large. We don't have to quantify what "large" means in order to run the hypothesis test. All we say is that the larger the value of $\hat{\theta}$ we observe, the stronger is the evidence against H_0. Of course in other situations we might choose smaller instead of larger values to represent stronger evidence. More complicated choices are possible too; see (15.26).

Having observed $\hat{\theta}$, the *achieved significance level* of the test, abbreviated ASL, is defined to be the probability of observing at least that large a value when the null hypothesis is true,

$$\text{ASL} = \text{Prob}_{H_0}\{\hat{\theta}^* \geq \hat{\theta}\}. \qquad (15.4)$$

The smaller the value of ASL, the stronger the evidence against H_0, as detailed below. The quantity $\hat{\theta}$ in (15.4) is fixed at its observed value; the random variable $\hat{\theta}^*$ has the null hypothesis distribution, the distribution of $\hat{\theta}$ if H_0 is true. As before, the star notation differentiates between the actual observation $\hat{\theta}$ and a hypothetical $\hat{\theta}^*$ generated according to H_0.

The hypothesis test of H_0 consists of computing ASL, and seeing if it is too small according to certain conventional thresholds.

Formally, we choose a small probability α, like .05 or .01, and *reject* H_0 if ASL is less than α. If ASL is greater than α, then we *accept* H_0, which amounts to saying that the experimental data does not decisively reject the null hypothesis (15.2) of absolutely no difference between F and G. Less formally, we observe ASL and rate the evidence against H_0 according to the following rough conventions:

$$
\begin{array}{ll}
\text{ASL} < .10 & \text{borderline evidence against } H_0 \\
\text{ASL} < .05 & \text{reasonably strong evidence against } H_0 \\
\text{ASL} < .025 & \text{strong evidence against } H_0 \\
\text{ASL} < .01 & \text{very strong evidence against } H_0
\end{array}
\tag{15.5}
$$

A traditional hypothesis test for the mouse data might begin with the assumption that F and G are normal distributions with possibly different means

$$
F = N(\mu_T, \sigma^2), \qquad G = N(\mu_C, \sigma^2). \tag{15.6}
$$

The null hypothesis is $H_0 : \mu_T = \mu_C$. Under H_0, $\hat{\theta} = \bar{z} - \bar{y}$ has a normal distribution with mean 0 and variance $\sigma^2[1/n + 1/m]$,

$$
H_0 : \hat{\theta} \sim N\left(0, \sigma^2(\frac{1}{n} + \frac{1}{m})\right); \tag{15.7}
$$

see Problem 3.4. Having observed $\hat{\theta}$, the ASL is the probability that a random variable $\hat{\theta}^*$ distributed as in (15.7) exceeds $\hat{\theta}$,

$$
\begin{aligned}
\text{ASL} &= \text{Prob}\left\{ Z > \frac{\hat{\theta}}{\sigma\sqrt{1/n + 1/m}} \right\} \\
&= 1 - \Phi\left(\frac{\hat{\theta}}{\sigma\sqrt{1/n + 1/m}} \right),
\end{aligned}
\tag{15.8}
$$

where Φ is the cumulative distribution function of the standard normal variate Z.

We don't know σ. A standard estimate based on (15.6) is

$$
\bar{\sigma} = \{[\sum_{i=1}^{n}(z_i - \bar{z})^2 + \sum_{j=1}^{m}(y_j - \bar{y})^2]/[n + m - 2]\}^{1/2}, \tag{15.9}
$$

which equals 54.21 for the mouse data. Substituting $\bar{\sigma}$ in (15.8) and remembering that $\hat{\theta} = 30.63$ gives

$$\text{ASL} = 1 - \Phi(\frac{30.63}{54.21\sqrt{1/9 + 1/7}}) = .131. \qquad (15.10)$$

This calculation treats $\bar{\sigma}$ as if it were a fixed constant. Student's t-test, which takes into account the randomness in $\bar{\sigma}$, gives

$$\text{ASL} = \text{Prob}\left\{t_{14} > \frac{30.63}{54.21\sqrt{1/9 + 1/7}}\right\} = .141, \qquad (15.11)$$

t_{14} indicating a t variate with 14 degrees of freedom. Student's test is based on the test statistic $\hat{\theta}/[\bar{\sigma}\sqrt{1/n + 1/m}]^{1/2}$, instead of $\hat{\theta}$. This statistic has a t_{n+m-2} distribution under the null hypothesis. In this case neither (15.10) nor (15.11) allows us to reject the null hypothesis H_0 according to (15.5), not even by the weakest standards of evidence.

The main practical difficulty with hypothesis tests comes in calculating the ASL, (15.4). We have written $\text{Prob}_{H_0}\{\hat{\theta}^* > \hat{\theta}\}$ as if the null hypothesis H_0 specifies a single distribution, from which we can calculate the probability of $\hat{\theta}^*$ exceeding $\hat{\theta}$. In most problems the null hypothesis (15.2), $F = G$, leave us with a family of possible null hypothesis distributions, rather than just one. In the normal case (15.6) for instance, the null hypothesis family (15.7) includes all normal distributions with expectation 0. In order to actually calculate the ASL, we had to either approximate the null hypothesis variance as in (15.10), or use Student's method (15.11). Student's method nicely solves the problem, but it only applies to the normal situation (15.6).

Fisher's permutation test is a clever way of calculating an ASL for the general null hypothesis $F = G$. Here is a simple description of it before we get into details. If the null hypothesis is correct, any of the survival times for any of the mice could have come equally well from either of the treatments. So we combine all the $m + n$ observations from both groups together, then take a sample of size m without replacement to represent the first group; the remaining n observations constitute the second group. We compute the difference between group means and then repeat this process a large number of times. If the original difference in sample means falls outside the middle 95% of the distribution of differences, the two-sided permutation test rejects the null hypothesis at a 5% level.

Permutation tests are based on the *order statistic representation* of the data $\mathbf{x} = (\mathbf{z}, \mathbf{y})$ from a two-sample problem. Table 15.1 shows

Table 15.1. *Order statistic representation for the mouse data of Table 2.1. All 16 data points have been combined and ordered from smallest to largest. The group code is "z" for Treatment and "y" for Control. For example, the 5th smallest of all 16 data points equals 31, and occurs in the Control group.*

group:	y	z	z	y	y	z	y	y
rank:	1	2	3	4	5	6	7	8
value:	10	16	23	27	31	38	40	46

group:	y	y	z	z	y	z	y	z
rank:	9	10	11	12	13	14	15	16
value :	50	52	94	99	104	141	146	197

the order statistic representation for the mouse data of Table 2.1. All 16 survival times have been combined and ranked from smallest to largest. The bottom line gives the ranked values, ranging from the smallest value 10 to the largest value 197. Which group each data point belongs to, "z" for Treatment or "y" for Control, is shown on the top line. The ranks 1 through 16 are shown on the second line. We see for instance that the 11th smallest value in the combined data set occurred in the Treatment group, and equaled 94. Table 15.1 contains the same information as Table 2.1, but arranged in a way that makes it easy to compare the relative sizes of the Treatment and Control values.

Let N equal the combined sample size $n + m$, and let $\mathbf{v} = (v_1, v_2, \cdots, v_N)$ be the combined and ordered vector of values; $N = 16$ and $\mathbf{v} = (10, 16, 23, \cdots, 197)$ for the mouse data. Also let $\mathbf{g} = (g_1, g_2, \cdots, g_N)$ be the vector that indicates which group each ordered observation belongs to, the top line in Table 15.1. [1] Together $\mathbf{v}$ and $\mathbf{g}$ convey the same information as $\mathbf{x} = (\mathbf{z}, \mathbf{y})$.

---

[1] It is convenient but not necessary to have $\mathbf{v}$ be the ordered elements of $(\mathbf{z}, \mathbf{y})$. Any other rule for listing the elements of $(\mathbf{z}, \mathbf{y})$ will do, as long as it doesn't involve the group identities. Suppose that the elements of $(\mathbf{z}, \mathbf{y})$ are vectors in R^2 for example. The $\mathbf{v}$ could be formed by ordering the members of $(\mathbf{z}, \mathbf{y})$ according to their first components and then breaking ties according to the order of their second components. If there are identical elements in $(\mathbf{z}, \mathbf{y})$, then the rule for forming v must include randomization, for instance "pairs of identical elements are ordered by the flip of a fair coin."

The vector $\mathbf{g}$ consists of n z's and m y's. There are

$$\binom{N}{n} = \frac{N!}{n!m!} \tag{15.12}$$

possible $\mathbf{g}$ vectors, corresponding to all possible ways of partitioning N elements into two subsets of size n and m. Permutation tests depend on the following important result:

<u>Permutation Lemma</u>. *Under $H_0 : F = G$, the vector $\mathbf{g}$ has probability $1/\binom{N}{n}$ of equaling any one of its possible values.*

In other words, all permutations of z's and y's are equally likely if $F = G$. We can think of a test statistic $\hat{\theta}$ as a function of $\mathbf{g}$ and $\mathbf{v}$, say

$$\hat{\theta} = S(\mathbf{g}, \mathbf{v}). \tag{15.13}$$

For instance, $\hat{\theta} = \bar{z} - \bar{y}$ can be expressed as

$$\hat{\theta} = \frac{1}{n} \sum_{g_i=z} v_i - \frac{1}{m} \sum_{g_i=y} v_i, \tag{15.14}$$

where $\sum_{g_i=z} v_i$ indicates the sum of the v_i over values of $i = 1, 2, \cdots, N$ having $g_i = z$.

Let $\mathbf{g}^*$ indicate any one of the $\binom{N}{n}$ possible vectors of n z's and m y's, and define the *permutation replication* of $\hat{\theta}$,

$$\hat{\theta}^* = \hat{\theta}(\mathbf{g}^*) = S(\mathbf{g}^*, \mathbf{v}). \tag{15.15}$$

There are $\binom{N}{n}$ permutation replications $\hat{\theta}^*$. The distribution that puts probability $1/\binom{N}{n}$ on each one of these is called the *permutation distribution of $\hat{\theta}$*, or of $\hat{\theta}^*$. The *permutation ASL* is defined to be the permutation probability that $\hat{\theta}^*$ exceeds $\hat{\theta}$,

$$\begin{aligned} \text{ASL}_{\text{perm}} &= \text{Prob}_{\text{perm}}\{\hat{\theta}^* \geq \hat{\theta}\} \\ &= \#\{\hat{\theta}^* \geq \hat{\theta}\}/\binom{N}{n}. \end{aligned} \tag{15.16}$$

The two definitions of ASL_{perm} in (15.16) are identical because of the Permutation Lemma.

In practice ASL_{perm} is usually approximated by Monte Carlo methods, according to Algorithm 15.1.

The permutation algorithm is quite similar to the bootstrap algorithm of Figure 6.1. The main difference is that sampling is carried out *without* replacement rather than with replacement.

Algorithm 15.1

Computation of the two-sample permutation test statistic

1. Choose B independent vectors $\mathbf{g}^*(1)$, $\mathbf{g}^*(2), \cdots, \mathbf{g}^*(B)$, each consisting of n z's and m y's and each being randomly selected from the set of all $\binom{N}{n}$ possible such vectors. [B will usually be at least 1000; see Table (15.3).]

2. Evaluate the permutation replications of $\hat{\theta}$ corresponding to each permutation vector,

$$\hat{\theta}^*(b) = S(\mathbf{g}^*(b), \mathbf{v}), \qquad b = 1, 2, \cdots, B. \qquad (15.17)$$

3. Approximate $\mathrm{ASL_{perm}}$ by

$$\widehat{\mathrm{ASL}}_{\mathrm{perm}} = \#\{\hat{\theta}^*(b) \geq \hat{\theta}\}/B. \qquad (15.18)$$

The top left panel of Figure 15.1 shows the histogram of $B = 1000$ permutation replications of the mean difference $\hat{\theta} = \bar{z} - \bar{y}$, (15.3); 132 of the 1000 $\hat{\theta}^*$ replications exceeded $\hat{\theta} = 30.63$, so this reinforces our previous conclusion that the data in Table 2.1 does not warrant rejection of the null hypothesis $F = G$:

$$\widehat{\mathrm{ASL}}_{\mathrm{perm}} = 132/1000 = .132. \qquad (15.19)$$

The permutation ASL is close to the t-test ASL, (15.11), even though there are no normality assumptions underlining $\mathrm{ASL_{perm}}$. This is no accident, though the very small difference between (15.19) and (15.11) is partly fortuitous. Fisher demonstrated a close theoretical connection between the permutation test based on $\bar{z} - \bar{y}$, and Student's test. See Problem 15.9. His main point in introducing permutation tests was to support the use of Student's test in non-normal applications.

How many permutation replications are required? For convenient notation let $A = \mathrm{ASL_{perm}}$ and $\hat{A} = \widehat{\mathrm{ASL}}_{\mathrm{perm}}$. Then $B \cdot \hat{A}$ equals the number of $\hat{\theta}^*(b)$ values exceeding the observed value $\hat{\theta}$, and so has a binomial distribution as in Problem 3.6,

$$B \cdot \hat{A} \sim \mathrm{Bi}(B, A); \quad \mathrm{E}(\hat{A}) = A; \quad \mathrm{var}(\hat{A}) = \frac{A(1-A)}{B}. \qquad (15.20)$$

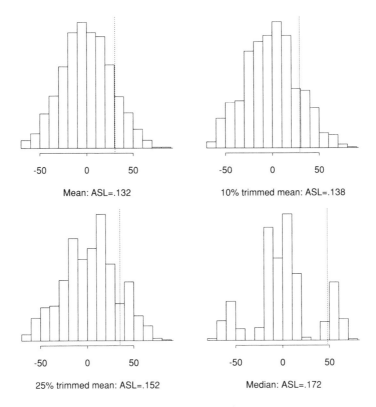

Mean: ASL=.132   10% trimmed mean: ASL=.138

25% trimmed mean: ASL=.152   Median: ASL=.172

Figure 15.1. *Permutation distributions for four different test statistics $\hat{\theta}$, mouse data, Table 2.1; dashed line indicates observed value of $\hat{\theta}$; $\widehat{ASL}_{\mathrm{perm}}$ leads to non-rejection of the null hypothesis for all four statistics. Top left: $\hat{\theta} = \bar{z} - \bar{y}$, difference of means, Treatment-Control groups. Top right: $\hat{\theta}$ equals the difference of 15% trimmed means. Bottom left: difference of 25% trimmed means. Bottom right: difference of medians.*

(Remember that $\hat{\theta}$ is a fixed quantity in (15.18), only $\hat{\theta}^*$ being random.) The coefficient of variation of $\hat{A}$ is

$$\mathrm{cv}_B(\hat{A}) = [\frac{(1 - A)/A}{B}]^{1/2}. \qquad (15.21)$$

The quantity $[(1 - A)/A]^{1/2}$ gets bigger as A gets smaller, as shown in Table 15.2.

Table 15.2. $[(1 - A)/A]^{1/2}$ *as a function of A.*

A:	.5	.25	.1	.05	.025:
$[(1 - A)/A]^{1/2}$:	1.00	1.73	3.00	4.36	6.24

Suppose we require $\mathrm{cv}_B(\hat{A})$ to be .10, meaning that we don't want Monte Carlo error to affect our estimate of $\mathrm{ASL}_{\mathrm{perm}}$ by more than 10%. Table 15.3 gives the number of permutation replications B required.

The reader may have been bothered by a peculiar feature of permutation testing: the permutation replications $\hat{\theta}^* = S(\mathbf{g}^*, \mathbf{v})$ change part of the original data but leave another part fixed. Why should we resample $\mathbf{g}$ but not $\mathbf{v}$? Some good theoretical reasons have been given in the statistics literature, but the main reason is practical. "Conditioning on $\mathbf{v}$", i.e. keeping $\mathbf{v}$ fixed in the permutation resampling process, reduces the two-sample situation to a *single* distribution, under the null hypothesis $F = G$. This is the essence of the Permutation Lemma. The quantity $\mathrm{ASL}_{\mathrm{perm}} = \mathrm{Prob}_{\mathrm{perm}}\{\hat{\theta}^* > \hat{\theta}\}$ is well-defined, though perhaps difficult to calculate, because $\mathrm{Prob}_{\mathrm{perm}}$ refers to a unique probability distribution. The quantity $\mathrm{ASL} = \mathrm{Prob}_{H_0}\{\hat{\theta}^* > \hat{\theta}\}$ is not well defined because there is no single distribution Prob_{H_0}.

The greatest virtue of permutation testing is its accuracy. If $H_0 : F = G$ is true, there is almost exactly a 5% chance that $\mathrm{ASL}_{\mathrm{perm}}$ will be less than .05. In general,

$$\mathrm{Prob}_{H_0}\{\mathrm{ASL}_{\mathrm{perm}} < \alpha\} = \alpha \qquad (15.22)$$

for any value of α between 0 and 1, except for small discrepancies caused by the discreteness of the permutation distribution. See Problem 15.6. This is important because the interpretive scale (15.5) is taken very literally in many fields of application.

15.3 Other test statistics

The permutation test's accuracy applies to any test statistic $\hat{\theta}$. The top right panel of Figure 15.1 refers to the difference of the

Table 15.3. *Number of permutations required to make* $cv(\widehat{ASL}) \leq .10$, *as a function of the achieved significance level.*

ASL_{perm}:	.5	.25	.1	.05	.025
B:	100	299	900	1901	3894

.15 trimmed means, [2]

$$\hat{\theta} = \bar{z}_{.15} - \bar{y}_{.15}. \tag{15.23}$$

The bottom left panel refers to the difference of the .25 trimmed means, and the bottom right panel to the difference of medians. The same $B = 1000$ permutation vectors $\mathbf{g}^*$ were used in all four panels, only the statistic $\hat{\theta}^* = S(\mathbf{g}^*, \mathbf{v})$ changing. The four values of $\widehat{ASL}_{perm}$, $.132, .138, .152$, and $.172$, are all consistent with acceptance of the null hypothesis $F = G$.

The fact that every $\hat{\theta}$ leads to an accurate ASL_{perm} does not mean that all $\hat{\theta}$'s are equally good test statistics. "Accuracy" means these ASL_{perm} won't tend to be misleadingly small when H_0 is true, as stated in (15.22). However if H_0 is false, if Treatment really is better than Control, then we *want* ASL_{perm} to be small. This property of a statistical test is called *power*. The penalty for choosing a poor test statistic $\hat{\theta}$ is low power – we don't get much probability of rejecting H_0 when it is false. We will say a little more about choosing $\hat{\theta}$ in the bootstrap discussion that concludes this chapter.

Looking at Table 2.1, the two groups appear to differ more in variance than in mean. The ratio of the estimated variances is nearly 2.5,

$$\hat{\sigma}_z^2 / \hat{\sigma}_y^2 = 2.48. \tag{15.24}$$

Is this difference genuine or just an artifact of the small sample sizes?

We can answer this question with a permutation test. Figure 15.2

---

[2] The $100 \cdot \alpha\%$ trimmed mean "$\bar{x}_\alpha$" of n numbers $x_1, x_2, \cdots, x_n$ is defined as follows: (i) order the numbers $x_{(1)} \leq x_{(2)}, \cdots, \leq x_{(n)}$; (ii) remove the $n - \alpha$ smallest and $n \cdot \alpha$ largest numbers; (iii) then $\bar{x}_\alpha$ equals the average of the remaining $n \cdot (1 - 2\alpha)$ numbers. Interpolation is necessary if $n \cdot \alpha$ is not an integer. In this notation, the mean is $\bar{x}_0$ and the median is $\bar{x}_{.50}$; $\bar{x}_{.25}$ is the average of the middle 50% of the data.

shows 1000 permutation replications of

$$\hat{\theta} = \log(\hat{\sigma}_z^2/\hat{\sigma}_y^2). \qquad (15.25)$$

(The logarithm doesn't affect the permutation results, see Problem 15.1). 152 of the 1000 $\hat{\theta}^*$ values exceeded $\hat{\theta} = \log(2.48) = .907$, giving $\widehat{\text{ASL}}_{\text{perm}} = .152$. Once again there are no grounds for rejecting the null hypothesis $F = G$. Notice that we *might* have rejected H_0 with this $\hat{\theta}$ even if we didn't reject it with $\hat{\theta} = \bar{z} - \bar{y}$. This $\hat{\theta}$ measures deviations from H_0 in a different way than do the $\hat{\theta}$'s of Figure 15.1.

The statistic $\log(\hat{\sigma}_z^2/\hat{\sigma}_y^2)$ differs from $\bar{z} - \bar{y}$ in an important way. The Treatment was designed to increase survival times, so we expect $\bar{z} - \bar{y}$ to be greater than zero if the Treatment works, i.e., if H_0 is false. On the other hand, we have no a priori reason for believing $\hat{\theta} = \log(\hat{\sigma}_z^2/\hat{\sigma}_y^2)$ will be greater than zero rather than less than zero if H_0 is false. To put it another way, we would have been just as interested in the outcome $\hat{\theta} = -\log(2.48)$ as in $\hat{\theta} = \log(2.48)$.

In this situation, it is common to compute a *two-sided* ASL, rather than the *one-sided* ASL (15.4). This is done by comparing the absolute value of $\hat{\theta}^*$ with the absolute value of $\hat{\theta}$,

$$\widehat{\text{ASL}}_{\text{perm}}(\text{two-sided}) = \#\{|\hat{\theta}^*(b)| > |\hat{\theta}|\}/B. \qquad (15.26)$$

Equivalently, we count the cases where either $\hat{\theta}^*$ or $-\hat{\theta}^*$ exceed $|\hat{\theta}|$. The two-sided ASL is always larger than the one-sided ASL, giving less reason for rejecting H_0. The two-sided test is inherently more conservative. For the mouse data, statistic (15.25) gave a two-sided ASL of .338.

The idea of a significance test can be stated as follows: we rank all possible data sets $\mathbf{x}$ according to how strongly they contradict the null hypothesis H_0; then we reject H_0 if $\mathbf{x}$ is among the 5% (or 10%, or 1% etc., as in (15.5)) of the data sets that most strongly contradict H_0. The definition of ASL in (15.4) amounts to measuring contradiction according to the size of $\hat{\theta}(\mathbf{x})$, large values of $\hat{\theta}$ implying greater evidence against H_0. Sometimes, though, we believe that large negative values of $\hat{\theta}$ are just as good as large positive values for discrediting H_0. This was the case in (15.25). If so, we need to take this into account when defining the 5% of the data sets that most strongly contradict H_0. That is the point of definition (15.26) for the two-sided ASL.

There are many other situations where we need to be careful

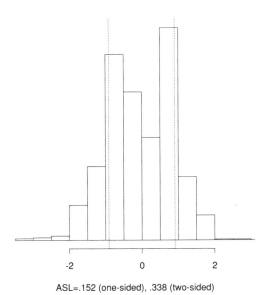

ASL=.152 (one-sided), .338 (two-sided)

Figure 15.2. $B = 1000$ *permutation replications of the log variance ratio*
$\hat{\theta} = \log(\hat{\sigma}_z^2/\hat{\sigma}_y^2)$ *for the mouse data of Table 2.1; 152 of the 1000 replica-*
tions gave $\hat{\theta}^*$ *greater than the observed value* $\hat{\theta} = .907$; 338 *of the 1000*
replications gave either $\hat{\theta}^*$ *or* $-\hat{\theta}^*$ *greater than .907. The dashed lines*
indicate $\hat{\theta}$ *and* $-\hat{\theta}$.

about ranking evidence against H_0. Suppose, for example, that we
run the four permutation tests of Figure 15.1, and decide to choose
the one with the smallest ASL, in this case $\widehat{\mathrm{ASL}} = .132$. Then we
are really ranking the evidence in **x** against H_0 according to the
statistic

$$\hat{\phi}(\mathbf{x}) = \min_k \{\widehat{\mathrm{ASL}}_k\}, \qquad (15.27)$$

where $\widehat{\mathrm{ASL}}_k$ is the permutation ASL for the kth statistic $\hat{\theta}_k$ $k =$
$1, 2, 3, 4$. Small values of $\hat{\phi}$ more strongly contradict H_0. It *isn't*
true that having observed $\hat{\theta} = .132$, the permutation ASL based
on $\hat{\phi}$ equals .132. More than 13.2% of the permutations will have

$\hat{\phi}^* < .132$ because of the minimization in definition (15.27).

Here is how to compute the correct permutation ASL for $\hat{\phi}$, using all 4000 permutation replications $\hat{\theta}_k^*(b)$ in Figure 15.1, $k = 1, 2, 3, 4$, $b = 1, 2, \cdots, 1000$. For each value of k and b define

$$A_k^*(b) = \frac{1}{1000} \sum_{i=1}^{B} I_{\{\theta_k^*(i) \geq \hat{\theta}_k^*(b)\}}, \tag{15.28}$$

where $I_{\{.\}}$ is the indicator function. So $A_k^*(b)$ is the proportion of the $\hat{\theta}_k^*$ values exceeding $\hat{\theta}_k^*(b)$. Then let

$$\hat{\phi}^*(b) = \min_k \{A_k^*(b)\}. \tag{15.29}$$

It is not obvious but it is true that the $\hat{\phi}^*(b)$ are genuine permutation replications of $\hat{\phi}$, (15.27), so the permutation ASL for $\hat{\phi}$ is

$$\widehat{\text{ASL}}_{\text{perm}} = \#\{\hat{\phi}^*(b) \leq \hat{\phi}\}/1000. \tag{15.30}$$

Figure 15.3 shows the histogram of the 1000 $\hat{\phi}^*(b)$ values. 167 of the 1000 values are less than $\hat{\phi} = .132$, giving permutation ASL = .167.

15.4 Relationship of hypothesis tests to confidence intervals and the bootstrap

There is an intimate connection between hypothesis testing and confidence intervals. Suppose $\hat{\theta}$, the observed value of the statistic of interest, is greater than zero. Choose α so that $\hat{\theta}_{\text{lo}}$, the lower end of the $1 - 2\alpha$ confidence interval for θ, exactly equals 0. Then $\text{Prob}_{\theta=0}\{\hat{\theta}^* \geq \hat{\theta}\} = \alpha$ according to (12.13). However if $\theta = 0$ is the null hypothesis, as in the mouse data example, then definition (15.4) gives ASL = α. For example, if the .94 confidence interval $[\hat{\theta}_{\text{lo}}, \hat{\theta}_{\text{up}}]$ has $\hat{\theta}_{\text{lo}} = 0$, then the ASL of the observed value $\hat{\theta}$ must equal .03 (since $.94 = 1 - 2 \cdot .03$).

In other words, we can use confidence intervals to calculate ASLs. With this in mind, Figure 15.4 gives the bootstrap distribution of two statistics we can use to form confidence intervals for the difference between the Treatment and Control groups, the mean difference $\hat{\theta}_0 = \bar{z} - \bar{y}$, left panel, and the .25 trimmed mean difference $\hat{\theta}_{.25} = \bar{z}_{.25} - \bar{y}_{.25}$, right panel. What value of α will make the lower end of the bootstrap confidence interval equal to zero? For the bootstrap percentile method applied to a statistic $\hat{\theta}^*$, the

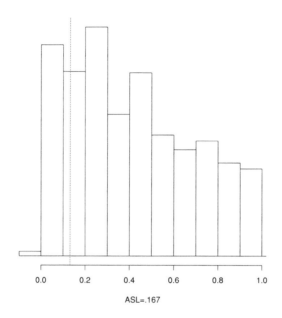

ASL=.167

Figure 15.3. *Permutation distribution for the minimum ASL statistic (15.27); based on the 1000 permutations used in Figure 15.1; dashed line indicates* $\hat{\phi} = .132$; *167 of the 1000* $\hat{\phi}^*$ *values are less than .132, so* $\widehat{ASL}_{perm} = .167$.

answer is

$$\alpha_0 = \#\{\hat{\theta}^*(b) < 0\}/B, \qquad (15.31)$$

the proportion of the bootstrap replications less than zero. (Then $\hat{\theta}_{lo} = \hat{\theta}^{*(\alpha_0)} = 0$ according to (13.5).) According to the previous paragraph, the ASL of $\hat{\theta}$ equals α_0, say

$$\widehat{ASL}_\% = \#\{\hat{\theta}^*(b) < 0\}/B. \qquad (15.32)$$

The $B = 1000$ bootstrap replications shown in Figure 15.4 gave

$$\widehat{ASL}_\%(\hat{\theta}_0) = .132 \qquad \text{and} \qquad \widehat{ASL}_\%(\hat{\theta}_{.25}) = .180. \qquad (15.33)$$

Notice how similar these results are to $\widehat{\mathrm{ASL}}_{\mathrm{perm}}(\hat{\theta}_0) = .132$, $\widehat{\mathrm{ASL}}_{\mathrm{perm}}(\hat{\theta}_{.25}) = .152$, Figure 15.1.

For the $\mathrm{BC_a}$ confidence intervals of Chapter 14, the ASL calculation gives

$$\widehat{\mathrm{ASL}}_{\mathrm{BC_a}} = \Phi^{-1}(\frac{w_0 - \hat{z}_0}{1 + \hat{a}(w_0 - \hat{z}_0)} - z_0), \qquad (15.34)$$

where

$$w_0 = \Phi^{-1}(\alpha_0) \qquad (15.35)$$

and the bias correction constant $\hat{z}_0$ is approximated according to formula (14.14). This formula gave $\hat{z}_0 = -.040$ for $\hat{\theta}_0$ and $\hat{z}_0 = .035$ for $\hat{\theta}_{.25}$.

The acceleration constant $\hat{a}$ is given by a two-sample version of (14.15). Let $\hat{\theta}_{z,(i)}$ be the value of $\hat{\theta}$ when we leave out z_i, and $\hat{\theta}_{y,(i)}$ be the value of $\hat{\theta}$ when we leave out y_i. Let $\hat{\theta}_{z,(\cdot)} = \sum_1^n \hat{\theta}_{z,(i)}/n$, $\hat{\theta}_{y,(\cdot)} = \sum_1^m \hat{\theta}_{y,(i)}/m$, $U_{z,i} = (n-1)(\hat{\theta}_{z,(\cdot)} - \hat{\theta}_{z,(i)})$, $U_{y,i} = (m-1)(\hat{\theta}_{y,(\cdot)} - \hat{\theta}_{y,(i)})$. Then

$$\hat{a} = \frac{1}{6} \frac{[\sum_{i=1}^n U_{z,i}^3/n^3 + \sum_{i=1}^m U_{y,i}^3/m^3]}{[\sum_{i=1}^n U_{z,i}^2/n^2 + \sum_{i=1}^m U_{y,i}^2/m^2]^{3/2}}, \qquad (15.36)$$

n and m being the lengths of $\mathbf{z}$ and $\mathbf{y}$. Formula (15.36) gives $\hat{a} = .06$ and $\hat{a} = -.01$ for $\hat{\theta}_0$ and $\hat{\theta}_{.25}$, respectively. Then

$$\widehat{\mathrm{ASL}}_{\mathrm{BC_a}}(\hat{\theta}_0) = .147 \qquad \text{and} \qquad \widehat{\mathrm{ASL}}_{\mathrm{BC_a}}(\hat{\theta}_{.25}) = .167 \quad (15.37)$$

according to (15.34).

Here are some points to keep in mind in comparing Figures 15.1 and 15.4:

- The permutation ASL is exact, while the bootstrap ASL is approximate. In practice, though, the two methods often give quite similar results, as is the case here.

- The bootstrap histograms are centered near $\hat{\theta}$, while the permutation histograms are centered near 0. In this sense, $\mathrm{ASL}_{\mathrm{perm}}$ measures how far the observed estimate $\hat{\theta}$ is from 0, while the bootstrap ASL measures how far 0 is from $\hat{\theta}$. The adjustments that the $\mathrm{BC_a}$ method makes to the percentile method, (15.34) compared to (15.31), are intended to reconcile these two ways of measuring statistical "distance."

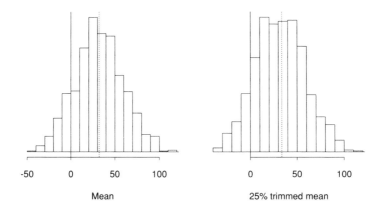

Figure 15.4. $B = 1000$ *bootstrap replications of the mean difference $\hat{\theta}$ for the mouse data, left panel, and the difference of the .25 trimmed means $\hat{\theta}_{.25}$, right panel; dashed lines are the observed estimates $\hat{\theta}_0 = 30.23$ and $\hat{\theta}_{.25} = 33.66$; 132 of the 1000 $\hat{\theta}_0^*$ values were less than zero; 180 of the 1000 $\hat{\theta}_{.25}^*$ values were less than zero.*

- The bootstrap ASL tests the null hypothesis $\theta = 0$ while the permutation ASL tests $F = G$. The latter is more special than the former, and can sometimes seem unrealistic. For the mouse data, we might wish to test the hypothesis that the means of the two groups were equal, $\theta_0 = 0$, without ever believing that the two distributions had the same variance, for instance. This is more of a theoretical objection than a practical one to permutation tests, which usually perform reasonably well even if $F = G$ is far from being a reasonable null hypothesis.

- The standard deviation of the permutation distribution is *not* a dependable estimate of standard error for $\hat{\theta}$ (it is not intended to be), while the bootstrap standard deviation is. Table 15.4 shows the standard deviations of the mouse data permutation and bootstrap distributions for $\hat{\theta}_0 = \bar{z} - \bar{y}$, $\hat{\theta}_{.15} = \bar{z}_{.15} - \bar{y}_{.15}$, $\hat{\theta}_{.25} = \bar{z}_{.25} - \bar{y}_{.25}$, and $\hat{\theta}_{.5} = \bar{z}_{.5} - \bar{y}_{.5}$, The bootstrap numbers show a faster increase in standard error as the trimming proportion increases from 0 to .5, and these are the numbers to be believed.

- The combination of a point estimate and a confidence interval is usually more informative than just a hypothesis test by itself. In the mouse experiment, the value 0.132 of $\widehat{\text{ASL}}_{\text{perm}}$ tells us

Table 15.4. *Standard deviations of the mouse data permutation and boot-strap distributions for* $\hat{\theta}_0 = \bar{z} - \bar{y}$, $\hat{\theta}_{.15} = \bar{z}_{.15} - \bar{y}_{.15}$, $\hat{\theta}_{.25} = \bar{z}_{.25} - \bar{y}_{.25}$, *and* $\hat{\theta}_{.5} = \bar{z}_{.5} - \bar{y}_{.5}$.

	$\hat{\theta}_0$	$\hat{\theta}_{.15}$	$\hat{\theta}_{.25}$	$\hat{\theta}_{.5}$
permutation:	27.9	28.6	30.8	33.5
bootstrap:	27.0	29.9	33.4	40.8

only that we can't rule out $\theta = 0$. The left panel of Figure 15.4 says that the true mean lies between -14.5 and 73.8 with confidence .90, BC_a method. In the authors' experience, hypothesis tests tend to be overused and confidence intervals underused in statistical applications.

Permutation methods tend to apply to only a narrow range of problems. However when they apply, as in testing $F = G$ in a two-sample problem, they give gratifyingly exact answers without parametric assumptions. The bootstrap distribution was originally called the "combination distribution." It was designed to extend the virtues of permutation testing to the great majority of statistical problems where there is nothing to permute. When there *is* something to permute, as in Figure 15.1, it is a good idea to do so, even if other methods like the bootstrap are also brought to bear. In the next chapter, we discuss problems for which the permutation method cannot be applied but a bootstrap hypothesis test can still be used.

15.5 Bibliographic notes

Permutation tests are described in many books, A comprehensive overview is given by Edgington (1987). Noreen (1989) gives an introduction to permutation tests, and relates them to the bootstrap.

15.6 Problems

15.1 Suppose that $\hat{\phi} = m(\hat{\theta})$, where $m(\cdot)$ is an increasing function. Show that the permutation ASL based on $\hat{\phi}$ is the same as the permutation ASL based on $\hat{\theta}$,

$$\widehat{\mathrm{ASL}}_{\mathrm{perm}}(\hat{\phi}) = \widehat{\mathrm{ASL}}_{\mathrm{perm}}(\hat{\theta}). \qquad (15.38)$$

15.2 Suppose that the N elements $(\mathbf{z}, \mathbf{y})$ are a random sample from a probability distribution having a continuous distribution on the real line. Prove the permutation lemma.

15.3 Suppose we take $\mathrm{ASL}_{\mathrm{perm}} = .01$ in (15.3). What is the entry for B?

15.4 (a) Assuming that (15.22) is exactly right, show that $\mathrm{ASL}_{\mathrm{perm}}$ has a uniform distribution on the interval $[0, 1]$.

(b) Draw a schematic picture suggesting the probability density of $\mathrm{ASL}_{\mathrm{perm}}$ when H_0 is true, overlaid with the probability density of $\mathrm{ASL}_{\mathrm{perm}}$ when H_0 is false.

15.5 Verify formula (15.34).

15.6 Formula (15.22) cannot be exactly true because of the discreteness of the permutation distribution. Let $M = \binom{N}{n}$. Show that

$$\mathrm{Prob}_{H_0}\{\mathrm{ASL}_{\mathrm{perm}} = \frac{k}{M}\} = \frac{1}{M} \qquad \text{for} \quad k = 1, 2, \cdots, M.$$

$$(15.39)$$

[You may make use of the permutation lemma, and assume that there are no ties among the M values $\hat{\theta}(\mathbf{g}^*, \mathbf{v})$.]

15.7 Define $\hat{\phi} = \widehat{\mathrm{ASL}}_{\mathrm{perm}}(\hat{\theta})$, for some statistic $\hat{\theta}$. Show that the ASL based on $\hat{\phi}$ is the same as the ASL based on $\hat{\theta}$,

$$\widehat{\mathrm{ASL}}_{\mathrm{perm}}(\hat{\phi}) = \widehat{\mathrm{ASL}}_{\mathrm{perm}}(\hat{\theta}) = \hat{\phi}. \qquad (15.40)$$

15.8 [†] Explain why (15.30) is true.

15.9 [†] With $\hat{\theta} = \bar{z} - \bar{y}$ as in (15.3), and $\hat{\sigma}$ defined as in (15.9) let $\hat{\phi}$ equal Student's t statistic $\hat{\theta}/[\hat{\sigma}\sqrt{1/n + 1/m}]$. Show that

$$\widehat{\mathrm{ASL}}_{\mathrm{perm}}(\hat{\phi}) = \widehat{\mathrm{ASL}}_{\mathrm{perm}}(\hat{\theta}). \qquad (15.41)$$

Hint: Use Problem 15.1.

† Indicates a difficult or more advanced problem.

## CHAPTER 16

Hypothesis testing with the bootstrap

16.1 Introduction

In Chapter 15 we describe the permutation test, a useful tool for hypothesis testing. At the end of that chapter we relate hypothesis tests to confidence intervals, and in particular showed how a bootstrap confidence interval could be used to provide a significance level for a hypothesis test. In this chapter we describe bootstrap methods that are designed directly for hypothesis testing. We will see that the bootstrap tests give similar results to permutation tests when both are available. The bootstrap tests are more widely applicable though less accurate.

16.2 The two-sample problem

We begin with the two-sample problem as described in the last chapter. We have samples $\mathbf{z}$ and $\mathbf{y}$ from possibly different probability distributions F and G, and we wish to test the null hypothesis $H_0 : F = G$. A bootstrap hypothesis test, like a permutation test, is based on a test statistic. In the previous chapter this was denoted by $\hat{\theta}$. To emphasize that a test statistic need not be an estimate of a parameter, we denote it here by $t(\mathbf{x})$. In the mouse data example, $t(\mathbf{x}) = \bar{z} - \bar{y}$, the difference of means with observed value 30.63. We seek an achieved significance level

$$\text{ASL} = \text{Prob}_{H_0}\{t(\mathbf{x}^*) \geq t(\mathbf{x})\} \qquad (16.1)$$

as in (15.4). The quantity $t(\mathbf{x})$ is fixed at its observed value and the random variable $\mathbf{x}^*$ has a distribution specified by the null hypothesis H_0. Call this distribution F_0. Now the question is, what is F_0? In the permutation test of the previous chapter, we fixed the

Algorithm 16.1

Computation of the bootstrap test statistic for testing $F = G$

---

1. Draw B samples of size $n + m$ with replacement from $\mathbf{x}$. Call the first n observations $\mathbf{z}^*$ and the remaining m observations $\mathbf{y}^*$.

2. Evaluate $t(\cdot)$ on each sample,

$$t(\mathbf{x}^{*b}) = \bar{\mathbf{z}}^* - \bar{\mathbf{y}}^*, \quad b = 1, 2, \cdots B. \tag{16.2}$$

3. Approximate $\mathrm{ASL}_{\mathrm{boot}}$ by

$$\widehat{\mathrm{ASL}}_{\mathrm{boot}} = \#\{t(\mathbf{x}^{*b}) \geq t_{obs}\}/B, \tag{16.3}$$

where $t_{obs} = t(\mathbf{x})$ the observed value of the statistic.

---

order statistics $\mathbf{v}$ and defined F_0 to be the distribution of possible orderings of the ranks $\mathbf{g}$. Bootstrap hypothesis testing, on the other hand, uses a "plug-in" style estimate for F_0. Denote the combined sample by $\mathbf{x}$ and let its empirical distribution be $\hat{F}_0$, putting probability $1/(n + m)$ on each member of $\mathbf{x}$. Under H_0, $\hat{F}_0$ provides a nonparametric estimate of the common population that gave rise to both $\mathbf{z}$ and $\mathbf{y}$. Algorithm 16.1 shows how ASL is computed.

Notice that the only difference between this algorithm and the permutation algorithm in equations (15.17) and (15.18) is that samples are drawn with replacement rather than without replacement. It is not surprising that it gives very similar results (left panel of Figure 16.1). One thousand bootstrap samples were generated, and 120 had $t(\mathbf{x}^*) > 30.63$. The value of $\mathrm{ASL}_{\mathrm{boot}}$ is $120/1000 = .120$ as compared to $.152$ from the permutation test.

More accurate testing can be obtained through the use of a studentized statistic. In the above test, instead of $t(\mathbf{x}) = \bar{z} - \bar{y}$ we could use

$$t(\mathbf{x}) = \frac{\bar{z} - \bar{y}}{\bar{\sigma}\sqrt{1/n + 1/m}}, \tag{16.4}$$

where $\bar{\sigma} = \{[\sum_{i=1}^{n}(z_i - \bar{z})^2 + \sum_{j=1}^{m}(y_j - \bar{y})^2]/[n + m - 2]\}^{1/2}$. This is the two-sample t statistic described in Chapter 15. The observed value of $t(\mathbf{x})$ was 1.12. Repeating the above bootstrap algorithm,

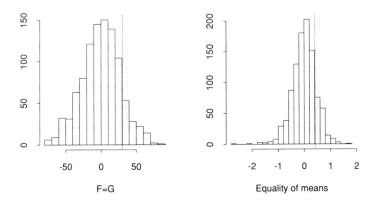

Figure 16.1. *Histograms of bootstrap replications for the mouse data example. The left panel is a histogram of bootstrap replications of $\bar{z} - \bar{y}$ for the test of $H_0 : F = G$, while the right panel is a histogram of bootstrap replications of the studentized statistic (16.5) for the test of equality of means. The dotted lines are drawn at the observed values (30.63 on the left, .416 on the right). In the left panel, $\widehat{ASL}_{boot}$ (the bootstrap estimate of the achieved significance level) equals .120, the proportion of values greater than 30.63. In the right panel, $\widehat{ASL}_{boot}$ equals .152.*

using $t(\mathbf{x}^*)$ defined by (16.4), produced 134 values out of 1000 larger than 1.12 and hence $\widehat{ASL}_{boot} = .134$. In this calculation we used exactly the same set of bootstrap samples that gave the value .120 for $\widehat{ASL}_{boot}$ based on $t(\mathbf{x}) = \bar{z} - \bar{y}$. Unlike in the permutation test, where we showed in Problem 15.9 that studentization does not affect the answer, studentization does produce a different value for $\widehat{ASL}_{boot}$. However, in this particular approach to bootstrapping the two-sample problem, the difference is typically quite small.

Algorithm 16.1 tests the null hypothesis that the two populations are identical, that is, $F = G$. What if we wanted to test only whether their means were equal? One approach would be to use the two-sample t statistic (16.4). Under the null hypothesis and assuming normal populations with equal variances, this has a Student's t distribution with $n + m - 2$ degrees of freedom. It uses the pooled estimate of standard error $\bar{\sigma}$. If we are not willing to assume that the variances in the two populations are equal, we could base the test on

$$t(\mathbf{x}) = \frac{\bar{z} - \bar{y}}{\sqrt{\bar{\sigma}_1^2/n + \bar{\sigma}_2^2/m}}, \tag{16.5}$$

where $\bar{\sigma}_1^2 = \sum_1^n (z_i - \bar{z})^2/(n-1)$, $\bar{\sigma}_2^2 = \sum_1^m (y_i - \bar{y})^2/(m-1)$. With normal populations, the quantity (16.5) no longer has a Student's t distribution and a number of approximate solutions have therefore been proposed. In the literature this is known as the Behrens-Fisher problem.

The equal variance assumption is attractive for the t-test because it simplifies the form of the resulting distribution. In considering a bootstrap hypothesis test for comparing the two means, there is no compelling reason to assume equal variances and hence we don't make that assumption. To proceed we need estimates of F and G that use only the assumption of a common mean. Letting $\bar{x}$ be the mean of the combined sample, we can translate both samples so that they have mean $\bar{x}$, and then resample each population separately. The procedure is shown in detail in Algorithm 16.2.

The results of this are shown in the right panel of Figure 16.1. The value of $\widehat{\text{ASL}}_{\text{boot}}$ was $152/1000 = .152$.

16.3 Relationship between the permutation test and the bootstrap

The preceding example illustrates some important differences between the permutation test and the bootstrap hypothesis test. A permutation test exploits special symmetry that exists under the null hypothesis to create a permutation distribution of the test statistic. For example, in the two-sample problem when testing $F = G$, all permutations of the order statistic of the combined sample are equally probable. As a result of this symmetry, the ASL from a permutation test is exact: in the two-sample problem, ASL_{perm} is the exact probability of obtaining a test statistic as extreme as the one observed, having fixed the data values of the combined sample.

In contrast, the bootstrap explicitly estimates the probability mechanism under the null hypothesis, and then samples from it to estimate the ASL. The estimate $\widehat{\text{ASL}}_{\text{boot}}$ has no interpretation as an exact probability, but like all bootstrap estimates is only guaranteed to be accurate as the sample size goes to infinity. On the other hand, the bootstrap hypothesis test does not require the

Algorithm 16.2

Computation of the bootstrap test statistic

for testing equality of means

1. Let $\hat{F}$ put equal probability on the points $\tilde{z}_i = z_i - \bar{z} + \bar{x}$, $i = 1, 2, \ldots n$, and $\hat{G}$ put equal probability on the points $\tilde{y}_i = y_i - \bar{y} + \bar{x}$, $i = 1, 2, \ldots m$, where $\bar{z}$ and $\bar{y}$ are the group means and $\bar{x}$ is the mean of the combined sample.

2. Form B bootstrap data sets $(\mathbf{z}^*, \mathbf{y}^*)$ where $\mathbf{z}^*$ is sampled with replacement from $\tilde{z}_1, \tilde{z}_2, \cdots \tilde{z}_n$ and $\mathbf{y}^*$ is sampled with replacement from $\tilde{y}_1, \tilde{y}_2, \ldots \tilde{y}_m$.

3. Evaluate $t(\cdot)$ defined by (16.5) on each data set,

$$t(\mathbf{x}^{*b}) = \frac{\bar{z}^* - \bar{y}^*}{\sqrt{\bar{\sigma}_1^{2*}/n + \bar{\sigma}_2^{2*}/m}}, \quad b = 1, 2, \cdots B. \quad (16.6)$$

4. Approximate $\mathrm{ASL}_{\mathrm{boot}}$ by

$$\widehat{\mathrm{ASL}}_{\mathrm{boot}} = \#\{t(\mathbf{x}^{*b}) \geq t_{obs}\}/B, \quad (16.7)$$

where $t_{obs} = t(\mathbf{x})$ is the observed value of the statistic.

---

special symmetry that is needed for a permutation test, and so can be applied much more generally. For instance in the two-sample problem, a permutation test can only test the null hypothesis $F = G$, while the bootstrap can test equal means and equal variances, or equal means with possibly unequal variances.

16.4 The one-sample problem

As our second example, consider a one-sample problem involving only the treated mice. Suppose that other investigators have run experiments similar to ours but with many more mice, and they observed a mean lifetime of 129.0 days for treated mice. We might want to test whether the mean of the treatment group in Table 2.1 was 129.0 as well:

$$H_0 : \mu_z = 129.0. \quad (16.8)$$

A one sample version of the normal test could be used. Assuming a normal population, under the null hypothesis

$$\bar{z} \sim N(129.0, \sigma^2/n), \tag{16.9}$$

where σ is the standard deviation of the treatment times. Having observed $\bar{z} = 86.9$, the ASL is the probability that a random variable $\bar{z}^*$ distributed accordingly to (16.9) is less than the observed value 86.9

$$\text{ASL} = \Phi(\frac{86.9 - 129.0}{\sigma/\sqrt{n}}), \tag{16.10}$$

where Φ is the cumulative distribution function of the standard normal.

Since σ is unknown, we insert the estimate

$$\bar{\sigma} = \{\sum_1^n (z_i - \bar{z})^2/(n-1)\}^{1/2} = 66.8 \tag{16.11}$$

into (16.10) giving

$$\text{ASL} = \Phi(\frac{-42.1}{66.8/\sqrt{7}}) = 0.05. \tag{16.12}$$

Student's t-test gives a somewhat larger ASL

$$\text{ASL} = \text{Prob}\{t_6 < \frac{-42.1}{66.8/\sqrt{7}}\} = 0.07. \tag{16.13}$$

So there is marginal evidence that the treated mice in our study have a mean survival time of less than 129.0 days. The two-sided ASLs are .10 and .14, respectively.

Notice that a two-sample permutation test cannot be used for this problem. If we had available all of the times for the treated mice (rather than just their mean of 129.0), we could carry out a two-sample permutation test of the equivalence of the two populations. However we do not have available all of the times but know only their mean; we wish to test $H_0 : \mu_z = 129.0$.

In contrast, the bootstrap can be used. We base the bootstrap hypothesis test on the distribution of the test statistic

$$t(\mathbf{z}) = \frac{\bar{z} - 129.0}{\bar{\sigma}/\sqrt{7}} \tag{16.14}$$

under the null hypothesis $\mu_z = 129.0$. The observed value is

$$\frac{86.9 - 129.0}{66.8/\sqrt{7}} = -1.67. \tag{16.15}$$

But what is the appropriate null distribution? We need a distribution $\hat{F}$ that estimates the population of treatment times *under H_0*. Note first that the empirical distribution $\hat{F}$ is not an appropriate estimate for F because it *does not obey H_0*. That is, the mean of $\hat{F}$ is not equal to the null value of 129.0. Somehow we need to obtain an estimate of the population that has mean 129.0. A simple way is to translate the empirical distribution $\hat{F}$ so that it has the desired mean. [1] In other words, we use as our estimated null distribution the empirical distribution on the values

$$\begin{aligned} \tilde{z}_i &= z_i - \bar{z} + 129.0 \\ &= z_i + 42.1 \end{aligned} \tag{16.16}$$

for $i = 1, 2, \cdots 7$. We sample $\tilde{z}_1^*, \ldots \tilde{z}_7^*$ with replacement from $\tilde{z}_1, \ldots \tilde{z}_7$, and for each bootstrap sample compute the statistic

$$t(\tilde{\mathbf{z}}^*) = \frac{\bar{\tilde{z}}^* - 129.0}{\bar{\tilde{\sigma}}^*/\sqrt{7}}, \tag{16.17}$$

where $\bar{\tilde{\sigma}}^*$ is the standard deviation of the bootstrap sample. A total of 100 out of 1000 samples had $t(\tilde{\mathbf{z}}^*)$ less than -1.67, and therefore the achieved significance level is $100/1000 = .10$, as compared to .05 and .07 for the normal and t tests, respectively.

Notice that our choice of null distribution assumes that the possible distributions for the treatment times, as the mean times vary, are just translated versions of one another. Such a family of distributions is called a *translation family*. This assumption is also present in the normal and t tests; but in those tests we assume further that the populations are normal. In either case, it might be sensible to take logarithms of the survival times before carrying out the analysis, because the logged lifetimes are more likely to satisfy a translation or normal family assumption (Problem 16.1).

There is a different but equivalent way of bootstrapping the one-sample problem. We draw with replacement from the (untranslated) data values $z_1, z_2, \ldots z_7$ and compute the statistic

$$t(\mathbf{z}^*) = \frac{\bar{z}^* - \bar{z}}{\bar{\sigma}^*/\sqrt{7}}, \tag{16.18}$$

---

[1] A different method is discussed in Problem 16.5.

where $\bar{\sigma}^*$ is the standard deviation of the bootstrap sample. This statistic is the same as (16.17) since

$$\bar{z}^* - 129.0 = (\bar{z}^* - \bar{z} + 129.0) - 129.0 = \bar{z}^* - \bar{z}$$

and the standard deviations are equal as well. This also shows the equivalence between the one-sample bootstrap hypothesis test and the bootstrap-t confidence interval described in Chapter 12. That interval is based on the percentiles of the statistic (16.18) under bootstrap sampling from $z_1, z_2, \ldots z_7$, exactly as above. Therefore the bootstrap-t confidence interval consists of those values μ_0 that are not rejected by the bootstrap hypothesis test described above. This general connection between confidence intervals and hypothesis tests is given in more detail in Section 12.3.

16.5 Testing multimodality of a population

Our second example is a much more exotic one. It is a case where a simple normal theory test does not exist and a permutation test cannot be used, but the bootstrap can be used effectively. The data are the thicknesses in millimeters of 485 stamps, printed in 1872. The stamp issue of that year was thought to be a "philatelic mixture", that is, printed on more than one type of paper. It is of historical interest to determine how many different types of paper were used.

A histogram of the data is shown in the top left panel of Figure 16.2. This sample is part of a large population of stamps from 1872, and we can imagine the distribution of thickness measurements for this population. We pose the statistical question: how many modes does this population have? A mode is defined to be a local maximum or "bump" of the population density. The number of modes is suggestive of the number of distinct types of paper used in the printing.

From the histogram in Figure 16.2, it appears that the population might have 2 or more modes. It is difficult to tell, however, because the histogram is not smooth. To obtain a smoother estimate, we can use a *Gaussian kernel density estimate*. Denoting the data by $x_1, \ldots x_n$, a Gaussian kernel density estimate is defined by

$$\hat{f}(t; h) = \frac{1}{nh} \sum_{1}^{n} \phi\left(\frac{t - x_i}{h}\right), \tag{16.19}$$

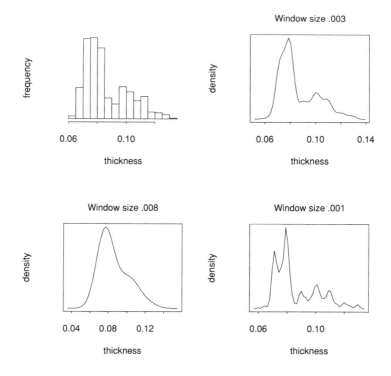

Figure 16.2. *Top left panel shows histogram of thicknesses of 485 stamps. Top right and bottom panels are Gaussian kernel density estimates for the same sample, using window size .003 (top right), .008 (bottom left) and .001 (bottom right).*

where $\phi(t)$ is the standard normal density $(1/\sqrt{2\pi})\exp(-t^2/2)$. The parameter h is called the *window size* and determines the amount of *smoothing* that is applied to the data. Larger values of h produce a smoother density estimate.

We can think of (16.19) as adding up n little Gaussian density curves centered at each point x_i, each having standard deviation h; Figure 16.3 illustrates this.

The top right panel of Figure 16.2 shows the resulting density estimate using $h = .003$; there are 2 or 3 modes. However by varying h, we can produce a greater or lesser number of modes. The

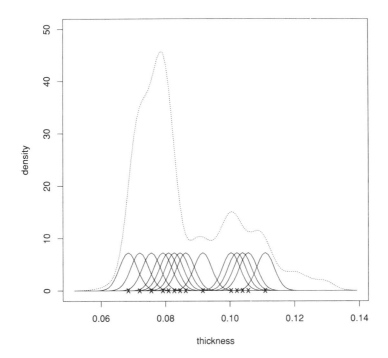

Figure 16.3. *Illustration of a Gaussian kernel density estimate. A small Gaussian density is centered at each data value (marked with an "x") and the density estimate (broken line) at each value is determined by adding up the values of all the Gaussian densities at that point. For the stamp data there are actually 485 little Gaussian densities used (one for each point); for clarity we have shown only a few.*

bottom left and right show the estimates obtained using $h = .008$ and $h = .001$, respectively. The former has one mode, while the latter has at least 7 modes! Clearly the inference that we draw from our data depends strongly on the value of h that we choose.

If we approach the problem in terms of hypothesis testing, there is a natural way to choose h. We will need the following important result, which we state without proof: as h increases, the number

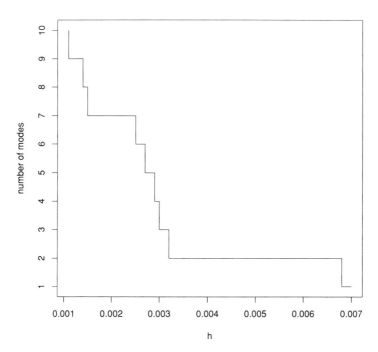

Figure 16.4. *Stamp data: number of modes in the Gaussian kernel density estimate as a function of the window size h.*

of modes in a Gaussian kernel density estimate is non-increasing. This is illustrated for the stamp data in Figure 16.4.

Now consider testing

$$H_0 : \text{number of modes} = 1 \qquad (16.20)$$

versus number of modes > 1. Since the number of modes decreases as h increases, there is a smallest value of h such that $\hat{f}(t;\ h)$ has one mode. Call this $\hat{h}_1$. Looking at Figure 16.4, $\hat{h}_1 \approx .0068$.

It seems reasonable to use $\hat{f}(t;\ \hat{h}_1)$ as the estimated null distribution for our test of H_0. In a sense, it is the density estimate closest to our data that is consistent with H_0. By "closest", we mean that it uses the least amount of smoothing (smallest value of h) among all estimates with one mode.

There is one small adjustment that we make to $\hat{f}(\cdot; \hat{h}_1)$. Formula (16.19) artificially increases the variance of the estimate (Problem 16.2), so we rescale it to have variance equal to the sample variance. Denote the rescaled estimate by $\hat{g}(\cdot; \hat{h}_1)$.

Finally, we need to select a test statistic. A natural choice is $\hat{h}_1$, the smallest window size producing a density estimate with one mode. A large value of $\hat{h}_1$ indicates that a great deal of smoothing must be done to create an estimate with one mode and is therefore evidence against H_0.

Putting all of this together, the bootstrap hypothesis test for H_0 : number of modes = 1 is based on the achieved significance level

$$\text{ASL}_{\text{boot}} = \text{Prob}_{\hat{g}(\cdot; \hat{h}_1)}\{\hat{h}_1^* > \hat{h}_1\}. \tag{16.21}$$

Here $\hat{h}_1$ is fixed at its observed value of .0068; the bootstrap sample $x_1^*, x_2^* \ldots x_n^*$ is drawn from $\hat{g}(\cdot; \hat{h}_1)$ and $\hat{h}_1^*$ is the smallest value of h producing a density estimate with one mode from the bootstrap data $x_1^*, x_2^* \ldots x_n^*$.

To approximate ASL_{boot} we need to draw bootstrap samples from the rescaled density estimate $\hat{g}(\cdot; \hat{h}_1)$. That is, rather than sampling with replacement from the data, we sample from a smooth estimate of the population. This is called *the smooth bootstrap*. Because of the convenient form of the Gaussian kernel estimate, drawing samples from $\hat{g}(\cdot; \hat{h}_1)$ is easy. We sample $y_1^*, y_2^*, \ldots y_n^*$ with replacement from $x_1, x_2, \ldots x_n$ and set

$$x_i^* = \bar{y}^* + (1 + \hat{h}_1^2/\hat{\sigma}^2)^{-1/2}(y_i^* - \bar{y}^* + \hat{h}_1 \epsilon_i); \quad i = 1, 2, \ldots n,$$
$$\tag{16.22}$$

where $\bar{y}^*$ is the mean of $y_1^*, y_2^*, \ldots y_n^*$, $\hat{\sigma}^2$ is the plug estimate of variance of the data and ϵ_i are standard normal random variables. The factor $(1 + \hat{h}_1^2/\hat{\sigma}^2)^{-1/2}$ scales the estimate so that its variance is approximately $\hat{\sigma}^2$ (Problem 16.3.) A summary of the steps is shown in Algorithm 16.3. (Actually a computational shortcut is possible for step 2; see Problem 16.3.)

We carried out this process with $B = 500$. Out of 500 bootstrap samples, none had $\hat{h}_1^* > .0068$, so $\widehat{\text{ASL}}_{\text{boot}} = 0$. We repeated this for H_0 : number of modes = 2, 3, ..., and Table 16.1 shows the resulting P-values. Interpreting these results in a sequential manner, starting with number of modes = 1, we reject the unimodal hypothesis but do not reject the hypothesis of 2 modes. This is

Algorithm 16.3

Computation of the bootstrap test statistic for multimodality

1. Draw B bootstrap samples of size n from $\hat{g}(\cdot;\,\hat{h}_1)$ using (16.22).

2. For each bootstrap sample compute $\hat{h}_1^*$ the smallest window width that produces a density estimate with one mode. Denote the B values of $\hat{h}_1^*$ by $\hat{h}_1^*(1),\ldots\hat{h}_1^*(B)$.

3. Approximate $\mathrm{ASL}_{\mathrm{boot}}$ by

$$\widehat{\mathrm{ASL}}_{\mathrm{boot}} = \#\{\hat{h}_1^*(b) \geq \hat{h}_1\}/B. \qquad (16.23)$$

---

where the inference process would end in many instances. If we were willing to entertain more exotic hypotheses, then from Table 16.1 there is also a suggestion that the population might have 7 modes.

16.6 Discussion

As the examples in this chapter illustrate, the two quantities that we must choose when carrying out a bootstrap hypothesis test are:

(a) A test statistic $t(\mathbf{x})$.

(b) A null distribution $\hat{F}_0$ for the data under H_0.

Given these, we generate B bootstrap values of $t(\mathbf{x}^*)$ under $\hat{F}_0$ and estimate the achieved significance level by

$$\widehat{\mathrm{ASL}}_{\mathrm{boot}} = \#\{t(\mathbf{x}^{*b}) \geq t(\mathbf{x})\}/B. \qquad (16.24)$$

As the stamp example shows, sometimes the choice of $t(\mathbf{x})$ and $\hat{F}_0$ are not obvious. The difficulty in choosing $\hat{F}_0$ is that, in most instances, H_0 is a composite hypothesis. In the stamp example, H_0 refers to all possible densities with one mode. A good choice for $\hat{F}_0$ is the distribution that obeys H_0 and is most reasonable for our data; this choice makes the test conservative, that is, the test is less likely to falsely reject the null hypothesis. In the stamp example, we tested for unimodality by generating samples from the unimodal distribution that is mostly nearly bimodal. In other

Table 16.1. *P-values for stamp example.*

number of modes(m)	$\hat{h}_m$	P-value
1	.0068	.00
2	0032	.29
3	.0030	.06
4	.0029	.00
5	.0027	.00
6	.0025	.00
7	.0015	.46
8	.0014	.17
9	.0011	.17

words, we used the smallest possible value for $\hat{h}_1$ and this makes the probability in (16.21) as large as possible.

The choice of test statistic $t(\mathbf{x})$ will determine the power of the test, that is, the chance that we reject H_0 when it is false. In the stamp example, if the actual population density is bimodal but the Gaussian kernel density does not approximate it accurately, then the test based on the window width $\hat{h}_1$ will not have high power.

Bootstrap tests are useful in situations where the alternative hypothesis is not well-specified. In cases where there is a parametric alternative hypothesis, likelihood or Bayesian methods might be preferable.

16.7 Bibliographic notes

Monte Carlo tests, related to the tests in this chapter, are proposed in Barnard (1963), Hope (1968), and Marriott (1979); see also Hall and Titterington (1989). Some theory of bootstrap hypothesis testing, and its relation to randomization tests, is given by Romano (1988, 1989). A discussion of practical issues appears in Hinkley (1988, 1989), Young (1988b), Noreen (1989), Fisher and Hall (1990), and Hall and Wilson (1991). See also Tibshirani (1992) for a comment on Hall and Wilson (1991). Young (1986) describe simulation-based hypothesis testing in the context of geometric statistics. Beran and Millar (1987) develop general asymptotic theory for stochastic minimum distance tests. In this work, the test statistic is the distance to a composite null hypothesis and a stochastic search procedure is used to approximate it. Besag and Clifford (1989) propose methods based on Markov chains for significance testing with dependent data. The two-sample problem with unequal variance has a long history: see, for example, Behrens (1929) and Welch (1947); Cox and Hinkley (1974) and Robinson (1982) give a more modern account. The use of the bootstrap for testing multimodality is proposed in Silverman (1981, 1983). It is applied to the stamp data in Izenman and Sommer (1988). Density estimation is described in many books, including Silverman (1986) and Scott (1992). The smooth bootstrap is studied by Silverman and Young (1987) and Hall, DiCiccio and Romano (1989).

16.8 Problems

16.1 Explain why the logarithm of survival times are more likely to be normally distributed than the times themselves.

16.2 (a) If y_i is sampled with replacement from $x_1, x_2, \ldots x_n$, ϵ_i has a standard normal distribution and $\hat{h}_1$ is considered fixed, show that

$$r_i = y_i^* + \hat{h}_1 \epsilon_i \qquad (16.25)$$

is distributed according to $\hat{f}(\cdot; \hat{h}_1)$, the Gaussian kernel density estimate defined by (16.19).

(b) Show that x_i^* given by (16.22) has the same mean as r_i^* but has variance approximately equal to $\hat{\sigma}^2$ rather than $\hat{\sigma}^2 + \hat{h}_1^2$ (the variance of r_i^*).

16.3 Denote by $\hat{h}_k$ the smallest window width producing a density estimate with k modes from our original data, and let $\hat{h}_k^*$ be the corresponding quantity for a bootstrap sample $\mathbf{x}^*$. Show that event

$$\{\hat{h}_k^* > \hat{h}_k\} \qquad (16.26)$$

is the same as the event

$$\{\hat{f}^*(\cdot; \hat{h}_k) \text{ has more than } k \text{ modes}\}, \qquad (16.27)$$

where $\hat{f}^*(\cdot; \hat{h}_k)$ is the Gaussian kernel density estimate based on the bootstrap sample $\mathbf{x}^*$. Hence it is not necessary to find $\hat{h}_k^*$ for each bootstrap sample; one need only check whether $\hat{f}(\cdot; \hat{h}_k)$ has more than k modes.

16.4 In the second example of this chapter, we tested whether the mean of the treatment group was equal to 129.0. We argued that one should not use the empirical distribution as the null distribution but rather should first translate it to have mean 129.0. In this problem we carry out a small simulation study to investigate this issue.

(a) Generate 100 samples $\mathbf{z}$ of size 7 from a normal population with mean 129.0 and standard deviation 66.8. For each sample, perform a bootstrap hypothesis test of $\mu_z = 129.0$ using the test statistic $\bar{z} - 129.0$ and using as the estimated null distribution 1) the empirical distribution, and 2) the empirical distribution translated to have mean 129.0.

Compute the average of ASL for each test, averaged over the 100 simulations.

(b) Repeat (a), but simulate from a normal population with a mean of 170. Discuss the results.

16.5 Suppose we have a sample $z_1, z_2, \ldots z_n$, and we want an estimate of the underlying population F restricted to have mean μ. One approach, used in Section 16.4, is to use the empirical distribution on the translated data values $z_i - \bar{z} + \mu$. A different approach is to leave the data values fixed, and instead change the probability p_i on each data value. Let $\mathbf{p} = (p_1, p_2, \ldots p_n)$ and let F_p be the distribution putting probability p_i on x_i for each i. Then it is reasonable to choose $\mathbf{p}$ so that the mean of $F_p = \sum p_i x_i = \mu$, and F_p is as close as possible to the empirical distribution $\hat{F}$. A convenient measure of closeness is the Kullback-Leibler distance

$$d_{F_p}(F_p, \hat{F}) = \sum_1^n p_i \log\left(\frac{1}{np_i}\right). \qquad (16.28)$$

(a) Using Lagrange multipliers, show that the probabilities that minimize expression (16.28) subject to $\sum p_i x_i = \mu$, $\sum p_i = 1$ are given by

$$p_i = \frac{\exp(tx_i)}{\sum_{i=1}^n \exp(tx_i)} \qquad (16.29)$$

where t is chosen so that $\sum p_i x_i = \mu$. This is sometimes called an *exponentially tilted* version of $\hat{F}$.

(b) Use this approach to carry out a test of $\mu = 129.0$ in the mouse data example of Section 16.4 and compare the results to those in that section.

CHAPTER 17

Cross-validation and other
estimates of prediction error

17.1 Introduction

In our discussion so far we have focused on a number of measures
of statistical accuracy: standard errors, biases, and confidence intervals. All of these are measures of accuracy for parameters of a
model. Prediction error is a different quantity that measures how
well a model predicts the response value of a future observation.
It is often used for model selection, since it is sensible to choose a
model that has the lowest prediction error among a set of candidates.

Cross-validation is a standard tool for estimating prediction error. It is an old idea (predating the bootstrap) that has enjoyed a
comeback in recent years with the increase in available computing
power and speed. In this chapter we discuss cross-validation, the
bootstrap and some other closely related techniques for estimation
of prediction error.

In regression models, prediction error refers to the expected
squared difference between a future response and its prediction
from the model:

$$PE = E(y - \hat{y})^2. \tag{17.1}$$

The expectation refers to repeated sampling from the true population. Prediction error also arises in the *classification* problem,
where the response falls into one of k unordered classes. For example, the possible responses might be Republican, Democrat, or
Independent in a political survey. In classification problems prediction error is commonly defined as the probability of an incorrect
classification

$$PE = Prob(\hat{y} \neq y), \tag{17.2}$$

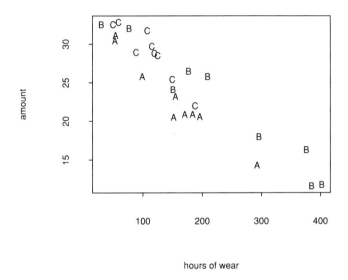

Figure 17.1. *Hormone data. Plot shows the amount of hormone remaining for a device versus the hours of wear. The symbol represents the lot number.*

also called the *misclassification rate*. The methods described in this chapter apply to both definitions of prediction error, and also to others. We begin with a intuitive description of the techniques, and then give a more detailed account in Section 17.6.2.

17.2 Example: hormone data

Let's look again at the hormone data example of chapter 9. Figure 17.1 redisplays the data for convenience. Recall that the response variable y_i is the amount of anti- inflammatory hormone remaining after z_i hours of wear, in 3 lots A, B, and C indicated by the plotting symbol in the figure. In Chapter 9 we fit regression lines to the data in each lot, with different intercepts but a common slope. The estimates are given in Table 9.3 on page 110.

Here we consider two questions: 1) "How well will the model predict the amount of hormone remaining for a new device?", and 2) "Does this model predict better (or worse) than a single regression line?" To answer the first question, we could look at the average residual squared error for all $n = 27$ responses,

$$\mathrm{RSE}/n = \sum_1^n (y_i - \hat{y}_i)^2/n = 2.20, \qquad (17.3)$$

but this will tend to be too "optimistic"; that is to say, it will probably underestimate the true prediction error. The reason is that we are using the same data to assess the model as were used to fit it, using parameter estimates that are fine-tuned to our particular data set. In other words the *test sample* is the same as the original sample, sometimes called the *training sample*. Estimates of prediction error obtained in this way are aptly called "apparent error" estimates.

A familiar method for improving on (17.3) is to divide by $n - p$ instead of n, where p is the number of predictor variables. This gives the usual unbiased estimate of residual variance $\hat{\sigma}^2 = \sum(y_i - \hat{y}_i)^2/(n - p)$. We will see that bigger corrections are necessary for the prediction problem.

17.3 Cross-validation

In order to get a more realistic estimate of prediction error, we would like to have a test sample that is separate from our training sample. Ideally this would come in the form of some new data from the same population that produced our original sample. In our example this would be hours of wear and hormone amount for some additional devices, say m of them. If we had these new data, say $(z_1^0, y_1^0), \ldots (z_m^0, y_m^0)$, we would work out the predicted values $\hat{y}_i^0$ from (9.3)

$$\hat{y}_i^0 = \hat{\beta}_j + \hat{\beta}_1 z_i^0 \qquad (17.4)$$

(where $j = A, B$, or C depending on the lot), and compute the average prediction sum of squares

$$\sum_1^m (y_i^0 - \hat{y}_i^0)^2/m. \qquad (17.5)$$

Algorithm 17.1

K-fold cross-validation

1. Split the data into K roughly equal-sized parts.
2. For the kth part, fit the model to the other $K - 1$ parts of the data, and calculate the prediction error of the fitted model when predicting the kth part of the data.
3. Do the above for $k = 1, 2, \ldots K$ and combine the K estimates of prediction error.

This quantity estimates how far, on the average, our prediction $\hat{y}_i^0$ differs from the actual value y_i^0.

Usually, additional data are not often available, for reasons of logistics or cost. To get around this, cross-validation uses part of the available data to fit the model, and a different part to test it. With large amounts of data, a common practice is to split the data into two equal parts. With smaller data sets like the hormone data, "K-fold" cross-validation makes more efficient use of the available information. The procedure is shown in Algorithm 17.1.

Here is K-fold cross-validation in more detail. Suppose we split the data into K parts. Let $k(i)$ be the part containing observation i. Denote by $\hat{y}_i^{-k(i)}$ the fitted value for observation i, computed with the $k(i)$th part of the data removed. Then the cross-validation estimate of prediction error is

$$\mathrm{CV} = \frac{1}{n} \sum_{i=1}^{n} (y_i - \hat{y}_i^{-k(i)})^2. \tag{17.6}$$

Often we choose $k = n$, resulting in "leave-one-out" cross-validation. For each observation i, we refit the model leaving that observation out of the data, and then compute the predicted value for the ith observation, denoted by $\hat{y}_i^{-i}$. We do this for each observation and then compute the average cross-validation sum of squares $\mathrm{CV} = \sum (y_i - \hat{y}_i^{-i})^2/n$.

We applied leave-one-out cross-validation to the hormone data: the value of CV turned out to be 3.09. By comparison, the average residual squared error (17.3) is 2.20 and so it underestimates the prediction error by about 29%. Figure 17.2 shows the usual residual

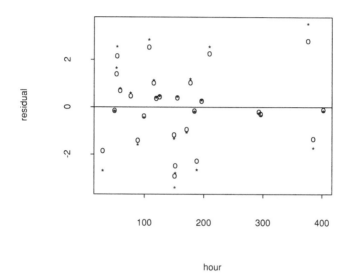

Figure 17.2. *Plot of residuals (circles) and cross-validated residuals (stars) for hormone data.*

$y_i - \hat{y}_i$ (circles) and the cross-validated residual $y_i - \hat{y}_i^{-i}$ (stars). Notice how the cross-validated residual is equal to or larger (in absolute value) than the usual residual for every case. (This turns out to be true in some generality: see Problems 17.1 and 18.1.)

We can look further at the breakdown of the CV by lot: the average values are 2.09, 4.76 and 2.43 for lots A, B and C, respectively. Hence the amounts for devices in lot B are more difficult to predict than those in lots A and C.

Cross-validation, as just described, requires refitting the complete model n times. In general this is unavoidable, but for least-squares fitting a handy shortcut is available (Problem 17.1).

17.4 C_p and other estimates of prediction error

There are other ways to estimate prediction error, and all are based on adjustments to the residual squared error RSE. The last part of this chapter describes a bootstrap approach. A simple analytic measure is the adjusted residual squared error

$$\text{RSE}/(n - 2p) \tag{17.7}$$

where p denotes the number of regressors in the model. This adjusts RSE/n upward to account for the fitting, the adjustment being larger as p increases. Note that $\text{RSE}/(n - 2p)$ is a more severe adjustment to RSE than the unbiased estimate of variance $\text{RSE}/(n - p)$.

Another estimate is (one form of) the "C_p" statistic

$$C_p = \text{RSE}/n + 2p\hat{\sigma}^2/n. \tag{17.8}$$

Here $\hat{\sigma}^2$ is an estimate of the residual variance; a reasonable choice for $\hat{\sigma}^2$ is $\text{RSE}/(n - p)$. (When computing the C_p statistic for a number of models, $\hat{\sigma}^2$ is computed once from the value of $\text{RSE}/(n - p)$ for some fixed large model.) The C_p statistic is a special case of Akaike's information criterion (AIC) for general models. It adjusts RSE/n so as to make it approximately unbiased for prediction error: $\text{E}(C_p) \approx \text{PE}$.

Implicitly these corrections account for the fact that the same data is being used to fit the model and to assess it through the residual squared error. The "p" in the denominator of the adjusted RSE and the second term of C_p are penalties to account for the amount of fitting. A simple argument shows that the adjusted residual squared error and C_p statistic are equivalent to a first order of approximation (Problem 17.4.)

Similar to C_p is Schwartz's criterion, or the BIC (Bayesian Information Criterion)

$$\text{BIC} = \text{RSE}/n + \log n \cdot p\hat{\sigma}^2/n \tag{17.9}$$

BIC replaces the "2" in C_p with $\log n$ and hence applies a more severe penalty than C_p, as long as $n > e^2$. As a result, when used for model comparisons, BIC will tend to favor more parsimonious models than C_p. One can show that BIC is a consistent criterion in the sense that it chooses the correct model as $n \to \infty$. This is not the case for the adjusted RSE or C_p.

In the hormone example, RSE $= 59.27$, $\hat{\sigma}^2 = 2.58$ and $p = 4$ and hence $RSE/(n - 2p) = 3.12$, $C_p = 2.96$, $BIC = 3.45$, as compared to the value of 3.09 for CV.

Why bother with cross-validation when simpler alternatives are available? The main reason is that for fitting problems more complicated than least squares, the number of parameters "p" is not known. The adjusted residual squared error, C_p and BIC statistics require knowledge of p, while cross-validation does not. Just like the bootstrap, cross-validation tends to give similar answers as standard methods in simple problems and its real power stems from its applicability in more complex situations. An example involving a classification tree is given below.

A second advantage of cross-validation is its robustness. The C_p and BIC statistics require a roughly correct working model to obtain the estimate $\hat{\sigma}^2$. Cross-validation does not require this and will work well even if the models being assessed are far from correct.

Finally, let's answer the second question raised above, regarding a comparison of the common slope, separate intercept model to a simpler model that specifies one common regression line for all lots. In the same manner as described above, we can compute the cross-validation sum of squares for the single regression line model. This value is 5.89 which is quite a bit larger than the value 3.27 for the model that allows a different intercept for each lot. This is not surprising given the statistically significant differences among the intercepts in Table 9.3. But cross-validation is useful because it gives a quantitative measure of the price the investigator would pay if he does not adjust for the lot number of a device.

17.5 Example: classification trees

For an example that illustrates the real power of cross-validation, let's switch gears and discuss a modern statistical procedure called "classification trees." In an experiment designed to provide information about the causes of duodenal ulcers (Giampaolo *et al.* 1988), a sample of 745 rats were each administered one of 56 model alkyl nucleophiles. Each rat was later autopsied for the development of duodenal ulcer and the outcome was classified as 1, 2 or 3 in increasing order of severity. There were 535 class 1, 90 class 2 and 120 class 3 outcomes. Sixty-seven characteristics of these compounds were measured, and the objective of the analysis was to ascertain which of the characteristics were associated with the development of duodenal ulcers.

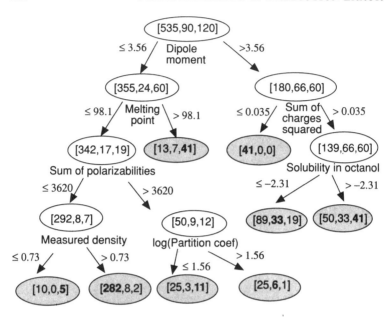

Figure 17.3. *CART tree. Classification tree from the CART analysis of data on duodenal ulcers. At each node of the tree a question is asked, and data points for which the answer is "yes" are assigned to the left branch and the others to the right branch. The shaded regions are the terminal nodes, or leaves, of the tree. The numbers in square brackets are the number of observations in each of the three classes present at each node. The bold number indicates the predicted class for the node. In this particular example, five penalty points are charged for misclassifying observations in true class 2 or 3, and one penalty point is charged for misclassifying observations in class 1. The predicted class is the one resulting in the fewest number of penalty points.*

The CART method (for Classification and Regression Trees) of Breiman, Friedman, Olshen and Stone (1984) is a computer-intensive approach to this problem that has become popular in scientific circles. When applied to these data, CART produced the classification tree shown in Figure 17.3.

At each node of the tree a yes-no question is asked, and data points for which the answer is "yes" are assigned to the left branch and the others to the right branch. The leaves of the tree shown in Figure 17.3 are called "terminal nodes." Each observation is assigned to one of the terminal nodes based on the answers to the questions. For example a rat that received a compound with Dipole moment $\leq$ 3.56 and melting point $>$ 98.1 would go left then right and end up in the terminal node marked "[13, 7, 41]." Triplets of numbers such as "[13, 7, 41]" below each terminal node number indicate the membership at that node, that is, there are 13 class 1, 7 class 2 and 41 class 3 observations at this terminal node.

Before discussing how the CART procedure built this tree, let's look at how it is used for classification. Each terminal node is assigned a class (1,2 or 3). The most obvious way to assign classes to the terminal nodes would be to use a majority rule and assign the class that is most numerous in the node. Using a majority rule, the node marked "[13,7,41]" would be assigned to class 3 and all of the other terminal nodes would be assigned to class 1. In this study, however, the investigators decided that it would be five times worse to misclassify an animal that actually had a severe ulcer or moderate ulcer than one with a milder ulcer. Hence, five penalty points were charged for misclassifying observations in true class 2 or 3, and one penalty point was charged for misclassifying observations in class 1. The predicted class is the one resulting in the fewest number of penalty points. In Figure 17.3 the predicted class is in boldface at each terminal node; for example, the node at the bottom left marked "[10,0,5]" has the "5" in boldface and hence is a class 3 node.

We can summarize the tree as follows. The top ("root") node was split on dipole moment. A high dipole moment indicates the presence of electronegative groups. This split separates the class 1 and 2 compounds: the ratio of class 2 to class 1 in the right split, 66/190, is more than 5 times as large as the ratio 24/355 in the left split. However, the class 3 compounds are divided equally, 60 on each side of the split. If in addition the sum of squared atomic charges is low, then CART finds that all compounds are class 1. Hence, ionization is a major determinant of biologic action in compounds with high dipole moments. Moving further down the right side of the tree, the solubility in octanol then (partially) separates class 3 from class 2 compounds. High octanol solubility probably reflects the ability to cross membranes and to enter the central nervous system.

On the left side of the root node, compounds with low dipole moment and high melting point were found to be class 3 severe. Compounds at this terminal node are related to cysteamine. Compounds with low melting points and high polarizability, all thiols in this study, were classified as class 2 or 3 with the partition coefficient separating these two classes. Of those chemicals with low polarizability, those of high density are class 1. These chemicals have high molecular weight and volume, and this terminal node contains the highest number of observations. The low density side of the split are all short chain amines.

In the terminology mentioned earlier, the data set of 745 observations is called the training sample. It is easy to work out the misclassification rate for each class when the tree of Figure 17.3 is applied to the training sample. Looking at the terminal nodes that predict classes 2 or 3, the number of errors for class 1 is $13 + 89 + 50 + 10 + 25 + 25 = 212$, so the apparent misclassification rate for class 1 is 212/535=39.6%. Similarly, the apparent misclassification rates for classes 2 and 3 are 56.7% and 18.3%. "Apparent" is an important qualifier here, since misclassification rates in the training sample can be badly biased downward, for the same reason that the residual squared error is overly optimistic in regression.

How does CART build a tree like that in Figure 17.3? CART is a fully automatic procedure that chooses the splitting variables and splitting points that best discriminate between the outcome classes. For example, "Dipole moment$\leq$ 3.56" is the split that was determined to best separate the data with respect to the outcome classes. CART chose both the splitting variable "Dipole moment" and the splitting value 3.56. Having found the first splitting rule, new splitting rules are selected for each of the two resulting groups, and this process is repeated.

Instead of stopping when the tree is some reasonable size, CART uses a more effective approach: a large tree is constructed and then pruned from the bottom. This latter approach is more effective in discovering interactions that involve several variables.

This brings up an important question: how large should the tree be? If we were to build a very large tree with only one observation in each terminal node, then the apparent misclassification rate would be 0%. However, this tree would probably do a poor job predicting the outcomes for a new sample. The reason is that the tree would be geared to the training sample; statistically speaking it is "overfit."

The best-sized tree would be the one that had the lowest misclassification rate for some new data. Thus if we had a second data set available (a test sample), we could apply trees of various sizes to it and then choose the one with lowest misclassification rate.

Of course in most situations we do not have extra data to work with, and this is where cross-validation comes in handy. Leave-one-out cross-validation doesn't work well here, because the resulting trees are not different enough from the original tree. Experience shows that it is much better to divide the data up into 10 groups of equal size, building a tree on 90% of the data, and then assessing its misclassification rate on the remaining 10% of the data. This is done for each of the 10 groups in turn, and the total misclassification rate is computed over the 10 runs. The best tree size is determined to be that tree size giving lowest misclassification rate. This is the size used in constructing the final tree from all of the data. The crucial feature of cross-validation is the separation of data for building and assessing the trees: each one-tenth of the data is acting as a test sample for the other 9 tenths. The precise details of the tree selection process are given in Problem 17.9.

The process of cross-validation not only provides an estimate of the best tree size, it also gives a realistic estimate of the misclassification rate of the final tree. The apparent error rates computed above are often unrealistically low because the training sample is used both for building and assessing the tree. For the tree of Figure 17.3, the cross-validated misclassification rates were about 10% higher than the apparent error rates. It is the cross-validated rates that provide an accurate assessment of how effective the tree will be in classifying a new sample.

17.6 Bootstrap estimates of prediction error

17.6.1 Overview

In the next two sections we investigate how the bootstrap can be used to estimate prediction error. A precise formulation will require some notation. Before jumping into that, we will convey the main ideas. The simplest bootstrap approach generates B bootstrap samples, estimates the model on each, and then applies each fitted model to the *original sample* to give B estimates of prediction error. The overall estimate of prediction error is the average of these B estimates. As an example, the left hand column of Table 17.1 shows 10 estimates of prediction error ("err") from 10 bootstrap samples, for the hormone data example described in Section 17.2. Their average is 2.52, as compared to the value of 2.20 for RSE/n.

Table 17.1. *Bootstrap estimates of prediction error for hormone data of Chapter 9. In each row of the table a bootstrap sample was generated by sampling with replacement from the hormone data, and the model specified in equation (9.21) was fit. The left column shows the resulting prediction error when this model is applied to the original data. The average of the left column (=2.52) is the simple bootstrap estimate of prediction error. The center column is the prediction error that results when the model is applied to the bootstrap sample, the so-called "apparent error." It is unrealistically low. The difference between the first and second columns is the "optimism" in the apparent error, given in the third column. The more refined bootstrap estimate adds the average optimism (=0.82) to the average residual squared error (=2.20), giving an estimate of 3.02.*

	$\mathrm{err}(\mathbf{x}^*, \widehat{F})$	$\mathrm{err}(\mathbf{x}^*, \widehat{F}^*)$	$\mathrm{err}(\mathbf{x}^*, \widehat{F}) - \mathrm{err}(\mathbf{x}^*, \widehat{F}^*)$
sample 1:	2.30	1.47	0.83
sample 2:	2.56	3.03	-0.47
sample 3:	2.30	1.65	0.65
sample 4:	2.43	1.76	0.67
sample 5:	2.44	2.00	0.44
sample 6:	2.67	1.17	1.50
sample 7:	2.68	1.23	1.45
sample 8:	2.39	1.55	0.84
sample 9:	2.86	1.76	1.10
sample 10:	2.54	1.37	1.17
AVERAGE:	2.52	1.70	0.82

This simple bootstrap approach turns out not to work very well, but fortunately, it is easy to improve upon. Take a look at the second column of Table 17.1: it shows the prediction error when the model estimated from the bootstrap sample is applied to the *bootstrap sample itself*. Not surprisingly, the values in the second column are lower on the average than those in the first column. The improved bootstrap estimate focuses on the difference between the first and second columns, called appropriately the "optimism"; it is the amount by which the average residual squared error (or "apparent error rate") underestimates the true prediction error. The overall estimate of optimism is the average of the B differences between the first and second columns, a value of 0.82 in this example.

Once an estimate of optimism is obtained, it is added to the apparent error rate to obtain an improved estimate of prediction error. Here we obtain 2.20+0.82=3.02. Of course 10 bootstrap samples are too few; repeating with 200 samples gave a value of 2.77 for the simple bootstrap estimate, and an estimate of .80 for the optimism leading to the value 2.20+0.80=3.00 for the improved estimate of prediction error. Essentially, we have added a bias-correction to the apparent error rate, in the same spirit as in Chapter 10.

17.6.2 Some details

The more refined bootstrap approach improves on the simpler approach by effectively removing the variability between the rows of Table 17.1, much like removing block effects in a two way analysis of variance. To understand further the justification for the bootstrap procedures, we need to think in terms of probability models for the data.

In Chapters 7 and 9, we describe two methods for bootstrapping regression models. The second method, which will be our focus here, treats the data $x_i = (c_i, y_i)$, $i = 1, 2, \ldots n$ as an i.i.d sample from the multi-dimensional distribution F. Recall that c_i might be a vector: in the hormone data, c_i would be the lot number and hours worn for the ith device. Call the entire sample x. A classification problem can be expressed in the same way, with y_i indicating the class membership of the ith observation. Our discussion below is quite general, covering both the regression and classification problems.

Suppose we estimate a model from our data, producing a predicted value of y at $c = c_0$ denoted by

$$\eta_x(c_0). \qquad (17.10)$$

We assume that $\eta_x(c_0)$ can be expressed as a plug-in statistic, that is $\eta_x(c_0) = \eta(c_0, \hat{F})$ for some function η, where $\hat{F}$ is the empirical distribution function of the data. If our problem is a regression problem as in the hormone example, then $\eta_x(c_0) = c_0\hat{\beta}$ where $\hat{\beta}$ is the least squares estimate of the regression parameter. In a classification problem, $\eta_x(c_0)$ is the predicted class for an observation with $c = c_0$.

Let $Q[y, \eta]$ denote a measure of error between the response y and the prediction η. In regression we often choose $Q[y, \eta] = (y - \eta)^2$;

in classification typically $Q[y, \eta] = I_{\{y \neq \eta\}}$, that is $Q[y, \eta] = 1$ if $y \neq \eta$ and 0 otherwise.

The prediction error for $\eta_{\mathbf{x}}(\mathbf{c}_0)$ is defined by

$$\text{err}(\mathbf{x}, F) \equiv \text{E}_{0F}\{Q[Y_0, \eta_{\mathbf{x}}(\mathbf{C}_0)]\}. \tag{17.11}$$

The notation E_{0F} indicates expectation over a new observation $(\mathbf{C}_0, Y_0)$ from the population F. Note that E_{0F} does not average over the data set $\mathbf{x}$, which is considered fixed. The apparent error rate is

$$\text{err}(\mathbf{x}, \hat{F}) = \text{E}_{0\hat{F}}\{Q[Y_0, \eta_{\mathbf{x}}(\mathbf{c}_i)]\} = \frac{1}{n}\sum_{1}^{n} Q[y_i, \eta_{\mathbf{x}}(\mathbf{c}_i)] \tag{17.12}$$

because "$\text{E}_{0\hat{F}}$" simply averages over the n observed cases $(\mathbf{c}_i, y_i)$. In regression with $Q[y, \eta] = (y - \eta)^2$, we have $\text{err}(\mathbf{x}, \hat{F}) = \sum_{1}^{n}[y_i - \eta_{\mathbf{x}}(\mathbf{c}_i)]^2/n$, while in classification with $Q[y, \eta] = I_{\{y \neq \eta\}}$, it equals $\#\{\eta_{\mathbf{x}}(\mathbf{c}_i) \neq y_i\}/n$ the misclassification rate over the original data set.

The K-fold cross-validation estimate of Section 17.3 can also be expressed in this framework. Let $k(i)$ denote the part containing observation i, and $\eta_{\mathbf{x}}^{-k(i)}(\mathbf{c})$ be the predicted value at $\mathbf{c}$, computed with the $k(i)$th part of the data removed. Then the cross-validation estimate of the true error rate is

$$\frac{1}{n}\sum_{i=1}^{n} Q[y_i, \eta_{\mathbf{x}}^{-k(i)}(\mathbf{c}_i)]. \tag{17.13}$$

To construct a bootstrap estimate of prediction error we apply the plug-in principle to equation (17.11). Let $\mathbf{x}^* = \{(\mathbf{c}_1^*, y_1^*), (\mathbf{c}_2^*, y_2^*), \ldots (\mathbf{c}_n^*, y_n^*)\}$ be a bootstrap sample. Then the plug-in estimate of $\text{err}(\mathbf{x}, F)$ is

$$\text{err}(\mathbf{x}^*, \hat{F}) = \frac{1}{n}\sum_{1}^{n} Q[y_i, \eta_{\mathbf{x}^*}(\mathbf{c}_i)] \tag{17.14}$$

In this expression $\eta_{\mathbf{x}^*}(\mathbf{c}_i)$ is the predicted value at $\mathbf{c} = \mathbf{c}_i$, based on the model estimated from the bootstrap data set $\mathbf{x}^*$.

We could use $\text{err}(\mathbf{x}^*, \hat{F})$ as our estimate, but it involves only a single bootstrap sample and hence is too variable. Instead, we must focus on the *average* prediction error

$$\text{E}_F[\text{err}(\mathbf{x}, F)], \tag{17.15}$$

with E_F indicating the expectation over data sets $\mathbf{x}$ with observations $\mathbf{x}_i \sim F$. The bootstrap estimate is

$$E_{\hat{F}}[\mathrm{err}(\mathbf{x}^*, \hat{F})] = E_{\hat{F}} \sum_1^n Q[y_i, \eta_{\mathbf{x}^*}(\mathbf{c}_i)]/n. \qquad (17.16)$$

Intuitively, the underlying idea is much the same as in Figure 8.3: in the "bootstrap world", the bootstrap sample is playing the role of the original sample, while the original sample is playing the role of the underlying population F.

Expression (17.16) is an ideal bootstrap estimate, corresponding to an infinite number of bootstrap samples. With a finite number B of bootstrap samples, we approximate this as follows. Let $\eta_{\mathbf{x}^{*b}}(\mathbf{c}_i)$ be the predicted value at $\mathbf{c}_i$, from the model estimated on bth bootstrap sample, $b = 1, 2, \ldots B$. Then our approximation to $E_{\hat{F}}[\mathrm{err}(\mathbf{x}^*, \hat{F})]$ is

$$\widehat{E}_{\hat{F}}[\mathrm{err}(\mathbf{x}^*, \hat{F})] = \frac{1}{B} \sum_{b=1}^B \sum_{i=1}^n Q[y_i, \eta_{\mathbf{x}^{*b}}(\mathbf{c}_i)]/n. \qquad (17.17)$$

In regression $\sum_1^n Q[y_i, \eta_{\mathbf{x}^{*b}}(\mathbf{c}_i)]/n = \sum_{i=1}^n [y_i - \eta_{\mathbf{x}^{*b}}(\mathbf{c}_i)]^2/n$; these are the values in the left hand column of Table 17.1, and their average (2.52) corresponds to the formula in equation (17.17).

The more refined bootstrap approach estimates the bias in $\mathrm{err}(\mathbf{x}, \hat{F})$ as an estimator of $\mathrm{err}(\mathbf{x}, F)$, and then corrects $\mathrm{err}(\mathbf{x}, \hat{F})$ by subtracting its estimated bias. We define the average optimism by

$$\omega(F) \equiv E_F[\mathrm{err}(\mathbf{x}, F) - \mathrm{err}(\mathbf{x}, \hat{F})]. \qquad (17.18)$$

This is the average difference between the true prediction error and the apparent error, over data sets $\mathbf{x}$ with observations $\mathbf{x}_i \sim F$. Note that $\omega(F)$ will tend to be positive because the apparent error rate tends to underestimate the prediction error. The bootstrap estimate of $\omega(F)$ is obtained through the plug-in principle:

$$\omega(\hat{F}) = E_{\hat{F}}[\mathrm{err}(\mathbf{x}^*, \hat{F}) - \mathrm{err}(\mathbf{x}^*, \hat{F}^*)]. \qquad (17.19)$$

Here $\hat{F}^*$ is the empirical distribution function of the bootstrap sample $\mathbf{x}^*$. The approximation to this ideal bootstrap quantity is

$$\widehat{\omega}(\hat{F}) = \frac{1}{B \cdot n} \left\{ \sum_{b=1}^{B} \sum_{i=1}^{n} Q[y_i, \eta_{\mathbf{X}^{*b}}(\mathbf{c}_i)] - \sum_{b=1}^{B} \sum_{i=1}^{n} Q[y_{ib}^*, \eta_{\mathbf{X}^{*b}}(\mathbf{c}_i^*)] \right\}.$$
(17.20)

In the above equation, $\eta_{\mathbf{X}^{*b}}(\mathbf{c}_i^*)$ is the predicted value at $\mathbf{c}_i^*$ from the model estimated on the bth bootstrap sample, $b = 1, 2, \ldots B$, and y_{ib}^* is the response value of the ith observation for the bth bootstrap sample. In Table 17.1, this is estimated by the average difference between the second and third columns, namely 0.82. The final estimate of prediction error is the apparent error plus the downward bias in the apparent error given by (17.20),

$$\text{err}(\mathbf{x}, \hat{F}) + \omega(\hat{F}) \tag{17.21}$$

which is approximated by $\frac{1}{n} \sum_{1}^{n} Q[y_i, \eta_{\mathbf{X}}(\mathbf{c}_i)] + \widehat{\omega}(\hat{F})$. This equals 2.20+0.82=3.02 in our example.

Both $\omega(\hat{F})$ and $\widehat{\mathrm{E}}[\text{err}(\mathbf{x}^*, \hat{F})]$ do not fix $\mathbf{x}$ (as specified in definition 17.11), but instead measure averages over data sets drawn from $\hat{F}$. The refined estimate in (17.21) is superior to the simple estimate (17.17) because it uses the observed $\mathbf{x}$ in the first term $\text{err}(\mathbf{x}, \hat{F})$; averaging only enters into the correction term $\omega(\hat{F})$.

17.7 The .632 bootstrap estimator

The simple bootstrap estimate in (17.17) can be written slightly differently

$$\widehat{\mathrm{E}}_{\hat{F}}[\text{err}(\mathbf{x}^*, \hat{F})] = \frac{1}{n} \sum_{i=1}^{n} \sum_{b=1}^{B} Q[y_i, \eta_{\mathbf{X}^{*b}}(\mathbf{c}_i)]/B. \tag{17.22}$$

We can view equation (17.22) as estimating the prediction error for each data point $(\mathbf{c}_i, y_i)$ and then averaging the error over $i = 1, 2, \ldots n$. Now for each data point $(\mathbf{c}_i, y_i)$, we can divide the bootstrap samples into those that contain $(\mathbf{c}_i, y_i)$ and those that do not. The prediction error for the data point $(\mathbf{c}_i, y_i)$ will likely be larger for a bootstrap sample *not* containing it, since such a bootstrap sample is "farther away" from $(\mathbf{c}_i, y_i)$ in some sense. The idea behind the .632 bootstrap estimator is to use the prediction error from just these cases to adjust the optimism in the apparent error rate.

Let ϵ_0 be the average error rate obtained from bootstrap data sets not containing the point being predicted (below we give details on the estimation of ϵ_0). As before, $\text{err}(\mathbf{x}, \hat{F})$ is the apparent error rate. It seems reasonable to use some multiple of $\epsilon_0 - \text{err}(\mathbf{x}, \hat{F})$ as an estimate of the optimism of $\text{err}(\mathbf{x}, \hat{F})$. The .632 bootstrap estimate of optimism is defined as

$$\hat{\omega}^{.632} = .632[\epsilon_0 - \text{err}(\mathbf{x}, \hat{F})]. \tag{17.23}$$

Adding this estimate to $\text{err}(\mathbf{x}, \hat{F})$ gives the .632 estimate of prediction error

$$\begin{aligned}\widehat{\text{err}}^{.632} &= \text{err}(\mathbf{x}, \hat{F}) + .632[\epsilon_0 - \text{err}(\mathbf{x}, \hat{F})] \\ &= .368 \cdot \text{err}(\mathbf{x}, \hat{F}) + .632 \cdot \epsilon_0. \end{aligned} \tag{17.24}$$

The factor ".632" comes from a theoretical argument showing that the bootstrap samples used in computing ϵ_0 are farther away on the average than a typical test sample, by roughly a factor of $1/.632$. The adjustment in (17.23) corrects for this, and makes $\widehat{\text{err}}^{.632}$ roughly unbiased for the true error rate. We will not give the theoretical argument here, but note that the value .632 arises because it is approximately the probability that a given observation appears in bootstrap sample of size n (Problem 17.7).

Given a set of B bootstrap samples, we estimate ϵ_0 by

$$\hat{\epsilon}_0 = \frac{1}{n} \sum_{i=1}^{n} \sum_{b \in C_i} Q[y_i, \eta_{\mathbf{X}^{*b}}(\mathbf{c}_i)]/B_i \tag{17.25}$$

where C_i is the set of indices of the bootstrap samples not containing the ith data point, and B_i is the number of such bootstrap samples. Table 17.2 shows the observation numbers appearing in each of the 10 bootstrap samples of Table 17.1. Observation #5, for example, does not appear in bootstrap samples 3,4,8, and 9. In the notation of equation (17.25), $C_i = (3, 4, 8, 9)$. So we would use only these four bootstrap samples in estimating the prediction error for observation $i = 5$ in equation (17.25).

In our example, $\hat{\epsilon}_0$ equals 3.63. Not surprisingly, this is larger than the apparent error 2.20, since it is the average prediction error for data points *not appearing* in the bootstrap sample used for their prediction. The .632 estimate of prediction error is therefore $.368 \cdot 2.20 + .632 \cdot 3.63 = 3.10$, close to the value of 3.00 obtained from the more refined bootstrap approach earlier.

Table 17.2. *The observation numbers appearing in each of the 10 bootstrap samples of Table 17.1.*

				Bootstrap	sample				
1	2	3	4	5	6	7	8	9	10
1	16	25	1	14	15	14	23	6	5
5	5	4	7	10	24	7	17	26	9
23	16	12	12	2	12	1	15	10	3
11	24	16	7	8	18	6	9	9	3
11	11	14	14	13	15	11	6	27	26
24	14	27	25	5	23	21	22	10	4
15	17	24	1	1	9	22	9	23	25
10	26	7	22	7	8	5	22	7	21
27	11	23	26	1	7	27	3	3	20
26	27	18	4	6	9	25	8	7	15
4	20	14	26	25	25	25	7	9	14
2	10	13	15	25	9	23	26	4	5
5	26	2	9	19	6	22	2	18	7
24	26	27	6	20	22	8	17	11	25
1	22	14	26	5	18	6	17	19	20
27	22	8	7	20	25	23	22	20	16
8	21	3	21	17	2	11	27	21	17
17	21	6	10	25	26	4	22	17	23
9	26	17	17	4	7	22	8	3	12
4	16	27	14	11	21	17	15	11	8
14	14	11	13	21	14	25	24	2	26
14	20	25	18	12	15	7	16	12	19
13	14	8	22	16	24	16	3	8	15
22	23	25	25	24	4	3	19	22	3
8	13	19	24	9	14	27	27	8	9
2	13	26	7	9	27	18	23	1	15
3	16	25	1	18	5	8	3	14	23

As a matter of interest, the average prediction error for data points that *did* appear in the bootstrap sample used for their prediction was 3.08; this value, however, is not used in the construction of the .632 estimator.

17.8 Discussion

All of the estimates of prediction error described in this chapter are significant improvements over the apparent error rate. Which is best among these competing methods is not clear. The methods are asymptotically the same, but can behave quite differently in small samples. Simulation experiments show that cross-validation is roughly unbiased but can show large variability. The simple bootstrap method has lower variability but can be severely biased downward; the more refined bootstrap approach is an improvement but still suffers from downward bias. In the few studies to date, the .632 estimator performed the best among all methods, but we need more evidence before making any solid recommendations.

S language functions for calculating cross-validation and bootstrap estimates of prediction error are described in the Appendix.

17.9 Bibliographic notes

Key references for cross-validation are Stone (1974, 1977) and Allen (1974). The AIC is proposed by Akaike (1973), while the BIC is introduced by Schwarz (1978). Stone (1977) shows that the AIC and leave one out cross-validation are asymptotically equivalent. The C_p statistic is proposed in Mallows (1973). Generalized cross-validation is described by Golub, Heath and Wahba (1979) and Wahba (1980); a further discussion of the topic may be found in the monograph by Wahba (1990). See also Hastie and Tibshirani (1990, chapter 3). Efron (1983) proposes a number of bootstrap estimates of prediction error, including the optimism and .632 estimates. Efron (1986) compares C_p, CV, GCV and bootstrap estimates of error rates, and argues that GCV is closer to C_p than CV. Linhart and Zucchini (1986) provide a survey of model selection techniques. The use of cross-validation and the bootstrap for model selection is studied by Breiman (1992), Breiman and Spector (1992), Shao (1993) and Zhang (1992). The CART (Classification and Regression Tree) methodology is due to Breiman *et al.* (1984). A study of cross-validation and bootstrap methods for these models is carried out by Crawford (1989). The CART tree example is taken from Giampaolo *et.al.* (1988).

17.10 Problems

17.1   (a) Let C be a regression design matrix as described on page 106 of Chapter 9. The projection or "hat" matrix that produces the fit is $H = C(C^T C)^{-1} C^T$. If h_{ii} denotes the iith element of H, show that the cross-validated residual can be written as

$$y_i - \hat{y}_i^{-i} = \frac{y_i - \hat{y}_i}{1 - h_{ii}}. \qquad (17.26)$$

(Hint: see the Sherman-Morrison-Woodbury formula in chapter 1 of Golub and Van Loan, 1983).

(b) Use this result to show that $y_i - \hat{y}_i^{-i} \geq y_i - \hat{y}_i$.

17.2 Find the explicit form of h_{ii} for the hormone data example.

17.3 Using the result of Problem 17.1 we can derive a simplified version of cross-validation, by replacing each h_{ii} by its average value $\bar{h} = \sum_1^n h_{ii}/n$. The resulting estimate is called "generalized cross-validation":

$$\text{GCV} = \frac{1}{n} \sum_1^n \left(\frac{y_i - \hat{y}_i}{1 - \bar{h}} \right)^2. \qquad (17.27)$$

Use a Taylor series approximation to show the close relationship between GCV and the C_p statistic.

17.4 Use a Taylor series approximation to show that the adjusted residual squared error (17.7) and the C_p statistic (17.8) are equal to first order, if RSE/n is used as an estimate of σ^2 in C_p.

17.5 Carry out a linear discriminant analysis of some classification data and use cross-validation to estimate the misclassification rate of the fitted model. Analyze the same data using the CART procedure and cross-validation, and compare the results.

17.6 Make explicit the quantities $\text{err}(\mathbf{x}, F)$, $\text{err}(\mathbf{x}, \hat{F})$ and their bootstrap counterparts, in a classification problem with prediction error equal to misclassification rate.

17.7 Given a data set of n distinct observations, show that the probability that an observation appears in a bootstrap sample of size n is $\rightarrow (1 - e^{-1}) \approx .632$ as $n \rightarrow \infty$.

17.8 (a) Carry out a bootstrap analysis for the hormone data, like the one in Table 17.1, using $B = 100$ bootstrap samples. In addition, calculate the average prediction error $\hat{\epsilon}_0$ for observations that do not appear in the bootstrap sample used for their prediction. Hence compute the .632 estimator for these data.

(b) Calculate the average prediction error $\hat{\epsilon}_j$ for observations that appear exactly j times in the bootstrap sample used for their prediction, for $j = 0, 1, 2, \ldots$. Graph $\hat{\epsilon}_j$ against j and give an explanation for the results.

17.9 *Tree selection in CART.* Let T be a classification tree and define the cost of a tree by

$$\text{cost}(T) = \text{mr}(T) + \lambda|T|, \qquad (17.28)$$

where $\text{mr}(T)$ denotes the (apparent) misclassification rate of T and $|T|$ is the number of terminal nodes in T. The parameter $\lambda \geq 0$ trades off the classification performance of the tree with its complexity. Denote by T_0 a fixed (large) tree, and consider all subtrees T of T_0, that is, all trees which can be obtained by pruning branches of T_0.

Let T_α be the subtree of T_0 with smallest cost. One can show that for each value $\alpha \geq 0$, a unique T_α exists (when more than one tree exists with the same cost, there is one tree that is a subtree of the others, and we choose that tree). Furthermore, if $\alpha_1 > \alpha_2$, then T_{α_1} is a subtree of T_{α_2}. The CART procedure derives an estimate $\hat{\alpha}$ of α by 10-fold cross-validation, and then the final tree chosen is $T_{\hat{\alpha}}$.

Here is how cross-validation is used. Let T_α^{-k} be the cost-minimizing tree for cost parameter α, when the kth part of the data is withheld ($k = 1, 2, \ldots 10$). Let $\text{mr}_k(T_\alpha^{-k})$ be the misclassification rate when T_α^{-k} is used to predict the kth part of the data.

For each fixed α, the misclassification rate is estimated by

$$\frac{1}{10} \sum_{k=1}^{10} \text{mr}_k(T_\alpha^{-k}). \qquad (17.29)$$

Finally, the value $\hat{\alpha}$ is chosen to minimize (17.29).

This procedure is an example of *adaptive estimation*, discussed in the next chapter. More details may be found in Breiman *et al.* (1984).

Write a computer program that grows and prunes classification trees. You may assume that the predictor variables are binary, to simplify the splitting process. Build in 10-fold cross-validation and try your program on a set of real data.

## CHAPTER 18

Adaptive estimation and calibration

18.1 Introduction

Consider a statistical estimator $\hat{\theta}_\lambda(\mathbf{x})$ depending on an adjustable parameter λ. For example, $\hat{\theta}_\lambda(\mathbf{x})$ might be a trimmed mean, with λ the trimming proportion. In order to apply the estimator to data, we need to choose a value for λ. In this chapter we use the bootstrap and related methods to assess the performance of $\hat{\theta}_\lambda(\mathbf{x})$ for each fixed λ. This idea is not much different from some of the ideas that are discussed in Chapters 6 and 17. However, here we take things further: based on this assessment, we choose the value $\hat{\lambda}$ that optimizes the performance of $\hat{\theta}_\lambda(\mathbf{x})$. Since the data themselves are telling us what procedure to use, this is called *adaptive estimation*. When this idea is applied to confidence interval procedures, it is sometimes referred to as *calibration*. We discuss two examples of adaptive estimation and calibration and then formalize the general idea.

18.2 Example: smoothing parameter selection for curve fitting

Our first example concerns choice of a smoothing parameter for a *curve fitting* or *nonparametric regression* estimator. Figure 18.1 shows a scatterplot of log C-peptide (a blood measurement) versus age (in years) for 43 diabetic children. The data are listed in Table 18.1. We are interested in predicting the log C-peptide values from the age of the child.

A smooth curve has been drawn through the scatterplot using a procedure called a cubic smoothing spline. Here's how it works. Denoting the data points by (z_i, y_i) for $i = 1, 2, \ldots n$, we seek a

Table 18.1. *Blood measurements on 43 diabetic children.*

obs #	age	log C-peptide	obs #	age	log C-peptide
1	5.2	4.8	23	11.3	5.1
2	8.8	4.1	24	1.0	3.9
3	10.5	5.2	25	14.5	5.7
4	10.6	5.5	26	11.9	5.1
5	10.4	5.0	27	8.1	5.2
6	1.8	3.4	28	13.8	3.7
7	12.7	3.4	29	15.5	4.9
8	15.6	4.9	30	9.8	4.8
9	5.8	5.6	31	11.0	4.4
10	1.9	3.7	32	12.4	5.2
11	2.2	3.9	33	11.1	5.1
12	4.8	4.5	34	5.1	4.6
13	7.9	4.8	35	4.8	3.9
14	5.2	4.9	36	4.2	5.1
15	0.9	3.0	37	6.9	5.1
16	11.8	4.6	38	13.2	6.0
17	7.9	4.8	39	9.9	4.9
18	11.5	5.5	40	12.5	4.1
19	10.6	4.5	41	13.2	4.6
20	8.5	5.3	42	8.9	4.9
21	11.1	4.7	43	10.8	5.1
22	12.8	6.6			

smooth function $f(z)$ that is close to the y values. That is, we require that $f(z)$ be smooth and that $f(z_i) \approx y_i$ for $i = 1, 2, \ldots n$. To formalize this objective, we define our solution $\hat{f}(z)$ to be the curve minimizing the criterion

$$J_\lambda(f) = \sum_1^n [y_i - f(z_i)]^2 + \lambda \int [f''(z)]^2 dx. \qquad (18.1)$$

The first term in $J_\lambda(f)$ measures the closeness of $f(z)$ to y, while the second term adds a penalty for the curvature of $f(z)$. (If you are unfamiliar with calculus, you can think of $\int [f''(z)]^2$ as

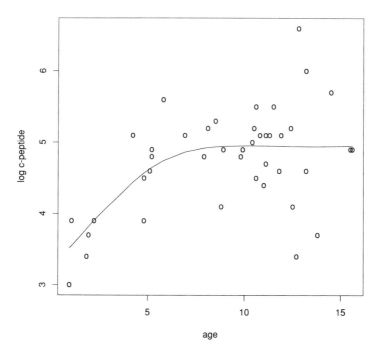

Figure 18.1. *Scatterplot of log C-peptide versus age for 43 diabetic children. The solid curve is a cubic smoothing spline that has been fit to the data.*

$\lambda \sum_2^{n-1} [f(z_{i+1}) - 2f(z_i) + f(z_{i-1})]^2$.) The penalty term will be small if $f(z)$ is smooth and large if $f(z)$ varies quickly. The *smoothing parameter* $\lambda \geq 0$ governs the tradeoff between the fit and smoothness of the curve. Small values of λ favor jagged curves that follow the data points closely: choosing $\lambda = 0$ means that we don't care about smoothness at all. Large values of λ favor smoother curves that don't adhere so closely to the data. For any fixed value of λ, the minimizer of $J_\lambda(f)$ can be shown to be a *cubic spline*: a set of piecewise cubic polynomials joined at the distinct values of z_i, called the "knots." Computer algorithms exist for computing a cubic spline, like the one shown in Figure 18.1.

What value of λ is best for our data? The left panel of Figure 18.2

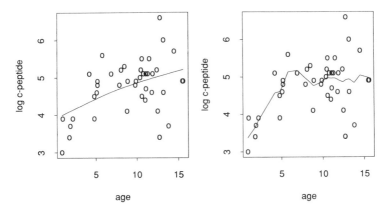

Figure 18.2. *As in Figure 18.1, but using a larger value of the smoothing parameter (left panel) and a smaller value of the smoothing parameter (right panel).*

shows the curve obtained for a larger value of λ: it is smoother than the curve in Figure 18.1 but doesn't seem to fit the data as well. In the right panel a smaller value of λ was used. Notice how the curve follows the data closely but is more jagged. Denote by $\hat{f}_\lambda(z)$ the function estimate based on our data set and the value λ. If we had new data (z', y'), it would be reasonable to choose λ to minimize the expected prediction error

$$\text{pse}(\lambda) = \text{E}[y' - \hat{f}_\lambda(z')]^2. \tag{18.2}$$

The expectation is over a new pair (z', y') from the distribution F that gave rise to our original sample. In lieu of new data, we can generate a bootstrap sample (z_i^*, y_i^*), $i = 1, 2, \ldots n$, and compute the curve estimate $\hat{f}_\lambda^*(z)$ based on this sample and a value λ. Then we find the error that $\hat{f}_\lambda^*(z)$ makes in predicting our original sample:

$$\text{pse}^*(\lambda) = \frac{1}{n} \sum_1^n [y_i - \hat{f}_\lambda^*(z_i)]^2. \tag{18.3}$$

Averaging this quantity over B bootstrap samples provides an estimate of the prediction error $\text{pse}(\lambda)$; denote this average by $\widehat{\text{pse}}(\lambda)$.

Why is (18.3) the appropriate formula, and not say $\sum_1^n (y_i^* - \hat{f}_\lambda(z_i))^2/n$, or $\sum_1^n (y_i^* - \hat{f}_\lambda^*(z_i^*))^2/n$? Formula (18.3) is obtained by applying the plug-in principle to the actual prediction error (18.2). To see this, it might be helpful to look back at Figure 8.3. The "real world" quantity pse(λ) involves $\hat{f}_\lambda(z')$, a function of our original data sample $\mathbf{x} = ((z_1, y_1), \ldots (z_n, y_n))$, and the new observation (z', y') which is distributed according to F, the true distribution of z and y. In the bootstrap world, we plug in $\hat{F}$ for F and generate bootstrap samples $\mathbf{x}^* \sim \hat{F}$. We calculate $\hat{f}_\lambda^*(z_i)$ from $\mathbf{x}^*$ in the same way that we calculated $\hat{f}_\lambda(z_i)$ from $\mathbf{x}$, while (z_i, y_i) plays the role of the "new" data (z', y').

We calculated $\widehat{\text{pse}}$ over a grid of λ values with $B = 100$ bootstrap samples, and obtained the pse(λ) estimate shown in the top left panel of Figure 18.3. The minimum occurs near $\lambda = .01$; this is the value of λ that was used in Figure 18.1. (In Figure 18.2 the values .44 and .0002 were used in the left and right panels, respectively). At the minimizing point we have drawn $\pm$se error bars to indicate the variability in the bootstrap estimate. The standard error is the sample standard error of the $B = 100$ individual estimates of pse(λ). Since smoother curves are usually preferred for aesthetic reasons, it is fairly common practice to choose the largest value $\tilde{\lambda} > \hat{\lambda}$ that produces not more than a one standard error increase in pse. In this case $\tilde{\lambda} = .03$; the resulting curve is very similar to the one in Figure 18.1.

The reader will recognize this procedure as an application of prediction error estimation, described in Chapter 17. Consequently, other techniques that are described in that chapter can be used in this problem as well. The top right panel shows a more refined bootstrap approach that focuses on the optimism of the apparent error rate. The bottom left and right panels use cross-validation and generalized cross-validation, respectively. Although the minimum of each of these three curves is close to $\lambda = .01$, the minimum is more clearly determined than in the top left panel.

A disadvantage of the bootstrap approaches is their computational cost. In contrast, the cross-validation and generalized cross-validation estimates can be computed very quickly in this context (Problem 18.1) and hence they are often the methods of choice for smoothing parameter selection of a cubic smoothing spline.

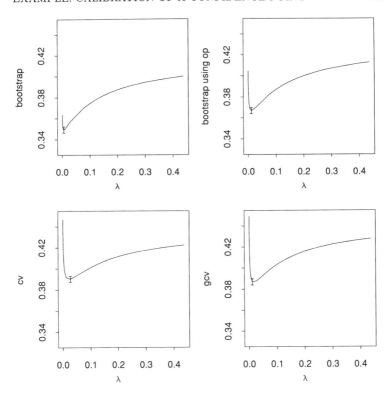

Figure 18.3. *Estimates of pse(λ). Top left panel uses the simple boot-strap approach; in top right, a more refined bootstrap approach is used, focusing on the optimism of the apparent error rate; the bottom left and right curves are obtained from cross-validation and generalized cross-validation, respectively. The minimum of each curve is indicated, with $\pm$se error bars.*

18.3 Example: calibration of a confidence point

As our second example, suppose $\hat{\theta}[\alpha]$ is an estimate of the lower αth confidence point for a parameter θ. That is, we intend to have

$$\text{Prob}\{\theta \leq \hat{\theta}[\alpha]\} = \alpha. \qquad (18.4)$$

The procedure $\hat{\theta}[\alpha]$ might be, for example, the standard normal point $\hat{\theta} - z^{(1-\alpha)}\hat{se}$, or the bootstrap percentile point described in Chapter 13. As we have seen in previous chapters, the actual coverage of a confidence procedure is rarely equal to the desired (nominal) coverage, and often is substantially different. One way to think about the coverage accuracy of a confidence procedure is in terms of its *calibration*: that is, for each α if (18.4) doesn't hold for $\hat{\theta}[\alpha]$, perhaps it will hold for $\hat{\theta}[\lambda]$ where $\lambda \neq \alpha$. For example, if we want the probability α in (18.4) to be 5%, perhaps we can achieve this by using the 3% confidence point. If we knew the mapping $\alpha \to \lambda$, we could construct a confidence procedure with exactly the desired coverage.

The bootstrap can be used to carry out the calibration. Here's how we do it. For convenient notation, denote the family of confidence points by

$$\hat{\theta}_\lambda = \hat{\theta}[\lambda]. \tag{18.5}$$

We seek a value $\hat{\lambda}$ such that

$$p(\hat{\lambda}) = \text{Prob}\{\theta \leq \hat{\theta}_{\hat{\lambda}}\} = \alpha. \tag{18.6}$$

Note that if the procedure is calibrated correctly, then (18.6) holds exactly with $\lambda = \alpha$.

Let

$$\hat{p}(\lambda) = \text{Prob}_*\{\hat{\theta} \leq \hat{\theta}_\lambda^*\}, \tag{18.7}$$

the bootstrap estimate of $p(\lambda)$. In (18.7), $\hat{\theta}$ is fixed and the "*" refers to bootstap sampling with replacement from the data. To approximate $\hat{p}(\lambda)$ we generate a number of bootstrap samples, compute $\hat{\theta}_\lambda^*$ for each one, and record the proportion of times that $\hat{\theta} \leq \hat{\theta}_\lambda^*$. This process is carried out simultaneously (using the same bootstrap samples) over a wide grid of λ values that includes the nominal value α.

Denoting by $\hat{\lambda}_\alpha$ the value of λ satisfying $\hat{p}(\lambda) = \alpha$, the calibrated confidence point is

$$\hat{\theta}[\hat{\lambda}_\alpha]. \tag{18.8}$$

Let's spell out the calibration process in more detail. Starting with a confidence limit $\hat{\theta}_\lambda$, the steps in calibrating $\hat{\theta}_\lambda$ are shown in Algorithm 18.1.

In many cases, the calculation of $\hat{\theta}_\lambda^*(b)$ in step (1a) above re-

Algorithm 18.1

Confidence point calibration via the bootstrap

---

1. Generate B bootstrap samples $\mathbf{x}^{*1}, \ldots \mathbf{x}^{*B}$. For each sample $b = 1, 2, \cdots B$:

   1a. Compute a λ-level confidence point $\hat{\theta}^*_\lambda(b)$ for a grid of values of λ. For example these might be the normal confidence points $\hat{\theta}^*(b) - z^{(1-\lambda)}\widehat{se}^*(b)$.

2. For each λ compute $\hat{p}(\lambda) = \#\{\hat{\theta} \leq \hat{\theta}^*_\lambda(b)\}/B$.

3. Find the value of λ satisfying $\hat{p}(\lambda) = \alpha$.

---

quires bootstrap sampling itself. This makes the overall calibration a nested computation, sometimes called a "double bootstrap." This is true if we use the normal confidence point for in step (1a) and there is no closed form expression for $\widehat{se}^*$ or if we use the percentile limit defined in Chapter 13.

As an example of bootstrap calibration, let's fix $\alpha = .05$ and consider a lower α confidence point for the correlation coefficient of the law school data (Table 3.1). Let $\hat{\theta}_\lambda$ be the bootstrap percentile interval based on $B = 200$ bootstrap samples. Using the same number of bootstrap samples in the calibration makes the total number $200 \cdot 200 = 40,000$. For reasonable accuracy the number 200 should probably be raised to at least 1000, but this would make the total number of bootstrap samples equal to $1,000,000$.[1] The estimate $\hat{p}(\lambda)$ is shown in the left panel of Figure 18.4. The 45° line is included for reference. The value $\hat{\lambda} = .01$ gives $\hat{p}(\hat{\lambda}) \approx .05$. The right panel of Figure 18.4 shows the corresponding plot for the upper 95% confidence point. The value $\hat{\lambda} = .93$ gives $\hat{p}(\hat{\lambda}) \approx .95$.

The calibrated percentile interval is constructed by selecting the 1% and 93% points (rather than the 5% and 95% points) of the bootstrap distribution of $\hat{\theta}$. This produces the interval

$$[.378, .938]. \qquad (18.9)$$

The percentile interval for these data is $[.524, .928]$ while the BC_a

---

[1] The more realistic calibration examples of Chapters 14 and 25 avoid most of the computational effort by use of the ABC approximation.

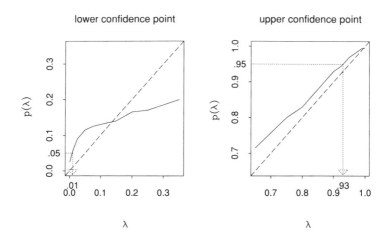

Figure 18.4. *Estimates of $p(\lambda)$ for the lower and upper confidence points (solid curves). The dotted line with the arrow indicates the calibration of the 5% and 95% points. The broken line is the 45° line for reference.*

interval from Chapter 14 is $[.410, .923]$. The calibration has moved the percentile point much closer to the BC_a point on the lower end and a little farther away from it on the upper end.

In fact, it is possible to carry out a nested calibration, that is, calibrate the calibration, and so on. Each calibration brings another order of accuracy, but at a formidable computational cost.

18.4 Some general considerations

The common theme in these two examples can be expressed in the following way. Given a statistical procedure $\hat{\theta}_\lambda(\mathbf{x})$ depending on a parameter λ, we require the value $\hat{\lambda}$ that minimizes some function $E[g(\hat{\theta}_\lambda)]$ or that achieves a specified value of $E[g(\hat{\theta}_\lambda)]$. Let $g^*(\hat{\theta}_\lambda^*)$ be $g(\hat{\theta}_\lambda)$, applied to a bootstrap data set. The bootstrap calibration procedure estimates $E[g(\hat{\theta}_\lambda)]$ by the bootstrap expectation of $g^*(\hat{\theta}_\lambda^*)$:

$$\widehat{E}[g(\hat{\theta}_\lambda)] = E_*[g^*(\hat{\theta}_\lambda^*)]. \tag{18.10}$$

Then we simply use $\widehat{E}[g(\hat{\theta}_\lambda)]$ in place of $E[g(\hat{\theta}_\lambda)]$: that is, find the value λ that minimizes $\widehat{E}[g(\hat{\theta}_\lambda)]$, or find the value of λ that achieves a specified value for $\widehat{E}[g(\hat{\theta}_\lambda)]$.

The form of the function $g(\cdot)$ depends on the problem at hand. In the scatterplot smoothing example $\hat{\theta}_\lambda = f_\lambda(\cdot)$ and

$$g(\hat{\theta}_\lambda) = (y' - \hat{f}_\lambda(z'))^2 \qquad (18.11)$$

and we seek the value of λ that minimizes $\text{pse}(\lambda) = E(y' - \hat{f}_\lambda(z'))^2$. Bootstrap calibration uses $\widehat{\text{pse}}(\lambda)$ the bootstrap expectation of $\text{pse}^*(\lambda)$, defined below (18.3), in place of $\text{pse}(\lambda)$ and then finds λ to minimize $\widehat{\text{pse}}(\lambda)$. For the confidence point application

$$g(\hat{\theta}_\lambda) = I_{\{\theta \le \hat{\theta}_\lambda\}} \qquad (18.12)$$

so that $E[g(\hat{\theta}_\lambda)] = \text{Prob}\{(\theta \le \hat{\theta}_\lambda\}$. We seek the value λ such that $E[g(\hat{\theta}_\lambda)] = \alpha$. Note that $g^*(\hat{\theta}_\lambda^*) = I_{\{\theta \le \hat{\theta}_\lambda^*\}}$, and $E_*[I_{\{\theta \le \hat{\theta}_\lambda^*\}}] = \text{Prob}_*\{\hat{\theta} \le \hat{\theta}_\lambda^*\}$. Therefore we find the value of λ such that $\text{Prob}_*\{\hat{\theta} \le \hat{\theta}_\lambda^*\} = \alpha$.

The bootstrap and related methods are potentially useful tools for adaptive estimation and calibration. However there are some problems that need to be tackled. One difficulty is the amount of computation required. For example, we have seen that the calibrated percentile interval requires $1000 \cdot 1000 = 1,000,000$ bootstrap samples, unless a computational shortcut can be found. Another more subtle issue is the procedure that we choose to calibrate. Let's look back at the confidence point problem where this issue is best understood. Rather than defining $\hat{\theta}_\lambda$ as in (18.5) we could use any one of the following definitions:

$$\hat{\theta}[\alpha] + \lambda, \qquad (18.13)$$

$$\hat{\theta} + \lambda, \qquad (18.14)$$

$$\hat{\theta} + \lambda \cdot \widehat{\text{se}}. \qquad (18.15)$$

The original definition (18.5) led to calibration of the nominal coverage probability of the confidence procedure $\hat{\theta}[\alpha]$, an approach also known as *pre-pivoting*. Using (18.13), we adjust the confidence point itself rather than the nominal coverage. That is, for each α we find the value of λ such that

$$p(\hat{\lambda}) = \text{Prob}\{\theta \le \hat{\theta}[\alpha] + \lambda\} = \alpha. \qquad (18.16)$$

In (18.14), we calibrate the distance from the point estimate $\hat{\theta}$, that is, we seek λ so that

$$p(\hat{\lambda}) = \text{Prob}\{\theta \leq \hat{\theta} + \lambda\} = \alpha. \tag{18.17}$$

Finally, in (18.15) we calibrate this distance standardized by an estimate of standard error $\hat{se}$, that is, we seek the value of λ so that

$$p(\hat{\lambda}) = \text{Prob}\{\theta \leq \hat{\theta} + \lambda \cdot \hat{se}\} = \alpha. \tag{18.18}$$

The differences in (18.5), (18.13)–(18.15) may seem subtle, but they turn out to be important. As long as $\hat{\theta}[\alpha]$ is a *first order accurate* confidence point as defined in Chapters 14 and 22 (such as the standard normal or bootstrap percentile points), the calibration process, using either (18.5) or (18.13), produces a *second order accurate* confidence point. (An example of this is given in Chapter 14). Hence the calibrated interval will enjoy the same accuracy as the BC_a procedure of Chapter 14.

Use of definition (18.15) also leads to second order accuracy. Interestingly, it is the same as the bootstrap-t interval described in Chapter 12 (Problem 18.4). However, definition (18.14) does not work, in the sense that the resulting confidence points are only first order accurate.

The question of how to choose the representation $\hat{\theta}_\lambda$ can arise in other problems, but is less clearly understood. In Problems 18.2 and 18.3 we investigate this issue for the estimation of a cubic smoothing spline.

18.5 Bibliographic notes

Curve fitting and smoothing parameter selection for curve fitting are discussed in Rice (1984), Silverman (1985), Hall and Titterington (1987), Eubank (1988), Härdle and Bowman (1988), Härdle, Hall, and Marron (1988), Härdle (1990), Hastie and Tibshirani (1990), and Wahba (1990). Hall (1992, chapter 4) gives an overview of bootstrap methods for this area, currently a very active one. The related problem of construction of confidence bands for curve estimates is studied by Hall and Titterington (1988), Hastie and Tibshirani (1990, chapter 3), Härdle (1990, chapter 4), and Hall (1992, chapter 4). Calibration of the bootstrap was first discussed by Hall (1986a, 1987), and Loh (1987, 1991). Hall and Martin (1988) give a general theory for bootstrap iteration and calibration. Prepivoting was suggested by Beran (1987, 1988). A summary of bootstrap iteration for confidence points is given in DiCiccio and Romano (1988). Adaptive estimation of the window width in a kernel density estimate is an interesting but difficult problem, and is studied in Taylor (1989), Romano (1988), Leger and Romano (1990), Faraway and Jhun (1990), and Hall (1990). The diabetes data are taken from Hastie and Tibshirani (1990).

18.6 Problems

18.1 The cubic smoothing spline fitting mechanism is a linear operation, that is, the vector of fitted values $\hat{\mathbf{y}}$ can be written as $\hat{\mathbf{y}} = \mathbf{S}\mathbf{y}$, where $\mathbf{y}$ is the vector of response values and $\mathbf{S}$ is an $n \times n$ matrix that depends on the x values and the smoothing parameter λ but not on $\mathbf{y}$. Give a simple argument to show that the deletion formula of Problem 17.1 holds for a cubic smoothing spline, with $\mathbf{S}$ replacing $\mathbf{H}$.

18.2 (a) In the diabetes data discussed in this chapter, suppose the ages were measured in weeks rather than years. Apply a cubic spline smoother with $\lambda = .01$, the value used in Figure 18.1. Does the curve estimate look the same? What has happened?

(b) Suggest how your finding in part (a) might effect the performance of the bootstrap and cross-validation for selection of λ.

18.3 (a) Generate 10 data sets of size 25 from the model

$$y = f(z) + \epsilon, \qquad (18.19)$$

where $f(z) = z^2$ and z is normally distributed with mean 0 and variance 3, and ϵ is normally distributed with mean 0 and variance 9. Apply a cubic smoothing spline to each simulated data set, choosing the smoothing parameter λ by the four methods described in the chapter. For each method, compute the average of mean squared error

$$\frac{1}{25} \sum_{i=1}^{25} [\hat{f}(z_i) - f(z_i)]^2 \qquad (18.20)$$

over the 10 simulations.

(b) Using the same simulated samples as in (a), repeat the exercise using $\beta = \lambda/\mathrm{sd}$ in place of λ, where sd is the standard deviation of the z values in the sample. Compare the results to those in part (a). Relate your findings to the previous exercise.

18.4 (a) Show that the confidence interval resulting from calibration of the endpoints based on equation (18.15) is the bootstrap-t interval described in Chapter 12.

(b) If the form (18.14) is used instead, show that the resulting interval corresponds to a bootstrap-t interval based on on $\hat{\theta} - \theta$ rather than $(\hat{\theta} - \theta)/\widehat{se}$.

(c) Give a reason why one might expect better behavior from (18.15) than (18.14). Draw an analogy to the results of Problems 18.2 and 18.3.

18.5 Suggest how to organize efficiently the computations in the bootstrap calibration procedure.

18.6 Explain in detail why expression (18.3) is the bootstrap analogue of the prediction error (18.2).

Assessing the error in bootstrap estimates

19.1 Introduction

So far in this book we have used the bootstrap and other methods to assess statistical accuracy. For the most part, we have ignored the fact that bootstrap estimates, like all statistics, are not exact but have inherent error. Typically bootstrap estimates are nearly unbiased, because of the way they are constructed (Problem 19.1); but they can have substantial variance. This comes from two distinct sources: sampling variability, due to the fact that we have only a sample of size n rather than the entire population, and bootstrap resampling variability, due to the fact that we take only B bootstrap samples rather than an infinite number. In this chapter we study these two components of variance, and also discuss the *jackknife-after-bootstrap*, a simple method for estimating the variability from a set of bootstrap estimates.

Figure 19.1 shows our setup. It is basically the same as Figure 6.1. We are in the one-sample situation with $\mathbf{x} = (x_1, x_2, \ldots x_n)$ generated from a population F. We have calculated from $\mathbf{x}$ our statistic of interest $s(\mathbf{x})$. We create B bootstrap samples $\mathbf{x}^{*1}, \ldots \mathbf{x}^{*B}$, each of size n, by sampling with replacement from $\mathbf{x}$ as in Figure 6.1; for each bootstrap sample $\mathbf{x}^{*b}$ we compute $s(\mathbf{x}^{*b})$, the bootstrap replication of the statistic. From the values $s(\mathbf{x}^{*b})$, we construct a bootstrap estimate of some feature of the distribution of $s(\mathbf{x})$, denoted by $\hat{\gamma}_B$. For example, $\hat{\gamma}_B$ might be the 95th percentile of the values $s(\mathbf{x}^{*b})$, intended to estimate the same percentile of the distribution of $s(\mathbf{x})$. Our objective in this chapter is to study how the variance of $\hat{\gamma}_B$ depends on the sample size n and the number of bootstrap samples B, and also how to estimate this variance from the bootstrap samples themselves.

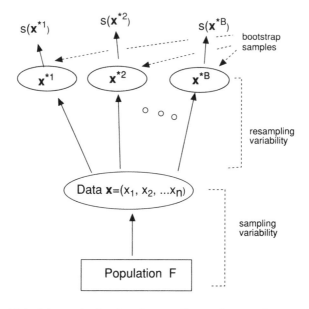

Figure 19.1. *Schematic showing the sampling and resampling components of variance*

19.2 Standard error estimation

Let's first focus on the bootstrap estimate of standard error for $s(\mathbf{x})$, where $\hat{\gamma}_B$ equals

$$\widehat{\mathrm{se}}_B = \{\frac{1}{B}\sum_{b=1}^{B}(s(\mathbf{x}^{*b}) - \bar{s})^2\}^{1/2}. \tag{19.1}$$

For convenience we divide by B rather than $B - 1$, as in Figure 6.1. Here $\bar{s} = \sum_{1}^{B} s(\mathbf{x}^{*b})/B$ is the mean of the bootstrap values.

The quantity $\widehat{\mathrm{se}}_B$, which measures the variability of the statistic $s(\mathbf{x})$, itself has a variance. It turns out that this variance has the approximate form

$$\mathrm{var}(\widehat{\mathrm{se}}_B) \doteq \frac{c_1}{n^2} + \frac{c_2}{nB}, \tag{19.2}$$

where c_1 and c_2 are constants depending on the underlying population F, but not on n or B. The derivation of equation (19.2) and other results is given in Section 19.5. The factor c_1/n^2 represents sampling variation, and it approaches zero as the sample size

n approaches infinity. The term c_2/nB represents the resampling variation, and it approaches 0 as $B \to \infty$ with n fixed. In Section 19.4, we describe jackknife-after-bootstrap method for estimating $\mathrm{var}(\widehat{\mathrm{se}}_B)$ from the data itself.

The variability of $\widehat{\mathrm{se}}_B$ can help in determining the necessary number of bootstrap replications B, and that is our focus here. As n and B change, so does $\mathrm{E}(\widehat{\mathrm{se}}_B)$. It is better to measure the size of $\widehat{\mathrm{se}}_B$ relative to $\mathrm{E}(\widehat{\mathrm{se}}_B)$, and hence we consider the coefficient of variation of $\widehat{\mathrm{se}}_B$:

$$\mathrm{cv}(\widehat{\mathrm{se}}_B) = \frac{\mathrm{var}(\widehat{\mathrm{se}}_B)^{1/2}}{\mathrm{E}(\widehat{\mathrm{se}}_B)}. \tag{19.3}$$

In section 19.5 we show that this equals

$$\mathrm{cv}(\widehat{\mathrm{se}}_B) = \left\{ \mathrm{cv}(\widehat{\mathrm{se}}_\infty)^2 + \frac{\mathrm{E}(\hat{\Delta}) + 2}{4B} \right\}^{1/2} \tag{19.4}$$

where $\hat{\Delta}$ is the kurtosis of the distribution of $\hat{\theta}$, and $\widehat{\mathrm{se}}_\infty$ is the ideal bootstrap estimate of standard error. This is the same as equation (6.9) of Chapter 6. Let's consider the case $\hat{\theta} = \bar{x}$ with $x_1, x_2, \ldots x_n$ normally distributed. Then (19.4) simplifies to

$$\mathrm{cv}(\widehat{\mathrm{se}}_B) = \left[\frac{1}{2n} + \frac{1}{2B} \right]^{1/2} \tag{19.5}$$

Figure 19.2 shows $\mathrm{cv}(\widehat{\mathrm{se}}_B)$ as a function n and B (solid curves). The figure caption gives the details. We see that that increasing B past 20 or 50 doesn't bring a substantial reduction in variance. The same conclusion was reached in Chapter 6.

19.3 Percentile estimation

Suppose now that our interest lies in a percentile. Let $\hat{q}_B^\alpha = \hat{G}_B^{-1}(\alpha)$, the estimated α-percentile of distribution of a statistic $\hat{\theta}$, based on B bootstrap samples. In other words

$$\hat{q}_B^\alpha = \{(\alpha \cdot B)\text{th largest of the } \theta^{*b}\} \tag{19.6}$$

(if $\alpha \cdot B$ is not an integer, we can use the convention given in Section 12.5, after equation (12.22) on page 160.

The variance of $\hat{q}_B^\alpha$ again has the form (19.2), but with different constants c_1 and c_2. As we did for standard error estimation, let's

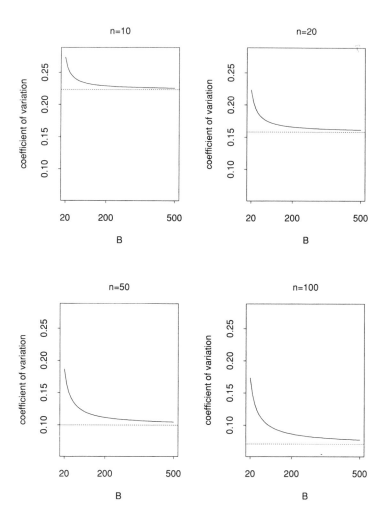

Figure 19.2. *Coefficient of variation of* $\widehat{se}_B(\bar{x})$ *where* $x_1, x_2, \ldots x_n$ *are drawn from a standard normal population. The solid curves in each panel show* $cv(\widehat{se}_B)$ *as a function of the sample size* n *and the number of bootstrap samples* B. *The dotted line is drawn at* $cv(\widehat{se}_\infty)$.

focus on the $\hat{\theta} = \bar{x}$ in the normal case. Then

$$\mathrm{cv}(\hat{q}_B^\alpha) \approx \frac{1}{G^{-1}(\alpha)} \left[\frac{\alpha(1-\alpha)}{g^2} \left(\frac{1}{n} + \frac{1}{B}\right) \right]^{1/2}. \tag{19.7}$$

In this expression, G is the cumulative distribution function of $\hat{\theta}$ and g is its derivative, evaluated at $G^{-1}(\alpha)$.

Figure 19.3 shows the analogue of Figure 19.2, for the upper 95% quantile of the distribution of a sample mean from a standard normal population. Although the curves decrease with B at the same rate as those in Figure 19.2, they are shifted upward. The results suggest that B should be ≥ 500 or 1000 in order to make the variability of $\hat{q}_B^\alpha$ acceptably low. More bootstrap samples are needed for estimating the 95th percentile than the standard error, because the percentile depends on the tail of the distribution where fewer samples occur. Generally speaking, bootstrap statistics $\hat{\gamma}_B$ that depend on the extreme tails of the distribution of $\hat{\theta}^*$ will require a larger number of bootstrap samples to achieve acceptable accuracy.

19.4 The jackknife-after-bootstrap

Suppose we have drawn B bootstrap samples and calculated $\widehat{se}_B$, a bootstrap estimate of the standard error of $s(\mathbf{x})$. We would like to have a measure of the uncertainty in $\widehat{se}_B$. The *jackknife-after-bootstrap* method provides a way of estimating $\mathrm{var}(\widehat{se}_B)$ using only information in our B bootstrap samples. Here is how it works. Suppose we had a large computer and set out to calculate the jackknife estimate of variance of $\widehat{se}_B$. This would involve the following steps:

- For $i = 1, 2, \ldots n$

    Leave out data point i and recompute $\widehat{se}_B$. Call the result $\widehat{se}_{B(i)}$.

- Define $\widehat{\mathrm{var}}_{\mathrm{jack}}(\widehat{se}_B) = [(n-1)/n] \sum_1^n (\widehat{se}_{B(i)} - \widehat{se}_{B(\cdot)})^2$ where $\widehat{se}_{B(\cdot)} = \sum_1^n \widehat{se}_{B(i)}/n$.

The difficulty with this endeavor is computation of $\widehat{se}_{B(i)}$: this requires a completely new set of bootstrap samples for each i. Fortunately there is a neat way to circumvent this problem. For each data point i, there are some bootstrap samples in which that data

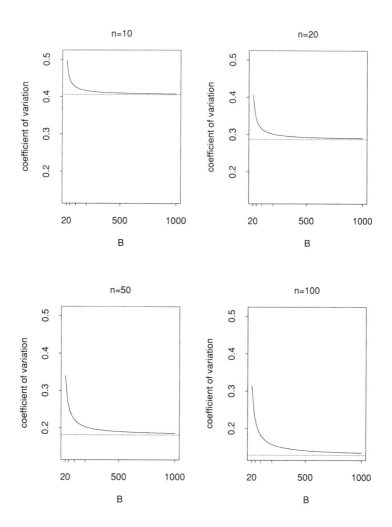

Figure 19.3. *Variability study of* $\hat{q}_B^\alpha$ *the estimated* α-*quantile of* $\bar{x}^*$, *where* $x_1, x_2, \ldots x_n$ *are drawn from a standard normal population. The solid curves in each panel show* $cv(\hat{q}_B^\alpha)$ *as a function of the sample size* n *and the number of bootstrap samples* B. *The dotted line is drawn at* $cv(\hat{q}_\infty^\alpha)$.

point *does not appear*, and we can use those samples to estimate $\widehat{\mathrm{se}}_{B(i)}$. In particular, we estimate $\widehat{\mathrm{se}}_{B(i)}$ by the sample standard deviation of $s(\mathbf{x}^{*b})$ over bootstrap samples $\mathbf{x}^{*b}$ that don't contain point i. Formally, if we let C_i denote the indices of the bootstrap samples that don't contain data point i, and there are B_i such samples, then

$$\widehat{\mathrm{se}}_{B(i)} = \left[\sum_{b \in C_i} (s(\mathbf{x}^{*b}) - \bar{s}_i)^2 / B_i \right]^{1/2}, \qquad (19.8)$$

where $\bar{s}_i = \sum_{b \in C_i} s(\mathbf{x}^{*b})/B_i$.

The reason that this shortcut works is the following fact.

Jackknife-after-bootstrap sampling lemma: *A bootstrap sample drawn with replacement from $x_1, x_2, \ldots x_{i-1}, x_{i+1}, \ldots x_n$ has the same distribution as a bootstrap sample drawn from $x_1, x_2, \ldots x_n$ in which none of the bootstrap values equals x_i.*

The proof of this lemma is straightforward.

As an example, consider the treatment times of the mouse data of Table 2.1: $(94, 197, 16, 38, 99, 141, 23)$. Table 19.1 shows 20 bootstrap samples[1] along with the bootstrap means $\bar{x}^{*b}$.

The bootstrap estimate of the standard error of the mean from these 20 samples is $\widehat{\mathrm{se}}_B = 23.4$. Here are the steps involved in computing $\widehat{\mathrm{var}}_{\mathrm{jack}}(\widehat{\mathrm{se}}_B)$. Consider the first data point $x = 94$. This point does not appear in bootstrap samples 1,3,5,6,7,9,12,13,15,18 and 19. Thus $\widehat{\mathrm{se}}_{B(1)}$ is the sample standard deviation of $\bar{x}^{*1}, \bar{x}^{*3}, \bar{x}^{*5}, \bar{x}^{*6}, \bar{x}^{*7}, \bar{x}^{*9}, \bar{x}^{*12}, \bar{x}^{*13}, \bar{x}^{*15}, \bar{x}^{*18}$ and $\bar{x}^{*19}$. This works out to be 28.6.

We carry out this calculation for data points $1, 2, 3, \cdots 7$ and obtain the 7 values for $\widehat{\mathrm{se}}_{B(i)}$: 28.6, 23.3, 16.9, 17.9, 24.5, 15.4 and 24.0. Finally, we take the sample variance of these 7 values to obtain $\widehat{\mathrm{var}}_{\mathrm{jack}}(\widehat{\mathrm{se}}_B) = 23.6$. Therefore $\widehat{\mathrm{se}}_{\mathrm{jack}}(\widehat{\mathrm{se}}_B) = 4.9$, which is about 20% of $\widehat{\mathrm{se}}_B$.

The jackknife-after-bootstrap can be applied to any bootstrap statistic, not just the standard error as above. For example, the bootstrap statistic might be the percentile $\hat{q}_B^{\alpha}$ discussed earlier. Then $\widehat{\mathrm{var}}_{\mathrm{jack}}(\hat{q}_B^{\alpha})$ is computed as above, except that we compute $\hat{q}_B^{\alpha}$ over all samples not containing data point i, rather that $\widehat{\mathrm{se}}_B$. Note that the jackknife-after-bootstrap runs into trouble if every

---

[1] $B = 20$ is used for this discussion but is really too small to provide needed accuracy for $\widehat{\mathrm{var}}_{\mathrm{jack}}(\widehat{\mathrm{se}}_B)$; $B = 200$ would be better as shown below.

Table 19.1. *20 Bootstrap samples, and bootstrap replicates of the mean, from the treatment group of the mouse data*

#			Bootstrap sample					Bootstrap mean
1.	16	23	38	16	141	99	197	75.71
2.	141	94	197	23	141	23	16	90.71
3.	197	38	16	23	23	197	197	98.71
4.	94	94	16	141	94	141	94	96.29
5.	99	23	99	141	38	99	23	74.57
6.	141	38	23	197	16	16	16	63.86
7.	197	38	38	38	197	16	16	77.14
8.	141	94	94	38	197	23	16	86.14
9.	141	16	197	23	16	141	141	96.43
10.	141	197	16	197	94	16	141	114.57
11.	94	99	141	23	141	197	16	101.57
12.	141	16	197	197	197	99	99	135.14
13.	16	141	197	197	99	197	99	135.14
14.	141	16	94	99	94	141	99	97.71
15.	197	99	38	16	23	197	141	101.57
16.	197	197	16	197	141	94	38	125.71
17.	94	38	94	99	16	99	94	76.28
18.	141	23	23	38	16	16	23	40.00
19.	99	197	99	38	23	141	99	99.43
20.	23	197	99	38	197	99	94	106.71

bootstrap sample contains a given point i. However this event is very rare if $n \geq 10$ and $B \geq 20$ (Problem 19.4).

How well does $\widehat{\text{var}}_{\text{jack}}(\hat{se}_B)$ estimate $\text{var}(\hat{se}_B)$? For convenience we focus on the square roots of these quantities, $\hat{se}_{\text{jack}}(\hat{se}_B) = [\widehat{\text{var}}_{\text{jack}}(\hat{se}_B)]^{1/2}$ and $se(\hat{se}_B) = [\text{var}(\hat{se}_B)]^{1/2}$. To investigate, we carried out a small simulation in the setting of Figure 19.2. Figure 19.4 shows $se(\hat{se}_B)$ (solid curve) along with the jackknife-after-bootstrap estimate (circles). These are the average values of $\hat{se}_{\text{jack}}(\hat{se}_B)$ over 50 simulated samples. Ideally these points should lie on the solid curves.

We see that $\hat{se}_{\text{jack}}(\hat{se}_B)$ overestimates $se(\hat{se}_B)$ by a large margin when B is as small as 20, but seems to improve as B gets up to 200.

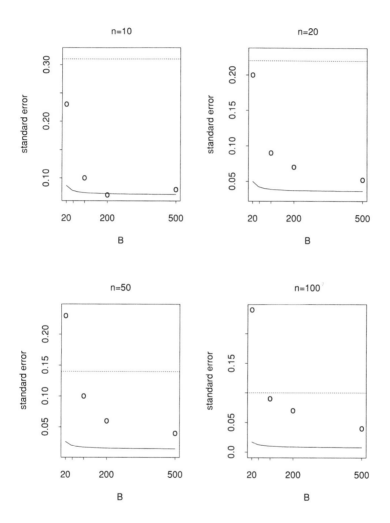

Figure 19.4. *Standard error of* $\widehat{se}_B$ *(solid curve) and the jackknife-after-bootstrap estimate* $\widehat{se}_{\text{jack}}(\widehat{se}_B)$ *(circles) averaged over 50 simulated samples. The dotted line shows the average of* $\widehat{se}_B$.

The reason that the jackknife-after-bootstrap is an overestimate, for small B, is somewhat subtle. It has to do with the fact that the same set of B samples is being used to estimate all of the n jackknife values, and hence the jackknife overestimates the resampling component of the variance. These results suggest that the jackknife-after-bootstrap method is only reliable when B is large.

19.5 Derivations

[1] To begin, let's see how (19.2) is obtained, and derive the form of the constants c_1 and c_2. Given the data $\mathbf{x}$, the quantity $\widehat{\text{se}}_B$ will have a certain expectation and variance when averaged over all possible bootstrap data sets for size B, say $\text{E}(\widehat{\text{se}}_B|\mathbf{x})$ and $\text{var}(\widehat{\text{se}}_B|\mathbf{x})$. The overall variance of $\widehat{\text{se}}_B$ is given by the formula

$$\text{var}(\widehat{\text{se}}_B) = \text{var}[\text{E}(\widehat{\text{se}}_B|\mathbf{x})] + \text{E}[\text{var}(\widehat{\text{se}}_B|\mathbf{x})]. \tag{19.9}$$

The outer variance and expectation in (19.9) refer to the random choice of the data $\mathbf{x}$. Let $\hat{m}_i$ be the ith moment of the bootstrap distribution of $s(\mathbf{x}^*)$ and $\hat{\Delta} = \hat{m}_4/\hat{m}_2^2 - 3$, the kurtosis of the bootstrap distribution of $s(\mathbf{x}^*)$. Both $\hat{m}_i$ and $\hat{\Delta}$ are functions of $\mathbf{x}$. Using standard formulas for the mean and variance of a sample standard deviation we obtain

$$\text{var}(\widehat{\text{se}}_B) \approx \text{var}(\hat{m}_2^{1/2}) + \text{E}[\frac{\hat{m}_2}{4B}(\hat{\Delta} + 2)]. \tag{19.10}$$

If we divide (19.10) by $\hat{m}_2$ and take its square root, we obtain expression (19.4) for the coefficient of variation of $\widehat{\text{se}}_B$ (since $\widehat{\text{se}}_\infty = \hat{m}_2^{1/2}$).

We can use (19.10) to derive (19.2) for most statistics $s(\mathbf{x})$. A particularly easy choice is the sample mean $s(\mathbf{x}) = \bar{x}$. Let σ^2 be the variance of F, μ_4 be the fourth moment and $\hat{\kappa}$ be the standardized kurtosis. Then $\hat{m}_2 = \hat{\sigma}^2/n$, $\hat{\Delta} \approx \hat{\kappa}/n$ and therefore

$$\begin{aligned} \text{var}(\widehat{\text{se}}_B) & \approx \text{var}(\frac{\hat{\sigma}}{\sqrt{n}}) + \text{E}[\frac{\hat{\sigma}^2}{4Bn}(\frac{\hat{\kappa}}{n} + 2)] \\ & \approx \frac{\mu_4/\mu_2 - \mu_2}{4n^2} + \frac{\sigma^2}{2nB} + \frac{\sigma^2\kappa}{4n^2B}, \end{aligned} \tag{19.11}$$

the leading terms having the form (19.2). In the case of Gaussian

---

[1] This section contains more advanced material.

F, $\mu_4 = 3\sigma^4$, $\kappa = 0$ and so (19.11) simplifies further to

$$\text{var}(\widehat{\text{se}}_B) = \frac{\sigma^2}{2n^2}(1 + \frac{n}{B}). \tag{19.12}$$

Taking the square root of this expression and dividing by $\sigma/\sqrt{n}$ gives the coefficient of variation (19.5). For the α-quantile, we have

$$\begin{aligned} \text{var}(\hat{q}_B^\alpha) = & \quad \text{var}(\text{E}(q_B^\alpha|\mathbf{x})) + \text{E}(\text{var}(q_B^\alpha|\mathbf{x})) \\ \approx & \quad \text{var}(\text{E}(\hat{q}_B^\alpha|\mathbf{x})) + \text{E}(\frac{\alpha(1-\alpha)}{B(g(G^{-1}(\alpha))^2)}). \end{aligned} \tag{19.13}$$

The approximation $\text{var}(q_B^\alpha|\mathbf{x}) \approx \alpha(1-\alpha)/B \cdot \hat{g}(\hat{G}^{-1}(\alpha))^2$ comes from the standard textbook approximation for the variance of a quantile. Then using $\text{E}(\hat{q}_B^\alpha|\mathbf{x}) \approx \hat{q}^\alpha$, $\text{var}(\hat{q}^\alpha) \approx \alpha(1-\alpha)/n(g(G^{-1}(\alpha)))^2$ we obtain formula (19.7).

19.6 Bibliographic notes

Formula for the mean and standard error of sample quantities can be found in Kendall and Stuart (1977, chapter 10). Efron (1987, section 6) studies the number of bootstrap replications necessary for achieving a given accuracy, and derived some of the formulae of this chapter. A different approach to this question is given by Hall (1986b). The jackknife-after-bootstrap technique in proposed in Efron (1992b).

19.7 Problems

19.1 Consider a statistic $s(\mathbf{x}) = t(\hat{F})$ based on an i.i.d sample $x_1, x_2, \ldots x_n$. Suppose we have a bootstrap estimate of some feature of the distribution of $s(\mathbf{x})$, denoted by $\hat{\gamma}_B$. Let $\gamma(\hat{F}) = \lim_{B \to \infty} \hat{\gamma}_B$. Show that $\hat{\gamma}_B$ is approximately unbiased for $\gamma(F)$. [Hint: use the relation $\text{E}(\cdot) = \text{E}_{\mathbf{X}}(\text{E}(\cdot|\mathbf{x}))$.]

19.2 Derive expression (19.10) for the variance of $\widehat{\text{se}}_B$ from relation (19.9).

19.3 Prove the jackknife-after-bootstrap sampling lemma.

19.4 (a) Given n distinct data items, show that the probability that a given data item does not appear in a bootstrap sample is $e_n = (1 - 1/n)^n$.

(b) Show that $e_n \to e^{-1} \approx .368$ as $n \to \infty$.

(c) Hence show that the probability that each of B bootstrap samples contains an item i is $(1-e_n)^B$. Evaluate this quantity for $n = 10, 20, 50, 100$ and $B = 10, 20, 50, 100$.

19.5 Verify the jackknife-after-bootstrap calculation for Table 19.1, leading to $\widehat{\text{var}}_{\text{jack}}(\widehat{\text{se}}_B) = 23.6$.

## CHAPTER 20

A geometrical representation for the bootstrap and jackknife

20.1 Introduction

[1] In this chapter we explore a different representation of a statistical estimator, for the purpose of studying the relationship among the bootstrap, jackknife, infinitesimal jackknife and delta methods. The representation is geometrical and as we will see, many of the results in the chapter can be nicely summarized in pictures.

Suppose we are in the simple one-sample situation of Chapter 6, having observed a random sample $\mathbf{x} = (x_1, x_2, \ldots x_n)$ from a population F. Consider a functional statistic

$$\hat{\theta} = t(\hat{F}), \tag{20.1}$$

where $\hat{F}$ denotes the empirical distribution function putting mass $1/n$ on each of our data points $x_1, x_2, \ldots x_n$. We turn to the *resampling representation* of t introduced in Section 10.4. Rather than thinking of $\hat{\theta}$ as a function of the values $x_1, x_2, \ldots x_n$, we fix $x_1, x_2, \ldots x_n$ and consider what happens when we vary the *amount* of probability mass that we put on each x_i. Let $\mathbf{P}^* = (P_1^*, \ldots P_n^*)^T$ be a vector of probabilities satisfying $0 \le P_i^* \le 1$ and $\sum_1^n P_i^* = 1$, and let $\hat{F}^* = \hat{F}(\mathbf{P}^*)$ be the distribution function putting mass P_i^* on x_i, $i = 1, 2, \ldots n$. We define $\hat{\theta}^*$ as a function of $\mathbf{P}^*$, say $T(\mathbf{P}^*)$, by

$$\hat{\theta}^* = T(\mathbf{P}^*) \equiv t(\hat{F}^*(\mathbf{P}^*)). \tag{20.2}$$

Notice the shift in emphasis in (20.2) from t, a function of $\hat{F}^*$ to T, a function of $\mathbf{P}^*$.

Henceforth we will work with $T(\mathbf{P}^*)$. This defines our statistic

---

[1] This chapter and the remaining chapters contain more advanced material.

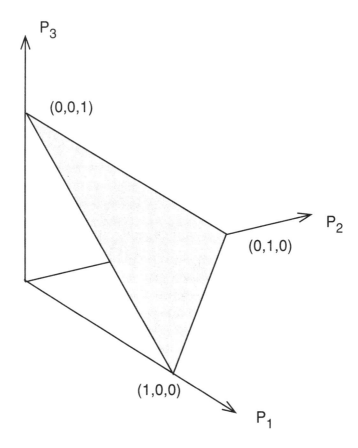

Figure 20.1. *The simplex for $n = 3$*

as a function whose domain is the set of vectors $\mathbf{P}^*$ satisfying $0 \leq P_i^* \leq 1$ and $\sum_1^n P_i^* = 1$. The set of such vectors is called an (n-dimensional) *simplex* and is denoted by S_n. For $n = 3$, the simplex is an equilateral triangle (Figure 20.1). Geometrically, let's focus on this case and lie the equilateral triangle flat on the page (Figure 20.2).

If we define

$$\mathbf{P}^0 = (\frac{1}{n}, \frac{1}{n}, \ldots, \frac{1}{n})^T, \qquad (20.3)$$

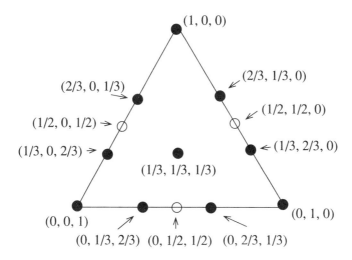

Figure 20.2. *Simplex for n = 3, laid flat on the page. The solid points indicate the support points of the bootstrap distribution while the open circles are the jackknife points.*

then $T(\mathbf{P}^0)$ is the observed value of the statistic, or in other words, t evaluated at $\hat{F}$. This is shown in the center of the simplex in Figure 20.2.

The jackknife values of the statistic are

$$\hat{\theta}_{(i)} = T(\mathbf{P}_{(i)}) \tag{20.4}$$

where

$$\mathbf{P}_{(i)} = \Big(\frac{1}{n-1}, \ldots, 0, \frac{1}{n-1}, \ldots \frac{1}{n-1}\Big)^T \quad (0 \text{ in } i\text{th place}). \tag{20.5}$$

These are also indicated in Figure 20.2.

The statistic $T(\mathbf{P}^*)$ can be thought of as a surface over its domain S_n as shown for $n = 3$ in Figure 20.3. Each point in the simplex at the bottom corresponds to a vector of probabilities $\mathbf{P}^*$; the value of the surface at $\mathbf{P}^*$ is $T(\mathbf{P}^*)$.

20.2 Bootstrap sampling

We can express bootstrap sampling in the framework described in the previous section. Sampling with replacement from $x_1, x_2, \ldots x_n$

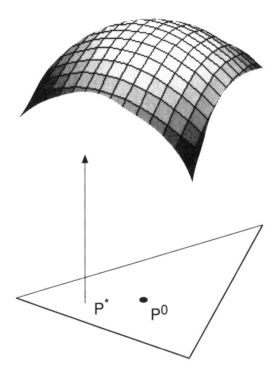

Figure 20.3. *The statistic* $T(\mathbf{P}^*)$ *viewed as a surface over the simplex.*

is equivalent to sampling $n\mathbf{P}^*$ from a multinomial distribution with n draws and equal class probabilities. Equivalently we can write

$$\mathbf{P}^* \sim \frac{1}{n}\text{Mult}(n, \mathbf{P}^0). \qquad (20.6)$$

The mean vector and covariance matrix of this distribution is

$$\mathbf{P}^* \sim \left(\mathbf{P}^0, [\frac{\mathbf{I}}{n^2} - \frac{\mathbf{P}^0\mathbf{P}^{0^T}}{n}]\right), \qquad (20.7)$$

where $\mathbf{I}$ is the $n \times n$ identity matrix.

The probability distribution (20.6) puts all of its support on vectors of the form $\mathbf{M}^*/n$, where $\mathbf{M}^*$ is an n-vector of nonnegative integers summing to n. The black dots in Figure 20.2 are the support points for $n = 3$. Problem 20.1 asks the reader to compute their associated probabilities under bootstrap sampling.

The correspondence between a bootstrap sample $x_1^*, \ldots x_n^*$ and the ith component of $\mathbf{P}^*$ is

$$P_i^* = \#\{x_j^* = x_i\}/n; \quad i = 1, 2, \ldots n, \tag{20.8}$$

the proportion of the bootstrap sample equaling x_i. As an example, consider the bootstrap sample x_2, x_1, x_1. This corresponds to $\mathbf{P}^* = (2/3, 1/3, 0)^T$ and according to (20.6) has probability

$$\binom{3}{2\;1\;0} \frac{1}{3}^2 \frac{1}{3}^1 \frac{1}{3}^0 = \frac{1}{9}, \tag{20.9}$$

where $\binom{a}{bcd}$ means $a!/(b! \cdot c! \cdot d!)$.

Note that the specific order x_2, x_1, x_1 is not important, and the factor $\binom{3}{210} = 3$ adds up the probabilities for the 3 possible orderings (x_2, x_1, x_1), (x_1, x_2, x_1), and (x_1, x_1, x_2).

The bootstrap estimate of variance for a statistic $T(\mathbf{P}^*)$ can be written as

$$\mathrm{var}_* T(\mathbf{P}^*), \tag{20.10}$$

where var_* indicates variance under the distribution (20.6). For the simple case $n = 3$, we could compute (20.10) exactly by adding up the 10 possible bootstrap samples weighted by their probabilities from (20.6) (see Problem 20.2). In this chapter we view the bootstrap estimate of variance as the "gold standard" and show how the jackknife and other estimators can be viewed as approximations to it.

20.3 The jackknife as an approximation to the bootstrap

A *linear* statistic $T(\mathbf{P}^*)$ has the form

$$T(\mathbf{P}^*) = c_0 + (\mathbf{P}^* - \mathbf{P}^0)^T \mathbf{U}, \tag{20.11}$$

where c_0 is a constant and $\mathbf{U} = (U_1, \ldots U_n)^T$ is a vector satisfying $\sum_1^n U_i = 0$. When viewed as a surface, a linear statistic defines a hyperplane over the simplex S_n. The mean $\bar{x}^* = \sum_1^n P_i^* x_i$ is a simple example of a linear statistic for which

$$c_0 = \bar{x}; \quad U_i = x_i - \bar{x} \tag{20.12}$$

(Problem 20.3).

The following result states that for any statistic, the jackknife estimate of variance for $T(\mathbf{P}^*)$ is almost the same as the bootstrap estimate of variance for a certain *linear approximation* to $T(\mathbf{P}^*)$:

Result 20.1 *The jackknife as an approximation to the bootstrap estimate of standard error.*

Let T^{LIN} be the unique hyperplane passing through the jackknife points $(\mathbf{P}_{(i)}, T(\mathbf{P}_{(i)}))$ for $i = 1, 2, \ldots n$. Then

$$var_* T^{\mathrm{LIN}} = \frac{n-1}{n} var_{\mathrm{jack}} \hat{\theta}, \tag{20.13}$$

where $var_{\mathrm{jack}} \hat{\theta}$ is the jackknife estimate of variance for $\hat{\theta}$:

$$var_{\mathrm{jack}} \hat{\theta} = \frac{n-1}{n} \sum_1^n (\hat{\theta}_{(i)} - \hat{\theta}_{(.)})^2 \tag{20.14}$$

and $\hat{\theta}_{(.)} = \sum_1^n \hat{\theta}_{(i)}/n$. In other words, the jackknife estimate of variance for $\hat{\theta} = t(\hat{F})$ equals $n/(n-1)$ times the bootstrap estimate of variance for T^{LIN}.

Proof:
By solving the set of n linear equations

$$\hat{\theta}_{(i)} = T^{\mathrm{LIN}}(\mathbf{P}_{(i)}) \tag{20.15}$$

for c_0 and $U_1, U_2, \ldots U_n$ we obtain

$$c_0 = \hat{\theta}; \quad U_i = (n-1)(\hat{\theta}_{(i)} - \hat{\theta}_{(.)}). \tag{20.16}$$

Using (20.7) and the fact that $\sum_1^n U_i = 0$,

$$\begin{aligned} var_* T^{\mathrm{LIN}}(\mathbf{P}^*) &= \mathbf{U}^T (var_* \mathbf{P}^*) \mathbf{U} = \tfrac{1}{n^2} \mathbf{U}^T \mathbf{U} \\ &= \tfrac{n-1}{n} \{ \tfrac{n-1}{n} \sum_1^n (\hat{\theta}_{(i)} - \hat{\theta}_{(.)})^2 \}. \Box \end{aligned} \tag{20.17}$$

The proof of this result can be approached differently: see Problem 11.6.

The "jackknife plane" T^{LIN} is shown in Figure 20.4. From Result 20.1 we see that the accuracy of the jackknife, as an approximation to the bootstrap, depends on how well T^{LIN} approximates $T(\mathbf{P}^*)$. In section 20.6 we examine the quality of this approximation in an example.

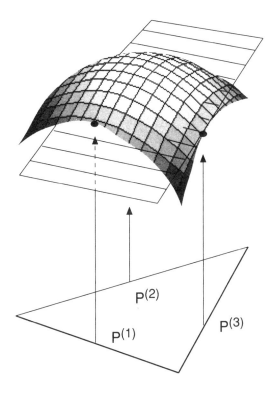

Figure 20.4. *The jackknife plane approximation to* $T(\mathbf{P}^*)$, *leading to the jackknife estimate of variance.*

20.4 Other jackknife approximations

The results of the previous section show that the bootstrap variance estimate arising from any approximation of the form (20.11) is

$$\frac{1}{n^2} \sum_1^n U_i^2. \qquad (20.18)$$

The jackknife uses the hyperplane passing through the jackknife points, and the resulting values $U_i = (n-1)(\hat{\theta}_{(i)} - \theta_{(.)})$. Another obvious choice would be the tangent plane approximation at $T(\mathbf{P}^0)$. This has the form

$$T^{\mathrm{TAN}}(\mathbf{P}^*) = T(\mathbf{P}^0) + (\mathbf{P}^* - \mathbf{P}^0)^T \mathbf{U}, \qquad (20.19)$$

where $\mathbf{U} = (U_1, \ldots U_n)$ is defined by

$$U_i = \lim_{\epsilon \to 0} \frac{T(\mathbf{P}^0 + \epsilon(\mathbf{e}_i - \mathbf{P}^0)) - T(\mathbf{P}^0)}{\epsilon}, \quad i = 1, 2, \ldots n, \quad (20.20)$$

and $\mathbf{e}_i = (0, 0, \ldots 0, 1, 0, \ldots 0)^T$ is the ith coordinate vector. The U_i are the *empirical influence values*, discussed in more detail in Chapter 21. This gives the variance estimate

$$\mathrm{var}^{IJ}\hat{\theta} = \frac{1}{n^2} \sum_1^n U_i^2 \qquad (20.21)$$

where U_i is defined by (20.20). This is called the *infinitesimal jack-knife* estimate of variance, and is also discussed in Chapter 21. Figure 20.5 shows the tangent plane approximation that leads to the infinitesimal jackknife estimate of variance.

The *positive jackknife*, yet another version of the jackknife, is based on

$$U_i = (n+1)(\hat{\theta}_{[i]} - \hat{\theta}), \qquad (20.22)$$

where $\theta_{[i]}$ denotes the value of $\hat{\theta}$ when x_i is repeated in the data set. It is discussed briefly in Section 21.3.

20.5 Estimates of bias

There is a similar relationship between the jackknife and bootstrap estimates of bias to that given for variances in Result 20.1 (page 288). For a linear statistic, both the jackknife and bootstrap estimates of bias are identically zero (Problem 20.4). We consider therefore an approximation involving *quadratic* statistics, defined by

$$T^{\mathrm{QUAD}}(\mathbf{P}^*) = c_0 + (\mathbf{P}^* - \mathbf{P}^0)^T \mathbf{U} + \frac{1}{2}(\mathbf{P}^* - \mathbf{P}^0)^T \mathbf{V}(\mathbf{P}^* - \mathbf{P}^0),$$

$$(20.23)$$

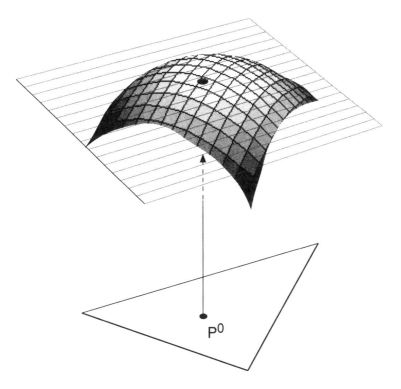

Figure 20.5. *The tangent plane approximation to $T(\mathbf{P}^*)$, leading to the infinitesimal jackknife estimate of variance.*

where $\mathbf{U}$ is an n-vector satisfying $\sum_1^n U_i = 0$ and $\mathbf{V}$ is an $n \times n$ symmetric matrix satisfying $\sum_i V_{ij} = \sum_j V_{ij} = 0$ for all i, j.

Result 20.2 *The jackknife as an approximation to the bootstrap estimate of bias.*

Let $T^{\mathrm{QUAD}}(\mathbf{P}^*)$ be a quadratic statistic passing through the jackknife points $(\mathbf{P}_{(i)}, T(\mathbf{P}_{(i)}))$ for $i = 1, 2, \dots n$, and the center point

$(\mathbf{P}^0, T(\mathbf{P}^0))$. *Then*

$$E_*(T^{\text{QUAD}}(\mathbf{P}^*) - \hat{\theta}) = \frac{n-1}{n}\widehat{\text{bias}}_{\text{jack}}(\hat{\theta}). \qquad (20.24)$$

Here $\widehat{\text{bias}}_{\text{jack}}(\hat{\theta})$ *is the jackknife estimate of bias for* $\hat{\theta}$:

$$\widehat{\text{bias}}_{\text{jack}}(\hat{\theta}) = (n-1)(\hat{\theta}_{(.)} - \hat{\theta}) \qquad (20.25)$$

and $\hat{\theta}_{(i)} = T(\mathbf{P}_{(i)})$, $\hat{\theta}_{(.)} = \sum_1^n \hat{\theta}_{(i)}/n$. *In other words, the jackknife estimate of bias for* $\hat{\theta} = t(\hat{F})$ *is* $n/(n-1)$ *times the bootstrap estimate of bias for the quadratic approximation* T^{QUAD}.

Proof:
Since T^{QUAD} passes through the points, $(\mathbf{P}_{(i)}, T(\mathbf{P}_{(i)}))$ for $i = 1, 2, \ldots n$, as well as $(\mathbf{P}^0, T(\mathbf{P}^0))$, $c_1, \mathbf{U}$, and $\mathbf{V}$ satisfy

$$\begin{aligned} c_0 &= T(\mathbf{P}^0) \\ \hat{\theta}_{(i)} &= c_0 + (\mathbf{P}_{(i)} - \mathbf{P}^0)^T\mathbf{U} + (\mathbf{P}_{(i)} - \mathbf{P}^0)^T\mathbf{V}(\mathbf{P}_{(i)} - \mathbf{P}^0) \end{aligned}$$
$$(20.26)$$

for $i = 1, 2, \ldots n$. Using (20.26) and the fact that $\sum_i V_{ij} = \sum_j V_{ij} = 0$ for all i, j, the jackknife estimate of bias is

$$\begin{aligned} (n-1)(\hat{\theta}_{(.)} - \hat{\theta}) &= \sum_1^n (\mathbf{P}_{(i)} - \mathbf{P}^0)^T\mathbf{U} + \\ & \quad \frac{1}{2}\sum_1^n (\mathbf{P}_{(i)} - \mathbf{P}^0)^T\mathbf{V}(\mathbf{P}_{(i)} - \mathbf{P}^0) \\ &= \frac{1}{2n^2}\sum_1^n V_{ii}. \end{aligned} \qquad (20.27)$$

Now for a general symmetric matrix $\mathbf{A}$, and a random vector $\mathbf{Y}$ with mean $\boldsymbol{\mu}$ and covariance matrix $\boldsymbol{\Sigma}$

$$\text{E}(\mathbf{Y}^T\mathbf{A}\mathbf{Y}) = \boldsymbol{\mu}^T\boldsymbol{\Sigma}\boldsymbol{\mu} + \text{tr}\boldsymbol{\Sigma}\mathbf{A}. \qquad (20.28)$$

Using this

$$\begin{aligned} \text{E}_*T^{\text{QUAD}}(\mathbf{P}^*) - T^{\text{QUAD}}(\mathbf{P}^0) &= \tfrac{1}{2}\text{tr}(\frac{\mathbf{I}}{n^2} - \frac{\mathbf{P}^0\mathbf{P}^{0T}}{n}) \\ &= \sum V_{ii}/2n^2 \quad \square \quad (20.29) \end{aligned}$$

In a similar fashion, suppose we approximate $T(\mathbf{P})$ by a two term Taylor series around $\mathbf{P}^0$ having the form (20.23). Then the bootstrap estimate of bias for this approximation equals $\sum V_{ii}/2n^2$, which is the same as the infinitesimal jackknife estimate of bias for T.

20.6 An example

The bootstrap surface of Figure 20.3 is difficult to view if n is larger than 3. It is possible to view "slices" of the surface and they can be quite informative. Consider for example the correlation coefficient applied to the law school data (Table 3.1). Figure 20.6 shows how the value of the correlation coefficient changes as the probability mass on each of the 15 data points is varied from 0 to 1 (solid curve). In each case, the remaining probability mass is spread evenly over the other 14 points. Notice how there is a large downward effect on the correlation coefficient as the amount of mass is increased on the 1st or 11th point. This make sense: these data points are $(576, 3.39)$ and $(653, 3.12)$, in the northwest and southeast part of the right panel of Figure 3.1, respectively.

The broken lines are the jackknife approximation to the surface. The approximation is generally quite good. When the mass on some data points is larger than .2, it starts to break down. However a probability mass greater than .2 corresponds to a data point appearing more than 3 times in a bootstrap sample, and this only occurs with probability approximately 1.5%. Furthermore, only about 20% of the samples will have at least one data value appearing more than 3 times. Therefore, the approximation is accurate where the bootstrap distribution puts most of its mass, and that is all that is needed for the jackknife to provide a reasonable approximation to the bootstrap estimate of standard error. The bootstrap and jackknife estimates of standard error are 0.127 and 0.142, respectively.

Notice how many of the curves are steeper between abscissa values 0 and 1/15 than they are past 1/15. In other words, the effect of deleting a data point is greater than the effect of doubling its probability. In this instance, the jackknife, which is based on slope estimates U_i between the abscissa values 0 and 1/15, will tend to give larger estimates of standard error than methods estimating the slope at 1/15 or beyond. The infinitesimal jackknife uses the tangent approximation at the observed data point $\mathbf{P}^0$, which corresponds to a tangent line to each curve through the dot at probability mass 1/15. It gives an estimate of standard error of 0.124, which is less than the jackknife value of 0.142 and close to

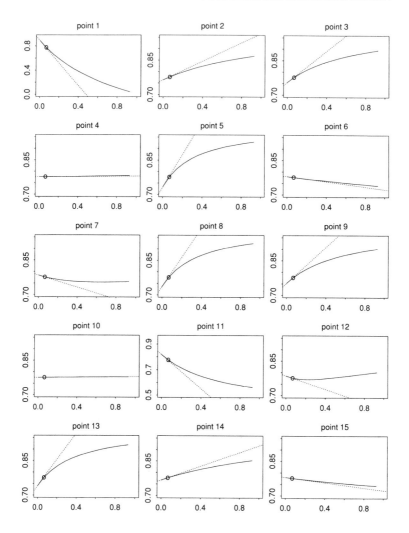

Figure 20.6. *Correlation coefficient for the law school data. Each plot shows the value of the correlation coefficient (solid curve) as the probability mass on the given data point is varied from 0 to 1 along the horizontal axis. The remaining probability is spread evenly among the other $n - 1$ points. Each plot therefore represents a slice of the resampling surface over the line running from the midpoint of a face of the simplex to the opposite vertex. The dot on each curve is the point (1/15, 0.776) corresponding to the original data set. The broken lines are the jackknife approximation. Note that the vertical scale is different on the 1st and 11th plots.*

the bootstrap value of 0.127. The positive jackknife uses the slope between the values 1/15 and 2/15, and gives an answer of 0.129 here.

This discussion suggests that for many statistics, the ordinary jackknife will tend to give larger estimates of standard error than the infinitesimal or positive jackknives. A result due to Efron and Stein (1981) says that we can expect the jackknife to give variance estimates with an upward bias. In the authors' experience, however, the jackknife gives more dependable estimates than the delta method or infinitesimal jackknife. It is the preferred variance estimate if bootstrap calculations are not done.

20.7 Bibliographic notes

The geometry of the bootstrap, and its relationship to the jack-knife and other estimates, appears in Efron (1979a, 1982). The idea of slicing the bootstrap surface is proposed in the dissertation of Therneau (1983).

20.8 Problems

20.1 Compute the probabilities of each of the support points of the bootstrap distribution in Figure 20.2.

20.2 Suppose our data values are $(1, 5, 4)$ and $\hat{\theta}$ is the sample mean.

(a) Work out the bootstrap estimate of variance by computing the probabilities of each of the 10 possible samples under the multinomial (20.6) and adding up the terms.

(b) Verify that the answer in (a) agrees with the closed form solution $\sum_1^n (x_i - \bar{x})^2 / n^2$ given in Chapter 5.

20.3 Show that the mean is a linear statistic of the form (20.11) with coefficients given by (20.12).

20.4 Show that for linear statistics, the jackknife and bootstrap estimates of bias are zero.

20.5 In this chapter we have discussed jackknife and other approximations to bootstrap bias and variance estimates. Suggest how one could obtain closed-form, jackknife-based approximations to higher moments of the bootstrap distribution.

## CHAPTER 21

An overview of nonparametric and parametric inference

21.1 Introduction

The objective of this chapter is to study the relationship of bootstrap and jackknife methodology to more traditional parametric approaches to statistical inference, specifically maximum likelihood estimation. Variance (or standard error) estimation is the focus for the comparison, and Figure 21.1 gives a summary of the possibilities. Exact or approximate inference is possible, using either a nonparametric or parametric specification for the population. We explore the relationships between these approaches, making clear the assumptions made by each.

Likelihood inference, based on construction of a parametric likelihood for a parameter, is discussed briefly in Section 21.4. We defer discussion of nonparametric likelihood inference until Chapter 24.

21.2 Distributions, densities and likelihood functions

Suppose we have a sample $x_1, x_2, \ldots x_n$ from a population. As in Chapter 3 we think of these values as independent realizations of a random variable X. The values of X may be real numbers or vectors of real numbers. A general way to describe the population that gives rise to X is through its *cumulative distribution function*

$$F(x) = \text{Prob}(X \leq x). \tag{21.1}$$

If the function $F(x)$ is differentiable, one can also describe the distribution of X through its *probability density function*:

$$f(x) = \frac{dF(x)}{dx}. \tag{21.2}$$

The probability that X lies in some set A can be obtained by

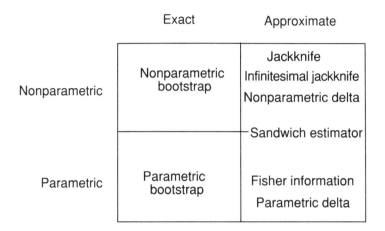

Figure 21.1. *A summary of the methods for variance estimation studied in this chapter.*

integration of the density function

$$\text{Prob}\{x \in A\} = \int_A f(x)dx. \tag{21.3}$$

Note that $f(x)$ is not a probability and can have a value greater than one. Assuming X is real-valued, for small $\Delta > 0$,

$$f(x)\Delta \doteq \text{Prob}(X \in [x, x + \Delta]). \tag{21.4}$$

As an example, if X has a standard normal distribution then

$$\begin{aligned} F(x) &= \int_{-\infty}^{x} \frac{1}{\sqrt{2\pi}} e^{-\frac{1}{2}t^2} dt \\ f(x) &= \frac{1}{\sqrt{2\pi}} e^{-\frac{1}{2}x^2}. \end{aligned} \tag{21.5}$$

A normal random variable takes on continuous values; recall that some random variables take on discrete values. A simple example is a binomial random variable with success probability say $1/3$. Then

$$f(x) = \binom{n}{x} (1/3)^x (1 - 1/3)^{n-x} \quad \text{for} \quad x = 0, 1, 2, \cdots, n. \tag{21.6}$$

In this discrete case $f(x)$ is often called a *probability mass function*.

Maximum likelihood is a popular approach to inference that is used when it can be assumed that X has a probability density (or probability mass function) depending on a finite number of unknown "parameters." This is discussed in section 21.4 and is called a *parametric* approach to inference. In the next section we focus on functional statistics for nonparametric inference, corresponding to the top half of Figure 21.1.

21.3 Functional statistics and influence functions

In Chapter 4 we discussed summary statistics of the form

$$\hat{\theta} = t(\hat{F}) \tag{21.7}$$

where $\hat{F}$ is the empirical distribution function. Such a statistic is the natural estimate of the population parameter

$$\theta = t(F). \tag{21.8}$$

For example, if $\theta = \mathrm{E}_F(X)$, then $\hat{\theta} = \mathrm{E}_{\hat{F}}(X) = \sum_1^n x_i/n$. Since $t(\hat{F})$ is a function of the distribution function $\hat{F}$, it is called a plug-in or *functional* statistic. Most estimates are functional statistics, but there are some exceptions. Consider for example the unbiased estimate of variance

$$s^2 = \frac{1}{n-1} \sum_1^n (x_i - \bar{x})^2. \tag{21.9}$$

Suppose we create a new data set of size $2n$ by duplicating each data point. Then $\hat{F}$ for the new data set puts mass $2/2n = 1/n$ on each x_i and hence is the same as it was for the original data set. But s^2 for the new data set equals

$$\frac{2}{2n-1} \sum_1^n (x - \bar{x})^2 \tag{21.10}$$

(Problem 21.1) which is not the same as (21.9). On the other hand, the plug-in-estimate of variance

$$\frac{1}{n} \sum_1^n (x_i - \bar{x})^2 \tag{21.11}$$

is the same in both cases; it is a functional statistic since it is obtained by substituting $\hat{F}$ for F in the formula for variance. [1] The difference between the unbiased and plug-in estimates of variance usually tends to be unimportant and creates no real difficulty.

More importantly, estimates which do not behave smoothly as functions of the sample size n are not functional statistics and cannot be studied in the manner described in this chapter. An example is

$$\hat{\theta} = \begin{cases} \text{median}(x_1, x_2, \ldots x_n) & \text{for } n \text{ odd ;} \\ \text{mean}(x_1, x_2, \ldots x_n) & \text{for } n \text{ even.} \end{cases} \qquad (21.12)$$

Such statistics seldom occur in real practice.

Suppose now that $t(\hat{F})$ is a functional statistic and consider an expansion of the form

$$t(\hat{F}) = t(F) + \frac{1}{n} \sum_1^n U(x_i, F) + O_p(n^{-1}). \qquad (21.13)$$

(The expression $O_p(n^{-1})$ reads "order n^{-1} in probability." A definition may be found in section 2.3 of Barndorff-Neilson and Cox, 1989.) Equation (21.13) is a kind of first order Taylor series expansion. As we will see, it is important to the understanding of many nonparametric and parametric estimates of variance of $t(\hat{F})$. The quantity $U(x_i, F)$ is called an *influence function* or *influence component* and is defined by

$$U(x, F) = \lim_{\epsilon \to 0} \frac{t[(1 - \epsilon)F + \epsilon \delta_x] - t(F)}{\epsilon}. \qquad (21.14)$$

The notation δ_x means a point mass of probability at x, and so $(1 - \epsilon)F + \epsilon \delta_x$ represents F with a small "contamination" at x. The function $U(x, F)$ measures the rate of change of $t(F)$ under this contamination; it is a kind of derivative. Two simple examples are the mean $t(F) = E_F(X)$ for which

$$U(x, F) = x - E_F X, \qquad (21.15)$$

and the median $t(F) = \text{median}(X) = F^{-1}(1/2)$ for which

$$U(x, F) = \frac{\text{sign}(x - \text{median}(X))}{2f_0}. \qquad (21.16)$$

---

[1] Plug-in estimates $\hat{\theta} = t(\hat{F})$ are always functional statistics, since $\hat{F}$ itself is a functional.

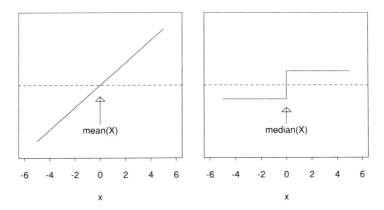

Figure 21.2. *Solid lines show the influence function for the mean (left panel) and the median (right panel), assuming both the mean and median are zero. Broken line is drawn at zero for reference.*

Here f_0 is the density of X evaluated its median, and $\text{sign}(x)$ denotes the sign of x: $\text{sign}(x) = 1, -1$, or 0 accordingly as $x > 0, x < 0$ or $x = 0$ (Problem 21.2). These are shown in Figure 21.2.

Notice that the effect of a contamination at x is proportional to $x - \text{E}_F(X)$ for the mean, but is bounded for the median. This reflects the fact that the median is resistant to outlying data values while the mean is not. Moving the point x further and further out on the x-axis has greater and greater effect on the mean, but not on the median.

The influence curve was originally proposed to study the resistance or robustness of a statistic. However, it is also useful for computing the approximate variance of a statistic. In particular (21.13) can be used to show that

$$\text{var}_F t(\hat{F}^*) \doteq \frac{1}{n}\text{var}_F U(x, F) = \frac{1}{n}\text{E}_F U^2(x, F). \qquad (21.17)$$

The final expression follows because $\text{E}_F U(x, F) = 0$ in general (Problem 21.3).

Formula (21.17) is the key to understanding many different variance estimates. By inserting an estimate of $U(x, F)$ into formula (21.17) we can derive the jackknife estimate of variance as well as many other variance estimates. Suppose we set $F = \hat{F}$ and rather than take the limit in the definition of $U(x, F)$, we set ϵ to

$-1/(n-1)$. Then we obtain the estimate

$$\left(\frac{n-1}{n}\right)^2 \sum_1^n (\hat{\theta}_{(i)} - \hat{\theta})^2 \tag{21.18}$$

where $\theta_{(i)}$ is the ith jackknife value. This is very similar to, but not exactly the same as, the jackknife estimate of variance $\widehat{\mathrm{var}}^{JK} t(\hat{F}) = [(n-1)/n] \sum_1^n (\hat{\theta}_{(i)} - \hat{\theta}_{(\cdot)})^2$ given in (20.14).

If instead we take $F = \hat{F}$ and go to the limit in the definition of $U(x, F)$, we obtain

$$\widehat{\mathrm{var}}^{IJ} t(\hat{F}) = \frac{1}{n^2} \sum_1^n U^2(x_i, \hat{F}) \tag{21.19}$$

which is called, appropriately, the *infinitesimal jackknife* estimate of variance. The quantity $U(x_i, \hat{F})$ is called an *empirical influence component*. Both the jackknife and infinitesimal jackknife estimates of variance are nonparametric since they use the nonparametric maximum likelihood estimate $\hat{F}$. They differ in the choice of ϵ: the infinitesimal jackknife takes the limit as $\epsilon \to 0$, while the jackknife uses the small negative value $-1/(n-1)$. There are other possibilities: the *positive jackknife* uses $\epsilon = 1/(n+1)$, giving $U_i = (n+1)(\theta_{[i]} - \hat{\theta})$ and the variance estimate

$$\left(\frac{n+1}{n}\right)^2 \sum_1^n (\hat{\theta}_{[i]} - \hat{\theta})^2 \tag{21.20}$$

where $\theta_{[i]}$ denotes the value of $\hat{\theta}$ when x_i is repeated in the data set. This not is usually a good estimate in small samples because it stresses the importance of any inflated data points. It can be badly biased downward, and is not commonly used.

Recall that the sample-based estimate of the left side of (21.17)

$$\mathrm{var}_{\hat{F}} t(\hat{F}^*) \tag{21.21}$$

is the bootstrap estimate of variance of $t(\hat{F})$. Here $\hat{F}^*$ is the empirical distribution corresponding to the bootstrap sample $\mathbf{x}^*$. Therefore the jackknife, positive jackknife and infinitesimal jackknife can all be viewed as approximations to the bootstrap estimate of variance, the approximation based on the first two terms in (21.13). This is why the bootstrap is labeled "exact" in Figure 21.1. If there is no error in approximation (21.13), $t(\hat{F})$ can be exactly represented in the form

$$t(\hat{F}) = t(F) + \frac{1}{n} \sum_{1}^{n} U(x_i, F), \qquad (21.22)$$

known as a *linear statistic*. (It is easy to check that the definition of linearity as defined (21.22) is the same as that given in equation (20.11) of Chapter 20.) In this case it is not surprising that the infinitesimal jackknife agrees with the bootstrap estimate of variance. The simplest example is the mean, for which both give the plug-in estimate of variance. Perhaps it is surprising that the jackknife estimate of variance also agrees with the bootstrap (except for the arbitrary factor $(n-1)/n$ included in the jackknife for historical reasons). The exact statements of these relationships are as follows.

RESULT 21.1. *Relationship between the nonparametric bootstrap, infinitesimal jackknife, and jackknife estimates of variance: If $t(\hat{F})$ is a linear statistic, then*

$$var_{\hat{F}} t(\hat{F}^*) = \widehat{var}^{IJ} t(\hat{F}) = \frac{n-1}{n} \widehat{var}^{JK} t(\hat{F}).$$

The proofs of these results are most easily expressed geometrically, using the *resampling representation*. They are given in sections 20.3 and 20.4 of Chapter 20.

21.4 Parametric maximum likelihood inference

In this section we describe the approaches to inference that fall in the bottom half of Figure 21.1. We begin by specifying a probability density or probability mass function for our observations

$$X \sim f_\theta(x). \qquad (21.23)$$

In this expression θ represents one or more unknown parameters that govern the distribution of X. This is called a *parametric model* for X. We denote the number of elements of θ by p. As an example, if X has a normal distribution with mean μ and variance σ^2, then

$$\theta = (\mu, \sigma^2), \qquad (21.24)$$

$p = 2$, and

$$f_\theta(x) = \frac{1}{\sqrt{2\pi\sigma^2}} e^{-\frac{1}{2}(\frac{x-\mu}{\sigma})^2}. \tag{21.25}$$

Maximum likelihood is based on the *likelihood function* defined by

$$L(\theta; \mathbf{x}) = \prod_1^n f_\theta(x_i). \tag{21.26}$$

The likelihood is defined only up to a positive multiplier, which we have taken to be one. We think of $L(\theta; \mathbf{x})$ as a function of θ with our data $\mathbf{x}$ fixed. In the discrete case, $L(\theta; \mathbf{x})$ is the probability of observing our sample. In the continuous case $L(\theta; \mathbf{x})\mathbf{\Delta}$ is approximately the probability of our sample lying in a small interval $[\mathbf{x}, \mathbf{x} + \mathbf{\Delta}]$, (21.4).

Denote the logarithm of $L(\theta; \mathbf{x})$ by

$$\ell(\theta; \mathbf{x}) = \sum_1^n \ell(\theta; x_i) \tag{21.27}$$

which we will sometimes abbreviate as $\ell(\theta)$. This expression is called the log-likelihood and each value $\ell(\theta; x_i) = \log f_\theta(x_i)$ is called a log-likelihood component.

The method of maximum likelihood chooses the value $\theta = \hat\theta$ to maximize $\ell(\theta; \mathbf{x})$. Consider for example the control group of the mouse data (Table 2.1, page 19). Let's assume the model

$$x_1, x_2, \ldots x_n \sim N(\theta, \sigma^2). \tag{21.28}$$

We set σ^2 to the value of the plug-in estimate

$$\hat\sigma^2 = 1799.2 = 42.42^2. \tag{21.29}$$

The left panel of Figure 21.3 shows the log-likelihood function $\ell(\theta; \mathbf{x})$ for the 9 data values.

The maximum occurs at $\hat\theta = 56.22$, which is also the sample mean $\bar{x}$. The explicit form of $\ell(\theta)$ in this example is

$$-n \log \hat\sigma \sqrt{2\pi} - \frac{1}{2}\sum_1^n (x_i - \theta)^2/\hat\sigma^2. \tag{21.30}$$

The log-likelihood is only defined up to an additive constant; for convenience, then, we have translated the curve in Figure 21.2 so

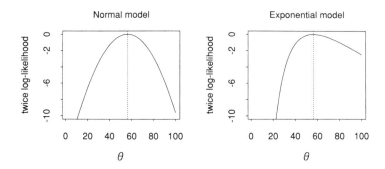

Figure 21.3. *Log-likelihood functions for the mean of the mouse data. Left panel is based on the normal model, while the right panel uses the exponential model. Dotted line is drawn at the maximum likelihood estimate* $\hat{\theta} = 56.22$.

that its maximum is at zero. The result is sometimes called the relative log-likelihood function.

As an alternative to normality we might assume that the observations come from an exponential distribution having density

$$f_\theta(x) = \frac{1}{\theta}e^{-x/\theta}, \quad x > 0. \tag{21.31}$$

The right panel of Figure 21.2 shows the log-likelihood for this model. The maximum also occurs at $\bar{x} = 56.22$, but the shape of the likelihood is quite different.

A different way to view maximum likelihood is to think of

$$\prod_{1}^{n} f_{\hat{\theta}}(x_i) \tag{21.32}$$

as the *maximum likelihood summary* of the data. The maximum likelihood summarizer is a probability density, not a number or vector, that summarizes the information in the data about our parametric model.

The likelihood function can be used to assess the precision of $\hat{\theta}$. We need a few more definitions. The *score function* is defined by

$$\dot{\ell}(\theta; \mathbf{x}) = \sum_{1}^{n} \dot{\ell}(\theta; x_i), \tag{21.33}$$

where $\dot{\ell}(\theta; x) = d\ell(\theta; x)/d\theta$. Assuming that the likelihood takes its maximum in the interior of the parameter space, $\dot{\ell}(\hat{\theta}; \mathbf{x}) = 0$. The *information* is

$$I(\theta) = -\sum_1^n \frac{d^2\ell(\theta; x_i)}{d\theta^2}. \tag{21.34}$$

When $I(\theta)$ is evaluated at $\theta = \hat{\theta}$, it is often called the *observed information*. The *Fisher information* (or expected information) is

$$i(\theta) = \mathrm{E}_\theta[I(\theta)]. \tag{21.35}$$

Finally, let θ_0 denote the true value of θ.

A standard result says that the maximum likelihood estimator has a limiting normal distribution

$$\hat{\theta} \to N(\theta_0, i(\theta_0)^{-1}). \tag{21.36}$$

Here we are independently sampling from $f_{\theta_0}(x)$ and the sample size $n \to \infty$. This suggests that the sampling distribution of $\hat{\theta}$ may be approximated by

$$N(\hat{\theta}, i(\hat{\theta})^{-1}). \tag{21.37}$$

Alternatively, $i(\hat{\theta})$ can be replaced by $I(\hat{\theta})$ to yield the approximation

$$N(\hat{\theta}, I(\hat{\theta})^{-1}). \tag{21.38}$$

The corresponding estimates for the standard error of $\hat{\theta}$ are

$$i(\hat{\theta})^{-1/2} \quad \text{and} \quad I(\hat{\theta})^{-1/2}. \tag{21.39}$$

Confidence points for θ can be constructed using approximations (21.37) or (21.38). The α confidence point has the form $\hat{\theta} - z^{(1-\alpha)} \cdot \{i(\hat{\theta})\}^{-1/2}$ or $\hat{\theta} - z^{(1-\alpha)} \cdot \{I(\hat{\theta})\}^{-1/2}$ respectively, where $z^{(1-\alpha)}$ is the $1 - \alpha$ percentile of the standard normal distribution.

Alternatively, a confidence interval can derived from the likelihood function, by using the approximation

$$2[\ell(\hat{\theta}) - \ell(\theta_0)] \sim \chi_1^2. \tag{21.40}$$

The resulting $1 - 2\alpha$ confidence interval is the set of all θ such that $2[\ell(\hat{\theta}) - \ell(\theta_0)] \leq \chi_1^{2(1-2\alpha)}$, where $\chi_1^{2(1-2\alpha)}$ is the $1 - 2\alpha$ percentile of the Chi-square distribution with one degree of freedom. It is also

Normal model

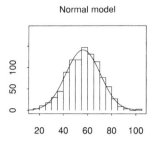

Exponential model

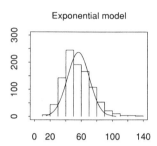

Figure 21.4. *Parametric bootstrap histograms 1000 replications of the mean $\hat{\theta}^*$. Left panel is based on the normal model, while the right panel uses the exponential model. Superimposed is the normal density curve based on (21.36).*

possible to carry out a nonparametric version of this, that is, to construct confidence intervals from a nonparametric likelihood for the parameter. Nonparametric likelihood is the subject of Chapter 24.

21.5 The parametric bootstrap

There is a more exact way of estimating the sampling distribution and variance of $\hat{\theta}$ in the parametric setting. We draw B samples of size n from the density $f_{\hat{\theta}}(x)$, and calculate the maximum likelihood estimate of θ for each one. The sample variance of these B values estimates the variance of $\hat{\theta}$. This process is called the parametric bootstrap method, and is described in Section 6.5 of Chapter 6. The only difference from the nonparametric bootstrap is that the samples are drawn from a parametric estimate of the population rather than the non-parametric estimate $\hat{F}$.

The left panel of Figure 21.4 shows a histogram of 500 parametric bootstrap values of $\hat{\theta}$. We drew 500 samples of size 9 from the normal model $N(\hat{\theta}, \widehat{se}^2)$ and computed the mean for each.

Superimposed on the histogram is the density $N(\hat{\theta}, i(\hat{\theta})^{-1})$ suggested by result (21.36). The agreement is very good, which is not at all surprising since the population is assumed to be normal and hence the asymptotic result (21.28) holds exactly for small samples. The parametric bootstrap estimate of variance is 183.7, while $1/i(\hat{\theta}) = 177.7$. In the normal model $1/i(\hat{\theta})$ is just the plug-in estimate of variance for the mean, so again this agreement is not surprising.

The right panel shows the results if we assume instead that the observations have an exponential distribution Now the large sample normal approximation is not very accurate. As the sample size n approaches infinity, the central limit theorem tells us that the histogram will start to look more and more like the normal density curve. In this instance, $n = 9$ is not close enough to infinity! However that the variance estimates are not very different: the parametric bootstrap estimate of variance based on 500 replicates is 359.5, while $1/i(\hat{\theta}) = \bar{x}^2/n = 351.2$.

21.6 Relation of parametric maximum likelihood, bootstrap and jackknife approaches

In order to relate the parametric maximum likelihood approach to the jackknife and other methods discussed earlier, we need to outline its multiparameter version. Suppose now that we have a vector of parameters $\boldsymbol{\eta}$ and we want to conduct inference for a real valued function $\theta = h(\boldsymbol{\eta})$. Let $\boldsymbol{\eta}_0$ be the true value of $\boldsymbol{\eta}$. If $\hat{\boldsymbol{\eta}}$ denotes the maximum likelihood estimate of $\boldsymbol{\eta}$, then the maximum likelihood estimate of θ is

$$\hat{\theta} = h(\hat{\boldsymbol{\eta}}). \tag{21.41}$$

Denote the parametric family of distribution functions of x by $F_{\boldsymbol{\eta}}$, with true value $F = F_{\eta_0}$.

As in the previous section let the score vector be $\dot{\ell}(\boldsymbol{\eta}; \mathbf{x})$, the information matrix be $I(\boldsymbol{\eta})$ and the expected information matrix be $i(\boldsymbol{\eta})$. These are the multiparameter analogues of the quantities introduced in the one parameter case. $\dot{\ell}(\boldsymbol{\eta}; \mathbf{x})$ is a vector of length p with ith element equal to $\partial\ell/\partial\eta_i$, $I(\boldsymbol{\eta})$ is a $p \times p$ matrix with ijth element $-\partial^2\ell/\partial\eta_i\partial\eta_j$, and $i(\boldsymbol{\eta})$ is a $p \times p$ matrix with ijth element $-\mathrm{E}_F(\partial^2\ell/\partial\eta_i\partial\eta_j)$. Denote by $\dot{h}(\boldsymbol{\eta})$ the gradient vector of $\theta = h(\boldsymbol{\eta})$

with respect to $\boldsymbol{\eta}$:

$$\dot{h}(\boldsymbol{\eta}) = \begin{pmatrix} \partial h(\boldsymbol{\eta})/\partial \eta_1 \\ \partial h(\boldsymbol{\eta})/\partial \eta_2 \\ \vdots \\ \partial h(\boldsymbol{\eta})/\partial \eta_p \end{pmatrix}. \tag{21.42}$$

An application of the chain rule shows that the inverse of the Fisher information for $h(\boldsymbol{\eta})$ is given by

$$i(h(\boldsymbol{\eta}))^{-1} = \dot{h}(\boldsymbol{\eta})^T i(\boldsymbol{\eta})^{-1}\dot{h}(\boldsymbol{\eta}). \tag{21.43}$$

The sample estimate replaces $\boldsymbol{\eta}$ with $\hat{\boldsymbol{\eta}}$ in the above equation. Furthermore, it can be shown that

$$h(\hat{\theta}) \to N(h(\theta_0), \dot{h}(\boldsymbol{\eta}_0)^T i(\boldsymbol{\eta}_0)^{-1}\dot{h}(\boldsymbol{\eta}_0)) \tag{21.44}$$

as $n \to \infty$, when sampling from $f_{\eta_0}(\cdot)$.

We can relate the Fisher information to the influence function method for obtaining variances by computing the influence component $U(x, F)$ for the maximum likelihood estimate $\hat{\theta}$:

$$U(x, F) = n \cdot \dot{h}(\boldsymbol{\eta})^T i(\boldsymbol{\eta})^{-1}\dot{\ell}(\boldsymbol{\eta}; x) \tag{21.45}$$

(Problem 21.5). If we evaluate $U(x, F)$ at $F = F_{\hat{\eta}}$, we see that $U(x, F_{\hat{\eta}})$ is a multiple of the score component $\dot{\ell}(\hat{\boldsymbol{\eta}}; x)$. This simple relationship between the score function and the influence function arises in the theory of "M-estimation" in robust statistical inference. In particular, the influence function of an M-estimate is a multiple of the "ψ" function that defines it.

Given result (21.45), the variance formula $\frac{1}{n}\mathrm{E}_F U^2(x, F)$ from (21.17) then leads to

$$\begin{aligned} \mathrm{var}_F \hat{\theta} &\doteq n \cdot \mathrm{E}_F \dot{h}(\boldsymbol{\eta})^T i(\boldsymbol{\eta})^{-1}[\dot{\ell}(\boldsymbol{\eta}; x)\dot{\ell}(\boldsymbol{\eta}; x)^T]i(\boldsymbol{\eta})^{-1}\dot{h}(\boldsymbol{\eta}) \\ &= n \cdot \dot{h}(\boldsymbol{\eta})^T i(\boldsymbol{\eta})^{-1}\Big\{\mathrm{E}_F[\dot{\ell}(\boldsymbol{\eta}; x)\dot{\ell}(\boldsymbol{\eta}; x)^T]\Big\}i(\boldsymbol{\eta})^{-1}\dot{h}(\boldsymbol{\eta}) \\ &= \dot{h}(\boldsymbol{\eta})^T i(\boldsymbol{\eta})^{-1}\dot{h}(\boldsymbol{\eta}), \end{aligned} \tag{21.46}$$

which is exactly the same as the inverse of the Fisher information (21.43) above. The last equation in (21.46) follows from the previous line by a basic identity relating the Fisher information to the covariance of the score function (Problem 21.6)

$$n \cdot \mathrm{E}_F\{\dot{\ell}(\boldsymbol{\eta}; x)\dot{\ell}(\boldsymbol{\eta}; x)^T\} = \mathrm{E}_F\{\dot{\ell}(\boldsymbol{\eta}; \mathbf{x})\dot{\ell}(\boldsymbol{\eta}; \mathbf{x})^T\} = -i(\boldsymbol{\eta}). \tag{21.47}$$

Hence the usual Fisher information method for estimation of variance can be thought of as an influence function-based estimate, using the model-based form of the influence function (21.45). We summarize for reference:

RESULT 21.2. *Relationship between the parametric bootstrap, infinitesimal jackknife and Fisher information-based estimates of variance:*
For a statistic $t(F_{\hat{\eta}})$,

$$\widehat{var}^{IJ} t(F_{\hat{\eta}}) = \dot{h}(\hat{\eta})^T i(\hat{\eta})^{-1} \dot{h}(\hat{\eta})$$

the right hand side being the inverse Fisher information for $h(\eta)$. *Furthermore if* $t(F_{\hat{\eta}})$ *is a linear statistic* $t(\hat{F}) + \frac{1}{n} \sum_{1}^{n} U(x_i, \hat{F})$ *then the infinitesimal jackknife and inverse Fisher information both agree with the parametric bootstrap estimate of variance for* $t(F_{\hat{\eta}})$.

21.6.1 Example: influence components for the mean

In the nonparametric functional approach, $U(x, \hat{F}) = x - \bar{x}$ from (21.15). Instead of operating nonparametrically, suppose we assume an exponential model. Then $\dot{\ell}(\hat{\eta}; x) = -1/\bar{x} + x/\bar{x}^2$, $i(\hat{\eta}) = n/\bar{x}^2$ and so

$$U(x, F_{\hat{\eta}}) = x - \bar{x}, \tag{21.48}$$

which again agrees with $U(x, \hat{F})$.

In general, the same value will be obtained for $U(x, F)$ whether we use formula (21.45), or treat $h(\hat{\eta})$ as a functional statistic and use the definition (21.14) directly. However, $U(x, F_{\hat{\eta}})$ and $U(x, \hat{F})$ may differ. A simple example where this occurs is the trimmed mean in the normal family. Thus, there are two potential differences between the parametric and nonparametric infinitesimal jackknife estimates of variance: the value of the influence curve U and the choice of distribution ($F_{\hat{\eta}}$ or $\hat{F}$) under which the expectation $E_F U^2$ is taken.

In this section we have assumed that the statistic $\hat{\theta}$ can be written as a functional of the empirical distribution function

$$\hat{\theta} = t(\hat{F}). \tag{21.49}$$

This implies that $t(\hat{F})$ and $\hat{\theta}$ are really estimating the same parameter, that is, $t(F_\eta) = h(\eta)$. For example in the normal family, if θ is the mean, then $t(\hat{F})$ would be the sample mean. We are not allowed to take $t(\hat{F})$ equal to the sample median, even though the mean θ of the normal distribution is also the median.

21.7 The empirical cdf as a maximum likelihood estimate

Suppose that we allow η to have an arbitrarily large number of components. Then the maximum likelihood estimate of the underlying population is the empirical distribution function $\hat{F}$. That is, it can be shown that $\hat{F}$ *is the nonparametric maximum likelihood estimate of F*. Here is the idea. We define the *nonparametric likelihood function* as

$$L(F) = \prod_1^n F(\{x_i\}), \qquad (21.50)$$

where $F(\{x_i\})$ is the probability of the set $\{x_i\}$ under F. Then it is easy to show that the empirical distribution function $\hat{F}$ maximizes $L(F)$ (Problem 21.4). As a result, the functional statistic $t(\hat{F})$ is the nonparametric maximum likelihood estimate of the parameter $t(F)$. In this sense, the nonparametric bootstrap carries out nonparametric maximum likelihood inference. Different approaches to nonparametric maximum likelihood inference are discussed in Chapter 24

21.8 The sandwich estimator

Note that the identity (21.47) holds only if the model is correct. A "semi-parametric" alternative to Fisher information uses the second expression on the right hand side of (21.46), estimating the quantity $E_F[\dot{\ell}(\eta; x)\dot{\ell}(\eta; x)^T]$ with the empirical covariance of the score function

$$\frac{1}{n} \sum_1^n \dot{\ell}(\hat{\eta}; x_i)\dot{\ell}(\hat{\eta}; x_i)^T. \qquad (21.51)$$

The resulting estimate of $\mathrm{var}_F \hat{\theta}$ is

$$\dot{h}(\hat{\eta})^T \left\{ i(\hat{\eta})^{-1} \sum_1^n \dot{\ell}(\hat{\eta}; x_i)\dot{\ell}(\hat{\eta}; x_i)^T i(\hat{\eta})^{-1} \right\} \dot{h}(\hat{\eta}). \qquad (21.52)$$

The quantity

$$i(\hat{\boldsymbol{\eta}})^{-1} \sum_{1}^{n} \dot{\ell}(\hat{\boldsymbol{\eta}}; x_i) \dot{\ell}(\hat{\boldsymbol{\eta}}; x_i)^T i(\hat{\boldsymbol{\eta}})^{-1} \qquad (21.53)$$

is sometimes called the "sandwich estimator", because the Fisher information sandwiches the empirical covariance of the score vector. Like bootstrap and jackknife estimates, the sandwich estimator is consistent for the true variance of $\hat{\theta}$ even if the parametric model does not hold. This is not the case for the observed information.

The sandwich estimator arises naturally in M-estimation and the theory of estimating equations. In the simple case $\hat{\theta} = \bar{x}$ in the normal model, it is easy to show that the sandwich estimator equals the maximum likelihood estimate of variance $\sum_{1}^{n}(x_i - \bar{x})^2/n^2$.

21.8.1 Example: Mouse data

Let's compare some of these methods for the mouse data of Chapter 2. Denote the times in the treatment group by X_i and those in the control group by Y_i. The quantity of interest is the difference in means

$$\theta = \mathrm{E}(X) - \mathrm{E}(Y). \qquad (21.54)$$

A nonparametric bootstrap approach to this problem allows different distribution functions F and G for the two groups, and resamples each group separately. In a parametric approach, we might specify a different normal distribution $N(\mu_i, \sigma_i^2)$, $i = 1, 2$ for each group, and then define

$$\boldsymbol{\eta} = (\mu_1, \sigma_1^2, \mu_2, \sigma_2^2) \qquad (21.55)$$

$$\theta = h(\boldsymbol{\eta}) = \mu_1 - \mu_2. \qquad (21.56)$$

Alternatively, we might assume an exponential distribution for each group with means μ_1 and μ_2. Then

$$\boldsymbol{\eta} = (\mu_1, \mu_2) \qquad (21.57)$$

$$\theta = h(\boldsymbol{\eta}) = \mu_1 - \mu_2 \qquad (21.58)$$

Figure 21.5 shows a number of different sampling distributions of the maximum likelihood estimator $\hat{\theta}$.

The estimates of the standard error of $\hat{\theta}$ are shown in Table 21.1: All of the estimates are similar except for those arising from the ex-

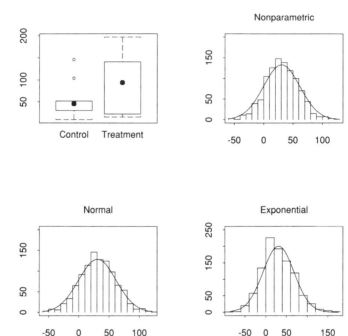

Figure 21.5. *Inference for the difference in means for the mouse data shown in top left. Bootstrap histograms of $\hat{\theta}$ are shown, from nonparametric bootstrap (top right), parametric bootstrap based on the normal distribution (bottom left) and the parametric bootstrap based on the exponential distribution (bottom right). Superimposed on each histogram is the normal density curve based on (21.36).*

ponential model, which are larger. The fact that the exponential-based standard errors are substantially larger than the nonparametric standard errors sheds doubt on the appropriateness of the exponential model for these data. Note that the sandwich estimator, which is exactly equal to the nonparametric bootstrap in this case, still performs well even under the exponential model assumption. Problem 21.7 asks the reader to compute these estimates.

Table 21.1. *Standard error estimates for the mouse data.*

Method	Formula	Value
Nonparametric		
bootstrap	$[\mathrm{var}_{\hat{F}}\, t(\hat{F}^*)]^{1/2}$	28.1
Jackknife	$[\frac{n-1}{n}\sum_1^n(\hat{\theta}_{(i)} - \hat{\theta}_{(\cdot)})]^{1/2}$	30.1
Infinitesimal jackknife	$[\frac{1}{n^2}\sum_1^n U^2(x_i, \hat{F})]^{1/2}$	28.9
Parametric bootstrap	$[\mathrm{var}_{F_{\hat{\eta}}}\hat{\theta}^*]^{1/2}$	
Normal		29.2
Exponential		37.7
Fisher information	$[\dot{h}(\hat{\eta})^T i(\hat{\eta})^{-1}\dot{h}(\hat{\eta})]^{1/2}$	
Normal		28.9
Exponential		37.8
Sandwich	$[\dot{h}(\hat{\eta})^T i(\hat{\eta})^{-1}V i(\hat{\eta})^{-1}\dot{h}(\hat{\eta})]^{1/2}$	
	where $V = \sum_1^n \dot{\ell}(\hat{\eta}; x_i)\dot{\ell}(\hat{\eta}; x_i)^T$	
Normal		28.1
Exponential		28.1

21.9 The delta method

The delta method is a special technique for variance estimation that is applicable to statistics that are functions of observed averages. Suppose that we can write

$$\hat{\theta}(X_1, X_2, \ldots X_n) = r(\bar{Q}_1, \bar{Q}_2, \ldots \bar{Q}_A), \qquad (21.59)$$

where $r(\cdot, \cdot, \ldots \cdot)$ is a known function and

$$\bar{Q}_a = \frac{1}{n}\sum_1^n Q_a(X_i). \qquad (21.60)$$

The simplest example is the mean, for which $Q_a(X_i) = X_i$; for the correlation we take

$$r(\bar{Q}_1, \bar{Q}_2, \bar{Q}_3, \bar{Q}_4, \bar{Q}_5) = \frac{\bar{Q}_4 - \bar{Q}_1\bar{Q}_2}{[\bar{Q}_3 - \bar{Q}_1^2]^{1/2}[\bar{Q}_5 - \bar{Q}_2^2]^{1/2}} \qquad (21.61)$$

with $X = (Y, Z), Q_1(X) = Y, Q_2(X) = Z, Q_3(X) = Y^2, Q_4(X) = YZ, Q_5(X) = Z^2$.

The idea behind the delta method is the following. Suppose we have a random variable U with mean μ and variance σ^2, and we seek the variance of a one-to-one function of U, say $g(U)$. By expanding $g(t)$ in a one term Taylor series about $t = \mu$ we have

$$g(t) \approx g(\mu) + (t - \mu)g'(\mu) \qquad (21.62)$$

and this gives

$$\mathrm{var}(g(U)) \approx [g'(\mu)]^2 \sigma^2. \qquad (21.63)$$

Now if U itself is a sample mean so that estimates of its mean μ and variance σ^2 are readily available, we can use this formula to obtain a simple estimate of $\mathrm{var}(g(U))$.

The delta method uses a multivariate version of this argument. Suppose $(Q_1(X), \ldots Q_A(X))$ has mean vector $\boldsymbol{\mu}_F$. A multivariate Taylor series has the form

$$r(q_1, q_2 \ldots q_A) \approx r(\mu_1, \mu_2, \ldots \mu_A) + \sum_{1}^{A} (q_i - \mu_u) \frac{\partial r}{\partial q_i} \Big|_{q_i = \mu_i}, \qquad (21.64)$$

or in convenient vector notation

$$r(\mathbf{q}) \approx r(\boldsymbol{\mu}) + \bigtriangledown r^T (\mathbf{q} - \boldsymbol{\mu}). \qquad (21.65)$$

This gives

$$\mathrm{var}(r(\bar{\mathbf{Q}})) \approx \frac{\bigtriangledown r_F^T \boldsymbol{\Sigma}_F \bigtriangledown r_F}{n} \qquad (21.66)$$

where $\boldsymbol{\Sigma}_F$ is the variance-covariance matrix of a single observation $X \sim F$. We have put the subscript F on the quantities in (21.66) to remind ourselves that they depend on the unknown distribution F. The *nonparametric delta method* substitutes $\hat{F}$ for F in (21.66); this simply entails estimation of the first and second moments of X by their sample (plug-in) estimates:

$$\mathrm{var}^{ND}(r(\bar{\mathbf{Q}})) = \frac{\bigtriangledown r_{\hat{F}}^T \boldsymbol{\Sigma}_{\hat{F}} \bigtriangledown r_{\hat{F}}}{n}. \qquad (21.67)$$

The *parametric delta method* uses a parametric estimate $F_{\hat{\eta}}$ for F:

$$\mathrm{var}^{PD}(r(\bar{\mathbf{q}})) = \frac{\bigtriangledown r_{\hat{\eta}}^T \boldsymbol{\Sigma}_{F_{\hat{\eta}}} \bigtriangledown r_{\hat{\eta}}}{n}. \qquad (21.68)$$

In both the nonparametric and parametric versions, the fact that the statistic $\hat{\theta}$ is a function of sample means is the key aspect of the delta method.

21.9.1 Example: delta method for the mean

Here $Q_1(X) = X$, $r(q) = q$, $\nabla r_F = 1$, $\Sigma_F = \text{var}(X)$. The non-parametric delta method gives $\Sigma_{\hat{F}} = \sum_1^n (x_i - \bar{x})^2/n$, the plug-in estimate of variance and finally

$$\text{var}^{ND} \bar{X} = \sum_1^n (x_i - \bar{x})^2/n^2, \tag{21.69}$$

which equals the bootstrap or plug-in estimate of variance of the mean.

If we use a parametric estimate $F_{\hat{\eta}}$ then parametric delta method estimate is $\sigma^2(F_{\hat{\eta}})/n$. For example, if we assume X has an exponential distribution (21.31), then $\hat{\theta} = \bar{x}$, $\sigma^2(F_{\hat{\eta}}) = 1/\bar{x}^2$ and $\text{var}^{PD}(\bar{X}) = 1/(n\bar{x}^2)$.

21.9.2 Example: delta method for the correlation coefficient

Application of the delta method to the correlation coefficient (21.61) shows how quickly the calculations can get complicated. Here $X = (Y, Z)$, $Q_1(X) = Y$, $Q_2(X) = Z$, $Q_3(X) = Y^2$, $Q_4(X) = YZ$, $Q_5(X) = Z^2$. Letting $\beta_{ab} = E_F[(Y - E_F Y)^a (Z - E_F Z)^b]$, after a long calculation, (21.68) gives

$$\begin{aligned}
\text{var}^{ND} r(\bar{\mathbf{Q}}) &= \frac{\hat{\theta}^2}{4n} \Big[\frac{\hat{\beta}_{40}}{\hat{\beta}_{20}^2} + \frac{\hat{\beta}_{04}}{\hat{\beta}_{02}^2} + \frac{2\hat{\beta}_{22}}{\hat{\beta}_{20}\hat{\beta}_{02}} + \frac{4\hat{\beta}_{22}}{\hat{\beta}_{11}^2} \\
&\qquad - \frac{4\hat{\beta}_{31}}{\hat{\beta}_{11}\hat{\beta}_{20}} - \frac{4\hat{\beta}_{13}}{\hat{\beta}_{11}\hat{\beta}_{02}} \Big]
\end{aligned} \tag{21.70}$$

where each $\hat{\beta}_{ab}$ is a (plug-in) sample moment, for example $\hat{\beta}_{13} = \sum (y_i - \bar{y})(z_i - \bar{z})^3/n$. The parametric delta method would use a (bivariate) parametric estimate $F_{\hat{\eta}}$ and then $\hat{\beta}_{ab}$ would be the estimated moments from $F_{\hat{\eta}}$.

21.10 Relationship between the delta method and infinitesimal jackknife

The infinitesimal jackknife applies to general functional statistics while the delta method works only for functions of means. Interestingly, when the infinitesimal jackknife is applied to a function of means, it gives the same answer as the delta method:

RESULT 21.3. *Relationship between the nonparametric delta method and the infinitesimal jackknife estimates of variance: If $t(\hat{F})$ is a function of sample means, as in (21.59), then*

$$var^{IJ}t(\hat{F}) = var^{ND}t(\hat{F}).$$

Using this, the relationship of the nonparametric delta method to the nonparametric bootstrap can be inferred from Result 21.1. An analogous result holds in the parametric case:

RESULT 21.4. *Relationship between the parametric delta method and the Fisher information: If $t(F_{\hat{\eta}})$ is a function of sample means, as in (21.59), then*

$$var^{PD}t(F_{\hat{\eta}}) = \dot{h}(\hat{\boldsymbol{\eta}})^T i(\hat{\boldsymbol{\eta}})^{-1}\dot{h}(\hat{\boldsymbol{\eta}})$$

which is the estimated inverse Fisher information from (21.43). This in turn equals the parametric infinitesimal jackknife estimate of variance by Result 21.2.

The proofs of these results are given in the next section. The underlying basis lies in the theory of exponential families.

21.11   Exponential families

In an exponential family, the variance of the vector of sufficient statistics equals the Fisher information for the natural parameter. This fact leads to simple proofs of Results 21.3 and 21.4 for exponential families, as we detail below.

A random variable X is said to have a density in the exponential family if

$$g_\eta(x) = h_0(x)e^{\eta^T \mathbf{q}(x) - \psi(\eta)}. \tag{21.71}$$

Here $\mathbf{q}(x) = (q_1(x), q_2(x), \ldots q_A(x))^T$ is a vector of *sufficient statistics*, $h_0(x)$ is a fixed density called the *base measure* and $\psi(\boldsymbol{\eta})$ is a function that adjusts $g_\eta(x)$ so that it integrates to one for each value of $\boldsymbol{\eta}$. We think of (21.71) as a family of distributions passing through $h_0(x)$, with the parameter vector $\boldsymbol{\eta}$ indexing the family members. $\boldsymbol{\eta}$ is called the *natural parameter* of the family.

The first two derivatives of $\psi(\boldsymbol{\eta})$ are related to the moments of $\mathbf{q}(X)$:

$$\mathrm{E}[\mathbf{q}(X)] \quad = \quad \psi'(\boldsymbol{\eta})$$

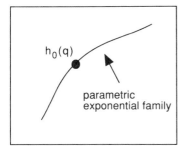 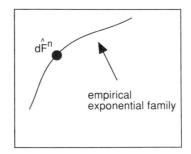

Figure 21.6. *Left panel: schematic of the exponential family for $\bar{\mathbf{q}}$. Right panel: empirical exponential family for $\bar{\mathbf{q}}$.*

$$\mathrm{var}[\mathbf{q}(X)] = \psi''(\boldsymbol{\eta}). \qquad (21.72)$$

As an example, if $X \sim N(\mu, 1)$, the reader is asked in Problem 21.8 to show that $g_\eta(x)$ can be written in the form (21.71) with

$$\boldsymbol{\eta} = \mu, \qquad (21.73)$$

$$h_0(x) = \frac{1}{\sqrt{2\pi}} e^{-\frac{1}{2}x^2}, \qquad (21.74)$$

$$q_1(x) = x, \qquad (21.75)$$

and

$$\psi(\boldsymbol{\eta}) = \frac{1}{2}\mu^2. \qquad (21.76)$$

If $X_1, X_2, \ldots X_n$ is a sample from an exponential family, the density of the sufficient statistics also has an exponential family form. Specifically, if $\bar{\mathbf{Q}} = (\sum_1^n q_1(X_i)/n, \sum_1^n q_2(X_i)/n, \ldots, \sum_1^n q_A(X_i)/n)^T$ then the density of $\bar{\mathbf{Q}}$ is

$$h_1(\bar{\mathbf{q}})e^{n[\boldsymbol{\eta}^T\bar{\mathbf{q}} - \psi(\boldsymbol{\eta})]} \qquad (21.77)$$

where $h_1(\bar{\mathbf{q}})$ is derived from $h_0(x)$ (Problem 21.9). This family is depicted in the left panel of Figure 21.6.

The maximum likelihood estimate of $\boldsymbol{\eta}$ satisfies the set of equations

$$\bar{\mathbf{q}} = \psi'(\hat{\boldsymbol{\eta}}) = \mathrm{E}_{\hat{\boldsymbol{\eta}}}(\bar{\mathbf{Q}}). \qquad (21.78)$$

In other words, the maximum likelihood estimate is the value of $\boldsymbol{\eta}$ that makes $\bar{\mathbf{q}}$ equal to its expectation under the model. The solution to these equations has the form

$$\hat{\boldsymbol{\eta}} = \mathbf{k}(\bar{\mathbf{q}}), \tag{21.79}$$

where $\mathbf{k}(\cdot)$ is the inverse of $\psi'(\hat{\boldsymbol{\eta}})$. Furthermore, the Fisher information for $\boldsymbol{\eta}$ is

$$i(\boldsymbol{\eta}) = n \cdot \psi''(\boldsymbol{\eta}) = n^2 \cdot \mathrm{var}(\bar{\mathbf{Q}}). \tag{21.80}$$

Usually, our interest is not in $\boldsymbol{\eta}$ but in the real-valued parameter $\theta = h(\boldsymbol{\eta})$. The maximum likelihood estimate of θ is

$$\hat{\theta} = h(\hat{\boldsymbol{\eta}}) = h(\mathbf{k}(\bar{\mathbf{q}})). \tag{21.81}$$

Finally, we get to the main point of this section. The inverse of the estimated Fisher information for θ is

$$\frac{\dot{h}^T [\psi''(\hat{\boldsymbol{\eta}})]^{-1} \dot{h}}{n}. \tag{21.82}$$

The parametric delta method, on the other hand, begins with $\bar{\mathbf{Q}}$ having variance $\psi''(\boldsymbol{\eta})/n$, and applies the transformation $h(\mathbf{k}(\cdot))$. Letting $\mathbf{K}$ be the matrix of derivatives of $\mathbf{k}$, the parametric delta method estimate of variance is equal to

$$
\begin{aligned}
\dot{h}^T \mathbf{K}^T \mathrm{var}(\bar{\mathbf{Q}}) \mathbf{K} \dot{h} &= \frac{\dot{h}^T \mathbf{K}^T \psi''(\hat{\boldsymbol{\eta}}) \mathbf{K} \dot{h}}{n} \\
&= \frac{\dot{h}^T [\psi''(\hat{\boldsymbol{\eta}})]^{-1} \dot{h}}{n}, \tag{21.83}
\end{aligned}
$$

since $\mathbf{K} = [\psi''(\hat{\boldsymbol{\eta}})]^{-1}$. Hence the parametric delta method estimate of variance equals the inverse of the Fisher information (Result 21.4).

In order to draw the same analogy in the nonparametric case, we need to define a family of distributions whose Fisher information for the sufficient statistics is the plug-in estimate. The appropriate family is called the *empirical exponential family*:

$$g_{\boldsymbol{\eta}}(\bar{\mathbf{q}}) = h_1(\bar{\mathbf{q}}) e^{\boldsymbol{\eta}^T \bar{\mathbf{q}} - \psi(\boldsymbol{\eta})} \tag{21.84}$$

defined with respect to the distribution $\hat{F}^n$, the product distribution of n independent copies of the empirical distribution $\hat{F}$. The parameter vector $\boldsymbol{\eta}$ corresponds to a distribution $F_{\boldsymbol{\eta}}$ that puts probability mass $g_{\boldsymbol{\eta}}(\mathbf{q}^*)/n^n$ on each of the n^n data sets $\mathbf{x}^* =$

$(x_1^*, x_2^*, \ldots x_n^*)^T$ where each x_i^* equals one of the original x_i data points, $\bar{\mathbf{q}}^* = \sum_1^n \mathbf{q}(x_j^*)/n$ and

$$dF_\eta(\mathbf{x}^*) = e^{\boldsymbol{\eta}^T \mathbf{q}^* - \psi(\boldsymbol{\eta})} d\hat{F}^n(\mathbf{x}^*). \qquad (21.85)$$

The value $\boldsymbol{\eta} = 0$ corresponds to the empirical distribution $\hat{F}^n$. The normalizing constant $\psi(\boldsymbol{\eta})$ is easily seen to equal $n \log(\sum_1^n e^{\boldsymbol{\eta}^T \mathbf{q}_k/n}/n)$ and hence the Fisher information for $\boldsymbol{\eta}$, evaluated at $\boldsymbol{\eta} = 0$ is

$$n \cdot \psi^{''}(0) = \sum_1^n (\mathbf{q}_k - \bar{\mathbf{q}})(\mathbf{q}_k - \bar{\mathbf{q}})^T/n^2 \qquad (21.86)$$

(Problem 21.10). Since (21.86) is the plug-in estimate of variance for the $\mathbf{q}_i$s, it is clear that the inverse Fisher information for a parameter $\theta = h(\boldsymbol{\eta})$ in this family, and the infinitesimal jackknife estimate of variance both equal the nonparametric delta estimate of variance. This proves result 21.3 in the exponential family case. A proof for the general case appears in Efron (1982, chapter 6).

21.12 Bibliographic notes

Functional statistics are fundamental to the theory of robust statistical inference, and are discussed in Huber (1981), Fernholz (1983), and Hampel *et al.* (1986). The infinitesimal jacknife was introduced by Jaeckel (1972), while the influence curve is proposed in Hampel (1974). The sandwich estimator is described in White (1981, 1982), Kent (1982), Royall (1986), and given its name by Lin and Wei (1989). A non-technical overview of maximum-likelihood inference is given by Silvey (1975). Lehmann (1983) gives a more mathematically sophisticated discussion. Cox and Hinkley (1974) provide a broad overview of inference. The delta method is discussed in chapter 6 of Efron (1982), where most of the results of this chapter are proven. Basic theory of exponential families is outlined in Lehmann (1983); their use in the bootstrap context may be found in Efron (1981, 1987). The justification of the empirical distribution function as a nonparametric maximum likelihood estimate was studied by Kiefer and Wolfowitz (1956) and Scholz (1980). The overview in this chapter was inspired by Stanford class notes developed by Andreas Buja.

21.13 Problems

21.1 Verify equation (21.10) and hence show that the unbiased estimate of variance is not a statistical functional.

21.2 Derive equations (21.15) and (21.16) for the influence functions of the mean and median.

21.3 Show that under appropriate regularity conditions $\mathrm{E}_F U(x, F) = 0$.

21.4 Prove that the empirical distribution function $\hat{F}$ maximizes $L(F) = \prod_1^n F(\{x_i\})$ and therefore, in this sense, is the nonparametric maximum likelihood estimate of F.

21.5 Derive equation (21.45) for the model-based form of the influence component.

21.6 Prove identity (21.47) relating the expected value of the squared score and the Fisher information.

21.7 Derive explicit expressions for the estimators in Table 21.1, and evaluate them for the mouse data. Verify the values in Table 21.1.

21.8 Show that the normal distribution has an exponential family form with components given by (21.76).

21.9 Show that the function $h_1(\bar{\mathbf{q}})$ in the exponential family (21.77) is the sum of $\prod_1^n h_0(x_i)$ over all $(x_1, x_2, \ldots x_n)$ for which $\sum_1^n q_1(x_i)/n = \bar{q}_1, \sum_1^n q_2(x_i)/n = \bar{q}_2, \ldots \sum_1^n q_A(x_i)/n = \bar{q}_A$.

21.10 Derive the form of $\psi(\boldsymbol{\eta})$ given above equation (21.86) and derive equation (21.86) for the Fisher information in the empirical exponential family.

## CHAPTER 22

Further topics in bootstrap confidence intervals

22.1 Introduction

Chapters 12–14 describe some methods for confidence interval construction using the bootstrap. In fact confidence intervals have received the most theoretical study of any topic in the bootstrap area. A full discussion of this theory would be beyond the scope and intent of this book. In this chapter we give the reader a heuristic description of some of the theory of confidence intervals, describe the underlying basis for the BC_a interval and discuss a computationally useful approximation to the BC_a interval called the "ABC" method.

22.2 Correctness and accuracy

Suppose we have a real-valued parameter of interest θ for which we would like a confidence interval. Rather than consider the two endpoints of the interval simultaneously, it is convenient to consider a single endpoint $\hat{\theta}[\alpha]$, with intended one-sided coverage α:

$$\text{Prob}(\theta \le \hat{\theta}[\alpha]) \approx \alpha \qquad (22.1)$$

for all α. First let's review some standard terminology. An approximate confidence point $\hat{\theta}[\alpha]$ is called *first order accurate* if

$$\text{Prob}(\theta \le \hat{\theta}[\alpha]) = \alpha + O(n^{-1/2}) \qquad (22.2)$$

and *second order accurate* if

$$\text{Prob}(\theta \le \hat{\theta}[\alpha]) = \alpha + O(n^{-1}), \qquad (22.3)$$

where the probabilities apply to the true population or distribution. Standard normal and Student's t intervals, described in Chap-

ter 12, are first order accurate but not second order accurate unless
the true distribution is normal. Some bootstrap methods provide
second order accurate intervals no matter what the true distribution may be.

Distinct from interval accuracy is the notion of *correctness*. This
refers to how closely a candidate confidence point matches an ideal
or exact confidence point. Let $\hat{\theta}_{\text{exact}}[\alpha]$ be an exact confidence point
that satisfies $\text{Prob}(\theta \le \hat{\theta}_{\text{exact}}[\alpha]) = \alpha$. A confidence point $\hat{\theta}[\alpha]$ is
called *first order correct* if

$$\hat{\theta}[\alpha] = \hat{\theta}_{\text{exact}}[\alpha] + O_p(n^{-1}) \tag{22.4}$$

and *second order correct* if

$$\hat{\theta}[\alpha] = \hat{\theta}_{\text{exact}}[\alpha] + O_p(n^{-3/2}). \tag{22.5}$$

Equivalently, a confidence point $\hat{\theta}[\alpha]$ is called *first order correct*
if

$$\hat{\theta}[\alpha] = \hat{\theta}_{\text{exact}}[\alpha] + O_p(n^{-1/2}) \cdot \hat{\sigma} \tag{22.6}$$

and *second order correct* if

$$\hat{\theta}[\alpha] = \hat{\theta}_{\text{exact}}[\alpha] + O_p(n^{-1}) \cdot \hat{\sigma} \tag{22.7}$$

where $\hat{\sigma}$ is any reasonable estimate of the standard error of $\hat{\theta}$. Since
$\hat{\sigma}$ itself is usually of order $n^{-1/2}$, (22.4) and (22.5) agree with (22.6)
and (22.7) respectively.

A fairly simple argument shows that correctness at a given order
implies accuracy at that order. In situations where exact endpoints
can be defined, standard normal and Student's t points are only
first order correct while some bootstrap methods produce second
order correct confidence points.

22.3 Confidence points based on approximate pivots

A convenient framework for studying bootstrap confidence points
is the "smooth function of means model." We assume that our
data are n independent and identically distributed random variables $X_1, X_2, \ldots X_n \sim F$. They may be real or vector-valued. Let
$E(X_i) = \mu$, and assume that our parameter of interest θ is some
smooth function of μ, that is, $\theta = f(\mu)$. If $\bar{X} = \sum_1^n X_i/n$, then
our estimate of θ is $\hat{\theta} = f(\bar{X})$. Letting $\text{var}(\hat{\theta}) = \tau^2/n$, we further
assume that $\tau^2 = g(\mu)$ for some smooth function g. The sample es-
timate of τ^2 is $\hat{\tau}^2 = g(\bar{X})$. This framework covers many commonly

occurring problems, including inference for the mean, variance and correlation, and exponential family models.

Our discussion will not be mathematically rigorous. Roughly speaking, we require appropriate regularity conditions to ensure that the central limit theorem can be applied to $\hat{\theta}$.

To begin, we consider four quantities:

$$P = \sqrt{n}(\hat{\theta} - \theta); \qquad Q = \sqrt{n}(\hat{\theta} - \theta)/\hat{\tau};$$
$$\hat{P} = \sqrt{n}(\hat{\theta}^* - \hat{\theta}); \qquad \hat{Q} = \sqrt{n}(\hat{\theta}^* - \hat{\theta})/\hat{\tau}^* \qquad (22.8)$$

Here $\hat{\theta}^*$ and $\hat{\tau}^*$ are $\hat{\theta}$ and $\hat{\tau}$ applied to a bootstrap sample. If F were known, exact confidence points could be based on the distribution of P or Q. Let $H(x)$ and $K(x)$ be the distribution functions of P and Q respectively, when sampling from F, and let $x^{(\alpha)} = H^{-1}(\alpha)$ and $y^{(\alpha)} = K^{-1}(\alpha)$ be the α-level quantiles of $H(x)$ and $K(x)$. Then the exact confidence points based on the pivoting argument for P

$$H(x) = \text{Prob}\left\{n^{1/2}(\hat{\theta} - \theta) \le x\right\} = \text{Prob}\left\{\theta \ge \hat{\theta} - n^{-1/2}x\right\} \quad (22.9)$$

(and similarly for Q) are

$$\hat{\theta}_{\text{uns}}[\alpha] = \hat{\theta} - n^{-1/2}x^{(1-\alpha)} \qquad (22.10)$$

$$\hat{\theta}_{\text{Stud}}[\alpha] = \hat{\theta} - n^{-1/2}\hat{\tau}y^{(1-\alpha)}. \qquad (22.11)$$

The first point is the "un-Studentized" point based on P, while the second is the "bootstrap-t" or Studentized point based on Q. Notice that a standard normal point has the form of $\hat{\theta}_{\text{Stud}}$ with the normal quantile $z^{(1-\alpha)}$ replacing $y^{(1-\alpha)}$, while the usual t interval uses the α-quantile of the Student's t distribution on $n-1$ degrees of freedom.

Of course F is usually unknown. The bootstrap uses $\hat{H}$ and $\hat{K}$, the distributions of $\hat{P}$ and $\hat{Q}$ under the estimated population $\hat{F}$, to estimate H and K. If $\hat{x}^{(\alpha)} = \hat{H}^{-1}(\alpha)$ and $\hat{y}^{(\alpha)} = \hat{K}^{-1}(\alpha)$, the estimated points are

$$\hat{\theta}_{\text{UNS}}[\alpha] = \hat{\theta} - n^{-1/2}\hat{x}^{(1-\alpha)} \qquad (22.12)$$

$$\hat{\theta}_{\text{STUD}}[\alpha] = \hat{\theta} - n^{-1/2}\hat{\tau}\hat{y}^{(1-\alpha)}. \qquad (22.13)$$

$\hat{\theta}_{\mathrm{STUD}}[\alpha]$ is the bootstrap-t endpoint discussed in Chapter 12. Some important results for these confidence points have been derived:

$$\hat{\theta}_{\mathrm{UNS}} = \hat{\theta}_{\mathrm{uns}} + O_p(n^{-1}); \quad \hat{H}(x) = H(x) + O_p(n^{-1/2}), \quad (22.14)$$

$$\hat{\theta}_{\mathrm{STUD}} = \hat{\theta}_{\mathrm{Stud}} + O_p(n^{-3/2}); \quad \hat{K}(x) = K(x) + O_p(n^{-1}). \quad (22.15)$$

In other words, the confidence point $\hat{\theta}_{\mathrm{STUD}}$ based on Q is second order accurate, and if we consider $\hat{\theta}_{\mathrm{Stud}}$ to be the "correct" interval, it is second order correct as well: it differs from $\hat{\theta}_{\mathrm{Stud}}$ by a term of size $O(n^{-3/2})$. The confidence point $\hat{\theta}_{\mathrm{UNS}}$ based on P is only first accurate. Interestingly, in order for the bootstrap to improve upon the standard normal procedure it should be based on a studentized quantity, at least when it is used in this simple way.

A fairly simple argument shows why studentization is important. Under the usual regularity conditions, the first four cumulants of P are

$$\begin{aligned}
\mathrm{E}(P) &= \frac{f_1(\theta)}{\sqrt{n}} + O(n^{-1}), \\
\mathrm{var}(P) &= f_2(\theta) + O(n^{-1}), \\
\mathrm{skew}(P) &= \frac{f_3(\theta)}{\sqrt{n}} + O(n^{-3/2}), \\
\mathrm{kurt}(P) &= O(n^{-1})
\end{aligned} \qquad (22.16)$$

while those of Q are

$$\begin{aligned}
\mathrm{E}(Q) &= \frac{f_4(\theta)}{\sqrt{n}} + O(n^{-1}), \\
\mathrm{var}(Q) &= 1 + O(n^{-1}), \\
\mathrm{skew}(Q) &= \frac{f_5(\theta)}{\sqrt{n}} + O(n^{-3/2}), \\
\mathrm{kurt}(Q) &= O(n^{-1}).
\end{aligned} \qquad (22.17)$$

The functions $f_1(\theta)$, $f_2(\theta)$, $f_3(\theta)$, $f_4(\theta)$ and $f_5(\theta)$ depend on θ but not n. In the above, "skew" and "kurt" are the standardized skewness and kurtosis $\mathrm{E}(\mu_3)/[\mathrm{E}(\mu_2)]^{3/2}$ and $\mathrm{E}(\mu_4)/[\mathrm{E}(\mu_2)]^2 - 3$, respectively, with μ_r the rth central moment. All other cumulants are $O(n^{-1})$ or smaller. Note that $\mathrm{var}(Q)$ does not involve any function $f_i(\theta)$. For details, see DiCiccio and Romano (1988) or Hall (1988a).

The use of $\hat{H}$ and $\hat{K}$ to estimate H and K is tantamount to substituting $\hat{\theta}$ for θ in these functions, and results in an error of

$O_p(n^{-1/2})$, that is $f_1(\hat\theta) = f_1(\theta) + O_p(n^{-1/2})$, $f_2(\hat\theta) = f_2(\theta) + O_p(n^{-1/2})$, etc. When these are substituted into the (22.16), the expectation, standardized skewness, and kurtosis of $\hat P$ are only $O(n^{-1})$ away from the corresponding cumulants of P, but

$$\text{var}(\hat P) = \text{var}(P) + O(n^{-1/2}). \qquad (22.18)$$

This causes the confidence point based on $\hat P$ to be only first order accurate. On the other hand, $\text{var}(Q) = 1 + O(n^{-1})$, $\text{var}(\hat Q) = 1 + O(n^{-1})$ so we do not incur an $O(n^{-1/2})$ error in estimating it. As a result, the confidence point based on $\hat Q$ is second order accurate.

22.4 The BC$_a$ interval

The α-level endpoint of the BC$_a$ interval, described in Chapter 14, is given by

$$\hat\theta_{\text{BC}_a}[\alpha] = \hat G^{-1}\left(\Phi(\hat z_0 + \frac{\hat z_0 + z^{(\alpha)}}{1 - \hat a(\hat z_0 + z^{(\alpha)})})\right), \qquad (22.19)$$

where $\hat G$ is the cumulative distribution function of the bootstrap replications $\hat\theta^*$, "$\hat z_0$" and "$\hat a$" are the bias and acceleration adjustments, and Φ is the cumulative distribution function of the standard normal distribution. It can be shown that the BC$_a$ interval is also second order accurate,

$$\text{Prob}(\theta \leq \hat\theta_{\text{BC}_a}[\alpha]) = \alpha + O(n^{-1}). \qquad (22.20)$$

In addition, the bootstrap-t endpoint and BC$_a$ endpoint agree to second order:

$$\hat\theta_{\text{BC}_a}[\alpha] = \hat\theta_{\text{STUD}}[\alpha] + O_p(n^{-3/2}), \qquad (22.21)$$

so that by the definition of correctness adopted in the previous section, $\hat\theta_{\text{BC}_a}$ is also second order correct. A proof of these facts is based on Edgeworth expansions of $H(x)$ and $K(x)$, and may be found in Hall (1988a).

Although the bootstrap-t and BC$_a$ procedures both produce second order valid intervals, a major advantage of the BC$_a$ procedure is its transformation-respecting property. The BC$_a$ interval for a parameter $\phi = m(\theta)$, based on $\hat\phi = m(\hat\theta)$ (where m is a monotone increasing mapping) is equal to $m(\cdot)$ applied to the endpoints of the BC$_a$ interval for θ based on $\hat\theta$. The bootstrap-t procedure is not transformation-respecting, and can work poorly if applied on the wrong scale. Generally speaking, the bootstrap-t works well for location parameters. The practical difficulty in applying it is to identify the transformation $h(\cdot)$ that maps the problem to a location form. One approach to this problem is the automatic variance stabilization technique described in chapter 12. The interval resulting from this technique is also second order correct and accurate.

22.5 The underlying basis for the BC_a interval

Suppose that we have our estimate $\hat{\theta}$ and have obtained an estimated standard error $\widehat{se}$ for $\hat{\theta}$, perhaps from bootstrap calculations. The BC_a interval is based on the following model. We assume that there is an increasing transformation such that $\phi = m(\theta), \hat{\phi} = m(\hat{\theta})$ gives

$$\frac{\hat{\phi} - \phi}{se_\phi} \sim N(-z_0, 1) \quad \text{with}$$

$$se_\phi = se_{\phi_0} \cdot [1 + a(\phi - \phi_0)]. \tag{22.22}$$

Here ϕ_0 is any convenient reference point on the scale of ϕ values. Notice that (22.22) is a generalization of the usual normal approximation

$$\frac{\hat{\theta} - \theta}{se} \sim N(0, 1). \tag{22.23}$$

The generalization involves three components that capture deviations from the ideal model (22.23): the *transformation* $m(\cdot)$, the *bias correction* z_0 and the *acceleration* a.

As described in Chapter 13, the percentile method generalizes the normal approximation (22.23) by allowing a transformation $m(\cdot)$ of θ and $\hat{\theta}$. The BC_a method adds the further adjustments z_0 and a, both of which are $O_p(n^{-1/2})$ in magnitude. The bias correction z_0 accounts for possible bias in $\hat{\phi}$ as an estimator of ϕ, while the acceleration constant a accounts for the possible change in the standard deviation of $\hat{\phi}$ as ϕ varies.

Why is model (22.22) a reasonable choice? It turns out that in a large class of problems, (22.22) holds to second order; that is, the error in the approximation (22.22) is typically $O_p(n^{-1})$. In contrast, the error in the normal approximation (22.23) is $O_p(n^{-1/2})$ in general. This implies that confidence intervals constructed using assumption (22.22) will typically be second order accurate and correct. All three components in (22.22) are needed to reduce the error to $O_p(n^{-1})$.

Now suppose that the model (22.22) holds exactly. Then an exact upper $1 - \alpha$ confidence point for ϕ can be shown to be

$$\phi[\alpha] = \hat{\phi} + \mathrm{se}_{\hat{\phi}} \frac{z_0 + z^{(\alpha)}}{1 - a(z_0 + z^{(\alpha)})}. \tag{22.24}$$

Let G be the cumulative distribution function of $\hat{\theta}$. Then if we map the endpoint $\phi[\alpha]$ back to the θ scale via the inverse transformation $m^{-1}(\cdot)$, we obtain

$$\theta[\alpha] = G^{-1}\left(\Phi\left(z_0 + \frac{z_0 + z^{(\alpha)}}{1 - a(z_0 + z^{(\alpha)})}\right)\right). \tag{22.25}$$

This is exactly the BC$_a$ endpoint defined in Chapter 14 and equation (22.19), except that it involves the theoretical quantities z_0, a and G rather than estimates.

The distribution G can be estimated by the bootstrap cumulative distribution function $\hat{G}$; depending on the situation, this would be obtained from either parametric or nonparametric bootstrap sampling. Letting Z be a standard normal variate, with cumulative distribution function Φ, the estimate of z_0 is obtained from

$$\mathrm{Prob}_\theta\{\hat{\theta} < \theta\} = \mathrm{Prob}_\phi\{\hat{\phi} < \phi\} = \mathrm{Prob}\{Z < z_0\} = \Phi(z_0). \tag{22.26}$$

Substituting $\theta = \hat{\theta}$ gives

$$\begin{aligned} \hat{z}_0 &= \Phi^{-1}(\mathrm{Prob}_{\hat{\theta}}\{\hat{\theta}^* < \hat{\theta}\}) \\ &= \Phi^{-1}\left(G(\hat{\theta})\right). \end{aligned} \tag{22.27}$$

This is the formula used in Chapter 14; notice that $\hat{z}_0$ measures the *median bias* of $\hat{\theta}$.

The acceleration constant "a" always has the meaning given in (22.22): it measures the rate of change of the standard error on a normalized scale. This sounds difficult to compute, but it is in fact easier to get a good estimate for "a" than for z_0. Here are some convenient formulas. In one-parameter models, a good approximation for a is

$$\hat{a} = \frac{1}{6}\mathrm{skew}_{\theta=\hat{\theta}}(\dot{\ell}_\theta), \tag{22.28}$$

where $\dot{\ell}_\theta$ is the score function. In one parameter models, it turns out that $\hat{a}$ and $\hat{z}_0$ are equal to second order, $\hat{a} = \hat{z}_0 + O_p(n^{-1})$.

In multiparameter models, a is estimated by reducing to the one-parameter *least favorable family* and then applying formula (22.28). We won't give details here, although we discuss least favorable families in Section 22.7. For the multinomial distribution, which corresponds to the nonparametric bootstrap, the resulting formula is

$$\hat{a} = \frac{\sum_{i=1}^n U_i^3}{6\{\sum_{i=1}^n U_i^2\}^{3/2}}. \tag{22.29}$$

where U_i is the ith infinitesimal jackknife value (or empirical influence component) $U(x_i, \hat{F})$ defined in (21.3). Alternatively, we may use the ith jackknife value, as in equation (14.15) of Chapter 14. This avoids having to explicitly define $\hat{\theta}$ as a functional statistic, and is done in the S function bcanon given in the Appendix. Note that $\hat{z}_0$ and $\hat{a}$ do not agree to second order in multiparameter models, as z_0 now includes a component that measures the curvature of the level surfaces of $\hat{\theta}$. Some more details on this point are given in the next section.

22.6 The ABC approximation

The computational burden for bootstrap intervals can be an obstacle, especially if the interval is to computed repeatedly. We describe next a useful approximation to the BC_a interval which replaces bootstrap sampling with numerical derivatives. It is called the "ABC" procedure for approximate bootstrap confidence interval or approximate BC_a interval and is applicable in exponential families and nonparametric problems using the multinomial distribution. We will define the ABC interval in the nonparametric case, and then show how it can be viewed as an approximation to the BC_a interval. S language programs for the ABC intervals appear in the Appendix.

Having observed $\mathbf{x} = (x_1, x_2, \cdots x_n)$, we assume a multinomial distribution with support on the observed data. Formally, if we denote the resampling vector by $\mathbf{P}^*$, we assume that $n\mathbf{P}^*$ has a multinomial distribution with success probabilities $\mathbf{P}^0 = (1/n, 1/n, \cdots, 1/n)^T$. Our statistic has the form

$$\hat{\theta} = T(\mathbf{P}). \tag{22.30}$$

The delta method approximation for the standard error of $\hat{\theta}$ (discussed in Chapter 21) is

$$\hat{\sigma} = \left(\sum_{i=1}^{n} \dot{T}_i^2 / n^2\right)^{1/2}, \tag{22.31}$$

where $\dot{T}$ is the empirical influence component

$$\dot{T}_i = \lim_{\epsilon \to 0} \frac{T((1-\epsilon)\mathbf{P}^0 + \epsilon\mathbf{e}_i) - T(\mathbf{P}^0)}{\epsilon}, \tag{22.32}$$

and $\mathbf{e}_i$ is the ith coordinate vector $(0, 0, \cdots, 0, 1, 0, \cdots, 0)^T$. This is the same definition as (20.20).

Let $\hat{\theta}[1-\alpha]$ indicate the endpoint of an approximate $100(1-\alpha)\%$ one-sided upper confidence interval for θ. Then $(\hat{\theta}[\alpha], \hat{\theta}[1-\alpha])$ is an approximate $100(1-2\alpha)\%$ two-sided interval.

The ABC confidence limit for θ, denoted $\hat{\theta}_{\mathrm{ABC}}[1-\alpha]$, is constructed as follows:

$$w \equiv \hat{z}_0 + z^{(1-\alpha)}, \qquad \lambda \equiv w/(1-\hat{a}w)^2, \qquad \hat{\delta} \equiv \dot{T}(\mathbf{P}^0),$$
$$\hat{\theta}_{\mathrm{ABC}}[1-\alpha] = T(\mathbf{P}^0 + \lambda\hat{\delta}/\hat{\sigma}). \tag{22.33}$$

The direction $\hat{\delta}$ is called the *least favorable direction* and is discussed in section 22.7 below. The big advantage of the ABC procedure is that the constants $\hat{z}_0$ and $\hat{a}$ can be computed in terms of numerical second derivatives, and hence no resampling is needed. The acceleration constant $\hat{a}$ is $1/6$ times the standardized skewness of the empirical influence components:

$$\hat{a} = \frac{1}{6} \frac{\sum_{i=1}^{n} \dot{T}_i^3}{(\sum_{i=1}^{n} \dot{T}_i^2)^{3/2}}. \tag{22.34}$$

This is the same as formula (22.29). The estimate of z_0 involves two quantities. The first is the bias $b = \mathrm{E}(\hat{\theta}) - \theta$. A quadratic Taylor series expansion of $\theta = T(\mathbf{P}^0)$ gives approximate bias $\hat{b}$,

$$\hat{b} = \sum_{i=1}^{n} \ddot{T}_i / (2n^2), \tag{22.35}$$

where $\ddot{T}_i$ is an element of the second order influence function,

$$\ddot{T}_i = \lim_{\epsilon \to 0} \frac{T((1-\epsilon)\mathbf{P}^0 + \epsilon\mathbf{e}_i) - 2T(\mathbf{P}^0) + T((1-\epsilon)\mathbf{P}^0 - \epsilon\mathbf{e}_i)}{\epsilon^2}. \tag{22.36}$$

The second quantity needed for z_0 is the quadratic coefficient $\hat{c}_q$,

$$\hat{c}_q =$$

$$\lim_{\epsilon \to 0} \frac{T[(1-\epsilon)\mathbf{P}^0 + \epsilon\dot{T}/(n^2\hat{\sigma})] - 2T(\mathbf{P}^0) + T[(1-\epsilon)\mathbf{P}^0 - \epsilon\dot{T}/(n^2\hat{\sigma})]}{\epsilon^2}.$$

$$(22.37)$$

This coefficient measures the nonlinearity of the function $\theta = T(\mathbf{P})$ as we move in the least favorable direction. Let $\theta(\lambda) \equiv T(\hat{\mathbf{P}} + \lambda\hat{\delta}/\hat{\sigma})$. A quadratic Taylor series expansion gives

$$\theta(\lambda) \doteq \hat{\theta} + \hat{\sigma}(\lambda + \hat{c}_q\lambda^2);\tag{22.38}$$

$\hat{c}_q$ measures the ratio of the quadratic term to the linear term in $\{\theta(\lambda) - \hat{\theta}\}/\hat{\sigma}$. The size of $\hat{c}_q$ does not affect the standard intervals, which treat every function $T(\mathbf{P})$ as if it were linear, but it has an important effect on more accurate confidence intervals.

The bias correction constant $\hat{z}_0$ is a function of $\hat{a}, \hat{b},$ and $\hat{c}_q$. These three constants are approximated by using a small value of ϵ in formulas (22.34), (22.36), and (22.37). Then we define

$$\hat{\gamma} = \hat{b}/\hat{\sigma} - \hat{c}_q,\tag{22.39}$$

and estimate z_0 by

$$\begin{aligned}\hat{z}_0 &\doteq \Phi^{-1}\{2 \cdot \Phi(\hat{a}) \cdot \Phi(-\hat{\gamma})\} \\ &\doteq \hat{a} - \hat{\gamma}.\end{aligned}\tag{22.40}$$

It can be shown that $\hat{\gamma}$ is the total curvature of the level surface $\{\mathbf{P} : T(\mathbf{P}) = \hat{\theta}\}$: the greater the curvature, the more biased is $\hat{\theta}$. In equation (22.27) we gave as the definition of $\hat{z}_0$,

$$\hat{z}_0 = \hat{G}^{-1}(\Phi(\hat{\theta})),\tag{22.41}$$

where $\hat{G}$ is the cumulative distribution function of $\hat{\theta}^*$. Either form of $\hat{z}_0$ approximates z_0 sufficiently well to preserve the second order accuracy of the BC_a formulas. The definition of z_0 is more like a median bias than a mean bias, which is why $\hat{z}_0$ involves quantities other than $\hat{b}$.

A further approximation gives a computationally more convenient form of the ABC endpoint. *The quadratic ABC confidence limit for θ*, denoted $\theta_{\text{ABC}_q}[1-\alpha]$, is constructed from $(\hat{\theta}, \hat{\sigma}, \hat{a}, \hat{z}_0, \hat{c}_q)$

and $z^{(\alpha)} = \Phi^{-1}(\alpha)$ as follows:

$$w \equiv z_0 + z^{(1-\alpha)}, \qquad \lambda \equiv w/(1 - \hat{a}w)^2, \qquad \xi \equiv \lambda + c_q\lambda^2,$$
$$\hat{\theta}_{\mathrm{ABC_q}}[1 - \alpha] = \hat{\theta} + \hat{\sigma}\xi. \qquad (22.42)$$

This definition follows from a quadratic Taylor series expansion for $T(\mathbf{P}^0 + \lambda\hat{\delta}/\hat{\sigma})$ (Problem 22.2).

The ABC interval can be derived as an approximation to the BC_a interval. The $1 - \alpha$ endpoint of the BC_a interval is defined in equation (22.19). A two-term Cornish Fisher expansion for $\hat{G}$ has the form

$$\hat{G}^{-1}(1 - \beta) = \hat{\theta} + \hat{b} + \hat{\sigma}[z^{(1-\beta)} + (\hat{a} + \hat{c}_q)((z^{(1-\beta)})^2 - 1)]. \qquad (22.43)$$

Applying approximation (22.43) to the BC_a interval in the least favorable family $\mathbf{P}^0 + \tau\hat{\delta}$ gives endpoints $\mathbf{P}^0 + \lambda\hat{\delta}/\hat{\sigma}$; (Problem 22.3); transforming these by the function $T(\cdot)$ gives definition (22.33).

Here is a summary of the computational effort required for the ABC intervals. The algorithm begins by numerically evaluating $\dot{T} = \dot{T}(\hat{\mathbf{P}})$. This requires $2n$ recomputations of $T(\cdot)$, 2 for each of the first derivatives $\partial T(\mathbf{P})/\partial \mathbf{P}_i|_{P=\hat{P}} \dot{=} \{T(\hat{\mathbf{P}} + \epsilon e_i) - T(\hat{\mathbf{P}} - \lambda e_i)\}/2\epsilon$, e_i being the ith coordinate vector. The vector $\dot{T}$ gives $\hat{\sigma} = \sum\{\dot{T}_i^2/n^2\}^{1/2}$. Then the $n + 2$ second derivatives in (22.35) and (22.36) are calculated, each requiring 2 recomputations of $T(\cdot)$. Altogether $4n + 4$ recomputations of $T(\cdot)$ are required to compute the quadratic ABC limits (22.42), compared with the $2n$ recomputations necessary for numerically evaluating the standard normal interval $\hat{\theta} \pm z^{(1-\alpha)}\hat{\sigma}$. In complicated situations the recomputations of $T(\cdot)$ dominate calculational expense, so it is fair to say that the $\mathrm{ABC_q}$ limits require less than three times as much numerical effort as the standard limits.

Like the BC_a interval, the ABC interval (22.33) is transformation-respecting. This is not true for the $\mathrm{ABC_q}$ limits, a disadvantage that can sometimes limit their accuracy.

22.7 Least favorable families

The least favorable family plays an important role in the ABC interval, and is implicit in the construction of the BC_a interval. In this section we describe the least favorable family in more detail. Denote the rescaled multinomial distribution with success proba-

332   BOOTSTRAP CONFIDENCE INTERVALS

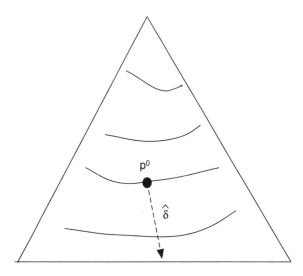

Figure 22.1. *Schematic drawing of the least favorable family for the multi-nomal distribution. The triangle depicts the simplex for $n = 3$. The solid curves are the level curves of constant value of the statistic $T(\mathbf{P})$. The least favorable direction $\hat{\delta}$ passes through $\mathbf{P}^0$ in the direction $\dot{T}(\mathbf{P}^0)$. From this, the least favorable family is defined by equation (22.45).*

bilities $\mathbf{P}$ by

$$g_P(\mathbf{P}^*). \qquad (22.44)$$

In other words, $g_P(\mathbf{P}^*)$ is the probability mass function of $\mathbf{X}/n$, where $\mathbf{X} = (X_1, X_2, \ldots X_n)$ has a multinomial distribution with success probabilities $\mathbf{P}$. The least favorable family for a parameter of interest $\theta = T(\mathbf{P})$ is defined as

$$h_\tau(\mathbf{P}^*) = g_{P^0 + \tau\hat{\delta}}(\mathbf{P}^*), \qquad (22.45)$$

where $\hat{\delta} = \dot{T}(\mathbf{P})$ evaluated at $\mathbf{P} = \mathbf{P}^0$. Figure 22.1 shows a schematic: h_τ is a one-dimensional family through the full n-dimensional family g_P passing through $\mathbf{P}^0$ and in the direction $\hat{\delta} = \dot{T}(\mathbf{P}^0)$.

This family is called least favorable because, at least asymptotically, inference for θ in h_τ is as difficult as it is in the full family g_P. Notice that in Figure 22.1, $\hat{\delta}$ is orthogonal to the level curves of $T(\mathbf{P})$ at $\mathbf{P}^0$; in general, $\hat{\delta}$ is orthogonal to the level curves in the metric of the Fisher information (Problem 22.1).

The specific property of the least favorable family is the following: the Fisher information for τ in h_τ at $\tau = 0$ equals

$$(\sum_{i=1}^{n} \dot{T}_i^2/n^2)^{-1}, \tag{22.46}$$

which is also the Fisher information for θ in g_P. Furthermore, any other one-dimensional subfamily has Fisher information at least as great as this.

In this sense, reduction from g_P to h_τ has not made the problem of inference for τ spuriously easier. Because of this, intervals for θ constructed from intervals for τ will have good coverage properties. Problem 22.1 gives the general definition of least favorable families and establishes the least favorable property. Problem 22.4 asks the reader to view the ABC interval as an approximate bootstrap-t interval constructed for τ and then mapped to the θ scale.

22.8 The ABC$_q$ method and transformations

The effect of the ABC$_q$ procedure may be examined by inverting the transformation given in (22.42):

$$
\begin{aligned}
\xi \equiv \frac{\theta - \hat{\theta}}{\hat{\sigma}} \to \lambda &\equiv \frac{2\xi}{1 + [1 + 4\hat{c}_q\xi]^{1/2}} \\
\to w &\equiv \frac{2\lambda}{(1 + 2\hat{a}\lambda) + (1 + 4\hat{a}\lambda)^{1/2}}
\end{aligned}
\tag{22.47}
$$

The ABC method amounts to taking $w - \hat{z}_0$, the transformation of the studentized pivot $\xi = (\theta - \hat{\theta})/\hat{\sigma}$, as standard normal. In fact, it can be shown that $w - \hat{z}_0$ is standard normal to second order. Figure 22.2 shows the estimated transformation $w - \hat{z}_0$ for the variance problem analyzed in Section 14.2 of Chapter 14. It appears logarithmic in shape, which seems reasonable since $\hat{\theta}$ is the variance.

The ABC$_q$ procedure is therefore similar to the bootstrap-t procedure, which estimates the distribution of $(\hat{\theta} - \theta)/\hat{\sigma}$ directly by bootstrap sampling. Although neither procedure is transformation-respecting, empirical evidence suggests that this is a less serious problem for the ABC$_q$ procedure. The original version, ABC, is transformation-respecting.

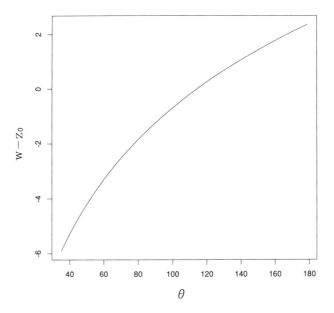

Figure 22.2. *Estimated transformation from* ABC_q *procedure, for the variance example of Section 14.2 of Chapter 14.*

22.9 Discussion

As we have seen, there are a number of different techniques that produce second-order accurate and correct confidence intervals. Through the use of bootstrap calibration (described in chapter 18), higher order accuracy can be achieved. By calibrating a second order accurate interval, we obtain a third order accurate interval having errors of order $O(n^{-3/2})$. A third order accurate interval can be calibrated, producing a fourth order accurate interval, and so on. A remaining challenge is to find computationally efficient methods for calibrating intervals. Calibration methods that start with a transformation-respecting interval and retain that property should also have better statistical behavior than procedures which are not transformation-respecting. Calibration of the ABC procedure is illustrated in the case history of Chapter 25.

22.10 Bibliographic notes

There has been a great deal written on the subject of bootstrap confidence intervals. Efron (1979a) proposes the percentile interval; the bias-corrected percentile and bootstrap-t intervals are described in Efron (1981). An early review of bootstrap confidence intervals appears in Tibshirani (1985). Buckland (1983, 1984, 1985) discusses algorithms for the percentile and bias-corrected percentile techniques. Tibshirani (1988) proposes automatic variance stabilization of the bootstrap-t procedure. The bias-corrected, accelerated interval (BC_a) is suggested in Efron (1987). See also Efron (1985). DiCiccio and Efron (1992) discuss the ABC (approximate bootstrap confidence) interval.

Singh (1981) was the first to establish second order accuracy of a bootstrap confidence interval, applying Edgeworth theory to the bootstrap-t interval. The theory of bootstrap confidence intervals is further developed in Swanepoel *et al.* (1983), Abramovitch and Singh (1985), Hartigan (1986), Hall (1986a), Bickel (1987), DiCiccio and Tibshirani (1987), Hall (1988a, 1988b), DiCiccio and Romano (1988, 1989, 1990) and Konishi (1991). Iteration for improving the coverage of bootstrap confidence intervals is described in Hall (1986a), Beran (1987, 1988), Loh (1987, 1991), Sheather (1987), Hall and Martin (1988), and Martin (1990). The material in section 22.3 is taken from Hall (1988a), Hartigan (1986) and DiCiccio (personal communication). A multiparametric version of the bootstrap-t method is proposed in Hall (1987).

Discussions of some of the issues concerning bootstrap confidence intervals appear in Schenker (1985), Robinson (1986, 1987), Peters and Freedman (1987), Hinkley (1988), and in the psychology literature, Lunneborg (1985), Rasmussen (1987), and Efron (1988).

General asymptotic theory for the bootstrap is developed in Bickel and Freedman (1981), Beran (1984), and Gine and Zinn (1989, 1990).

The least favorable family is due to Stein (1956).

22.11 Problems

22.1 (a) Consider a parametric family with parameter vector $\boldsymbol{\eta}$. Denote the families of densities by g_η, and let the maximum likelihood estimate of $\boldsymbol{\eta}$ be $\hat{\boldsymbol{\eta}}$. Suppose our parameter of interest is $\theta = t(\boldsymbol{\eta})$. Let $I(\hat{\boldsymbol{\eta}}) = -d^2 \log g_\eta / d\boldsymbol{\eta}\boldsymbol{\eta}^T|_{\hat{\eta}}$, the observed information for $\boldsymbol{\eta}$, evaluated at $\boldsymbol{\eta} = \hat{\boldsymbol{\eta}}$. The least favorable direction is defined as

$$\hat{\delta} = I(\hat{\boldsymbol{\eta}})^{-1}\dot{t}(\hat{\boldsymbol{\eta}}), \qquad (22.48)$$

where $\dot{t}(\hat{\boldsymbol{\eta}}) = dt(\boldsymbol{\eta})/d\boldsymbol{\eta}|_{\eta=\hat{\eta}}$. The least favorable family for θ is defined to be

$$h_\tau = g_{\hat{\eta}+\tau\hat{\delta}}. \qquad (22.49)$$

Show that the observed information for τ in h_τ is

$$1/\dot{t}(\hat{\boldsymbol{\eta}})^T I(\hat{\boldsymbol{\eta}})^{-1}\dot{t}(\hat{\boldsymbol{\eta}}), \qquad (22.50)$$

and show that this is also the observed information for θ in g_η.

(b) Show that any other subfamily $g_{\hat{\eta}+\tau d}$ (where d is a vector) has observed information for τ greater than or equal to (22.50).

(c) Verify that (22.45) is the least favorable family for $T(\mathbf{P})$ in the multinomial distribution.

22.2 Derive expression (22.42) from a quadratic Taylor series expansion for $T(\mathbf{P}^0 + \lambda\hat{\delta}/\hat{\sigma})$.

22.3 Show that the Cornish-Fisher expansion (22.43), applied to the BC_a interval in the least favorable family $\mathbf{P}^0 + \tau\hat{\delta}$ gives endpoints $\mathbf{P}^0 + \lambda\hat{\delta}/\hat{\sigma}$.

22.4 The maximum likelihood estimate of the parameter τ indexing the least favorable family is 0, with an estimated standard error of $\hat{\sigma}$. Therefore an α-level bootstrap-t endpoint τ has the form $0 + k(\alpha)\hat{\sigma}$ for some constant $k(\alpha)$. Show that the ABC endpoint can be viewed as a bootstrap-t interval constructed for τ, and then mapped to the θ scale, and give the corresponding value for $k(\alpha)$.

22.5 Let $\hat{\theta} = t(\hat{F})$ be the sample correlation coefficient (4.6) between y and z, for the data set $\mathbf{x} = ((z_1, y_1), (z_2, y_2), \cdots, (z_n, y_n))$.

(a)   Show that

$$\hat{\theta} = \frac{E_{\hat{F}}(zy) - E_{\hat{F}}(z) \cdot E_{\hat{F}}(y)}{[(E_{\hat{F}}(z^2) - E_{\hat{F}}(z)^2) \cdot (E_{\hat{F}}(y^2) - E_{\hat{F}}(y)^2)]^{1/2}}.$$

    (b) Describe how one could compute the empirical influence component (22.32).

22.6 Show that under model (22.22), the exact confidence points for ϕ are given by (22.24).

## CHAPTER 23

Efficient bootstrap computations

23.1 Introduction

In this chapter we investigate computational techniques designed to improve the accuracy and reduce the cost of bootstrap calculations. Consider for example an independent and identically distributed sample $\mathbf{x} = (x_1, x_2, \ldots x_n)$ from a population F and a statistic of interest $s(\mathbf{x})$. The ideal bootstrap estimate of the expectation of $s(\mathbf{x})$ is

$$\hat{e} = \mathrm{E}_{\hat{F}} s(\mathbf{x}^*), \tag{23.1}$$

where $\hat{F}$ is the empirical distribution function. Unless $s(\mathbf{x})$ is the mean or some other simple statistic, it is not easy to compute $\hat{e}$ exactly, so we approximate the ideal estimate by

$$\hat{e}_B = \frac{1}{B} \sum_{b=1}^{B} s(\mathbf{x}^{*b}), \tag{23.2}$$

where each $\mathbf{x}^{*b}$ is a sample of size n drawn with replacement from $\mathbf{x}$. [1]

Formula (23.2) is an example of a *Monte Carlo* estimate of the expectation $\mathrm{E}_{\hat{F}} s(\mathbf{x}^*)$. Monte Carlo estimates of expectations (or integrals) are defined as follows. Suppose $f(z)$ is a real-valued function of a possibly vector-valued argument z and $G(z)$ is the probability measure of z. We wish to estimate the expectation $\mathrm{E}_G[f(z)]$ which can also be written as

$$e = \int f(z) dG(z). \tag{23.3}$$

---

[1] Expectations are a natural starting point for our discussion, since most bootstrap quantities of interest can be written as functions of them.

A simple Monte Carlo estimate of e is

$$\hat{e} = \frac{1}{B} \sum_{b=1}^{B} f(z_b), \qquad (23.4)$$

where the z_b are realizations drawn from $G(z)$. Note that $\hat{e} \to e$ as $B \to \infty$ according to the law of large numbers; furthermore $E(\hat{e}) = e$ and $\text{var}(\hat{e}-e) = c/B$ so that the error (standard deviation of $\hat{e} - e$) goes to zero at the rate $1/\sqrt{B}$.

The bootstrap estimate $\hat{e}_B$ is a special case in which $f(z) = s(\mathbf{x})$ and $G(z)$ is the product measure $\hat{F} \times \hat{F} \cdots \times \hat{F}$. That is, $G(z)$ specifies that each of n random variables is independent and identically distributed from $\hat{F}$.

Simple Monte Carlo sampling is only one of many methods for multidimensional numerical integration. A number of more sophisticated methods have been proposed that can, in some cases, achieve smaller error for a given number of function evaluations B, or equivalently, require a smaller value of B to achieve a specified accuracy.

By viewing bootstrap sampling as a Monte Carlo integration method, we can exploit these ideas to construct more efficient methods for obtaining bootstrap estimates. The methods that we describe in this chapter can be divided roughly into two kinds: purely post-sampling adjustments, and combined pre- and post-sampling adjustments. The first type uses the usual bootstrap sampling but makes post-sampling adjustments to the bootstrap estimates. The post-sampling adjustments are variations on the "control function" method for integration. These are useful for bias and variance estimation. The second type uses a sampling scheme other than sampling with replacement and then makes post-sampling adjustments to account for the change. The two specific methods that we discuss are balanced bootstrap sampling for bias and variance estimation, and importance sampling for estimation of tail probabilities.

In considering the use of any of these methods, one must weigh the potential gains in efficiency with the ease of use. For example suppose a variance reduction method provides a five-fold savings but might require many hours to implement. Then it might be more cost effective for the statistician to use simple bootstrap sampling and let the computer run 5 times longer. The methods in this chapter are likely to be useful in situations where high accuracy is required, or in problems where an estimate will be recomputed many times rather than on a one-time basis.

23.2 Post-sampling adjustments

Control functions are a standard tool for numerical integration. They form the basis for the methods described in this section. We begin with a general description of control functions, and then illustrate how they can be used in the bootstrap context.

Our goal is to estimate the integral of a function with respect to a measure G

$$e = \int f(z)dG. \tag{23.5}$$

Suppose that we have a function $g(z)$ that approximates $f(z)$, and whose integral with respect to G is known. Then we can write

$$\int f(z)dG = \int g(z)dG + \int (f(z) - g(z))dG. \tag{23.6}$$

The value of the first integral on the right side of this expression is known. The idea is to use Monte Carlo sampling to estimate the integral of $f(z) - g(z)$ rather than $f(z)$ itself. The function $g(z)$ is called the control function for $f(z)$. To proceed, we generate B random samples from G and construct the estimate

$$\hat{e}_1 = \int g(z)dG + \frac{1}{B}\sum_1^B [f(z_i) - g(z_i)]. \tag{23.7}$$

The variance of this estimate is

$$\mathrm{var}(\hat{e}_1) = \frac{1}{B}\mathrm{var}[f(z) - g(z)]. \tag{23.8}$$

where the variance is taken with respect to $z \sim G$. By comparison, the simple estimate

$$\hat{e}_0 = \frac{1}{B}\sum_{b=1}^B f(z_b) \tag{23.9}$$

has variance $\mathrm{var}[f(z)]/B$. If $g(z)$ is a good approximation to $f(z)$, then $\mathrm{var}[f(z) - g(z)] < \mathrm{var}[f(z)]$ and as a result the control function will produce an estimate with lower variance for the same number of samples B.

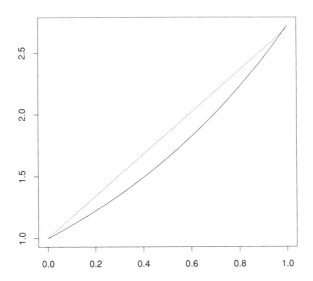

Figure 23.1. *Linear approximation (dotted line) to the exponential func-tion (solid line).*

As a simple example, suppose we want to estimate by Monte Carlo methods the integral of $f(z) = \exp(z)$ over the unit interval with respect to the uniform distribution. As shown in Figure 23.1, the function $g(z) = 1.0 + 1.7z$ provides a good approximation to $\exp(z)$, and the integral of $1.0 + 1.7z$ is $1.0z + .85z^2$.

To estimate the integral of $\exp(z)$, we draw B random numbers z_b and compute the quantity

$$\hat{e}_1 = 1.0 + .85 + \frac{1}{B} \sum_{b=1}^{B} [\exp(z_b) - (1.0 + 1.7z_b)]. \qquad (23.10)$$

If we try to integrate $\exp(z)$ by the direct Monte Carlo method, our estimate is

$$\hat{e}_0 = \frac{1}{B}\sum_{b=1}^{B} \exp(z_b). \tag{23.11}$$

In this simple case we can compute the reduction in variance achieved by importance sampling. A simple calculation shows that $\mathrm{var}(f) = (-1/2)e^2 + 2e - 3/2$, $\mathrm{var}(f - g) \approx (-1/2)e^2 + 3.7e - 6.36$ and hence

$$\frac{\mathrm{var}(\hat{e}_0)}{\mathrm{var}(\hat{e}_1)} \approx \frac{(-1/2)e^2 + 2e - 3/2}{(-1/2)e^2 + 3.7e - 6.36} \approx 78. \tag{23.12}$$

For the same number of samples B, the control function estimate has $1/78$ times the variance of the simple estimate.

The integral (23.5) is an expectation $\mathrm{E}_G f(z)$. In the bootstrap context expectations arise in the estimate of bias. Another quantity of special interest for the bootstrap is the variance

$$\mathrm{var}(f(z)) = \int f^2(z)\mathrm{var}(f(z)) = \int f(z)^2. \tag{23.13}$$

One could apply possibly different control functions to each of the two components separately, but in the bootstrap context it turns out to be better to use a single control function applied to $f(z)$. For this purpose, we require a function $g(z)$ with known variance, such that $f(z) \approx g(z)$. Note that

$$\mathrm{var}(f) \;=\; \mathrm{var}(g(z)) + \mathrm{var}(f(z) - g(z)) + 2 \cdot \mathrm{cov}(g(z), f(z) - g(z)) \tag{23.14}$$

so that in general we need to estimate the covariance between $g(z)$ and $f(z) - g(z)$ as well as $\mathrm{var}(f(z) - g(z))$ by Monte Carlo sampling. In section 23.4 we will see that it is possible to choose $g(z)$ orthogonal to $f(z)$ in the bootstrap setting, so that the covariance term vanishes.

23.3 Application to bootstrap bias estimation

Assume that we have an independent and identically distributed sample $\mathbf{x} = (x_1, x_2, \ldots x_n)$ from a population F and a statistic of interest $s(\mathbf{x})$. Rather than working with $s(\mathbf{x})$, it is convenient to use the resampling representation described in Chapter 20. Let $\mathbf{P}^* = (P_1^*, \ldots P_n^*)^T$ be a vector of probabilities satisfying $0 \leq P_i^* \leq$

1 and $\sum_1^n P_i^* = 1$, and let $\hat{F}^* = \hat{F}(\mathbf{P}^*)$ be the distribution function putting mass P_i^* on x_i, $i = 1, 2, \ldots n$. Assuming $s(\mathbf{x})$ is a functional statistic, if $\mathbf{x}^*$ is a bootstrap sample we can express $s(\mathbf{x}^*)$ as $T(\mathbf{P}^*)$, where each P_i^* is the proportion of the sample containing x_i, for $i = 1, 2, \cdots n$.

An effective and convenient form for the control function is a linear function

$$a_0 + \mathbf{a}^T \mathbf{P}^*. \tag{23.15}$$

This is convenient because its mean and variance under multinomial sampling

$$\mathbf{P}^* \sim \frac{1}{n} \text{Mult}(n, \mathbf{P}^0) \tag{23.16}$$

have the simple forms $a_0 + \mathbf{a}^T \mathbf{P}^0$ and $\mathbf{a}^T \Sigma \mathbf{a}$ respectively, where

$$\mathbf{P}^0 = (1/n, 1/n \cdots 1/n)^T \tag{23.17}$$

and

$$\Sigma = \frac{\mathbf{I}}{n^2} - \frac{\mathbf{P}^0 \mathbf{P}^{0T}}{n}. \tag{23.18}$$

If we use $a_0 + \mathbf{a}^T \mathbf{P}^*$ as a control function for estimating $\mathrm{E}_* T(\mathbf{P}^*)$, then our estimate has the form

$$
\begin{aligned}
\hat{e}_1 &= \mathrm{E}_*(a_0 + \mathbf{a}^T \mathbf{P}^*) + \frac{1}{B} \sum_{b=1}^B (T(\mathbf{P}^{*b}) - a_0 - \mathbf{a}^T \mathbf{P}^{*b}) \\
&= a_0 + \mathbf{a}^T \mathbf{P}^0 + \frac{1}{B} \sum_{b=1}^B T(\mathbf{P}^{*b}) - (a_0 + \mathbf{a}^T \bar{\mathbf{P}}^*). \quad (23.19)
\end{aligned}
$$

where $\bar{\mathbf{P}}^* = \frac{1}{B} \sum_{b=1}^B \mathbf{P}^{*b}$ the mean of the B resampling vectors $\mathbf{P}^{*b}, b = 1, 2, \ldots B$.

Which values of a_0 and $\mathbf{a}$ should we use? As we will see for estimation of variance, there is a number of good choices. For estimation of the expectation, however, it turns out that variance reduction can be achieved without having to choose a_0 and $\mathbf{a}$. Suppose $a_0 + \mathbf{a}^T \mathbf{P}^*$ is any linear function agreeing with $T(\mathbf{P}^*)$ at $\mathbf{P}^* = \mathbf{P}^0$, that is, $a_0 + \mathbf{a}^T \mathbf{P}^0 = T(\mathbf{P}^0)$. Then if we replace $a_0 + \mathbf{a}^T \bar{\mathbf{P}}^*$ in

(23.19) by $T(\bar{\mathbf{P}}^*)$, the new estimate is

$$\tilde{e}_1 = \frac{1}{B} \sum_{b=1}^{B} T(\mathbf{P}^{*b}) + T(\mathbf{P}^0) - T(\bar{\mathbf{P}}^*). \qquad (23.20)$$

Note that the new estimate $\tilde{e}_1$ adjusts the simple estimate $\sum_{b=1}^{B} T(\mathbf{P}^{*b})/B$ by a term that accounts for difference between $\bar{\mathbf{P}}^*$ and its theoretical expectation $\mathbf{P}^0$. It is not unreasonable to replace $a_0 + \mathbf{a}^T \bar{\mathbf{P}}^*$ by $T(\bar{\mathbf{P}}^*)$ since both quantities approach $T(\mathbf{P}^0)$ as $B \to \infty$.

How does this change the bootstrap estimate of bias? The usual estimate is

$$\overline{\text{bias}} = \frac{1}{B} \sum_{1}^{B} T(\mathbf{P}^{*b}) - T(\mathbf{P}^0). \qquad (23.21)$$

The new estimate is

$$\begin{aligned} \widehat{\text{bias}} &= \frac{1}{B} \sum_{b=1}^{B} [T(\mathbf{P}^{*b}) + T(\mathbf{P}^0) - T(\bar{\mathbf{P}}^*)] - T(\mathbf{P}^0) \\ &= \frac{1}{B} \sum_{b=1}^{B} T(\mathbf{P}^{*b}) - T(\bar{\mathbf{P}}^*). \qquad (23.22) \end{aligned}$$

This procedure is sometimes called *re-centering*.

There is another way to motivate $\widehat{\text{bias}}$. Suppose first that $T(\mathbf{P}^*)$ is a linear statistic defined in Section 20.3:

$$T^{\text{LIN}} = c_0 + (\mathbf{P}^* - \mathbf{P}^0)^T \mathbf{U} \qquad (23.23)$$

where $\mathbf{U}$ is an n-vector satisfying $\sum_{1}^{n} U_i = 0$. Then it is easy to show that $\widehat{\text{bias}} = 0$ (the true bias) but $\overline{\text{bias}} = (\bar{\mathbf{P}}^* - \mathbf{P}^0)^T \mathbf{U}$ which may not equal zero (Problem 23.2).

Suppose instead that $T(\mathbf{P}^*)$ is a quadratic statistic as defined in section 20.5:

$$T^{\text{QUAD}} = c_0 + (\mathbf{P}^* - \mathbf{P}^0)^T \mathbf{U} + \frac{1}{2} (\mathbf{P}^* - \mathbf{P}^0)^T \mathbf{V} (\mathbf{P}^* - \mathbf{P}^0) \qquad (23.24)$$

where $\mathbf{U}$ is an n-vector satisfying $\sum_{1}^{n} U_i = 0$ and $\mathbf{V}$ is an $n \times n$ symmetric matrix satisfying $\sum_i V_{ij} = \sum_j V_{ij} = 0$ for all i, j. Then the true bias of $T(\mathbf{P}^*)$ is easily shown to be

$$\text{bias}_\infty = \frac{1}{2}\text{tr}\mathbf{V}\Sigma, \qquad (23.25)$$

where Σ is given by expression (23.18).

The usual and new estimates can be written as

$$\overline{\text{bias}} = \frac{1}{2}\text{tr}V\hat{\Sigma} + (\bar{\mathbf{P}}^* - \mathbf{P}^0)\mathbf{U} + \frac{1}{2}(\bar{\mathbf{P}}^* - \mathbf{P}^0)^T\mathbf{V}(\bar{\mathbf{P}}^* - \mathbf{P}^0)$$

$$(23.26)$$

$$\widehat{\text{bias}} = \frac{1}{2}\text{tr}\mathbf{V}\hat{\Sigma} \qquad (23.27)$$

where $\hat{\Sigma}$ is the maximum likelihood estimate

$$\bar{\Sigma} = \frac{1}{B}\sum_{1}^{B}(\mathbf{P}^{*b} - \bar{\mathbf{P}}^*)(\mathbf{P}^{*b} - \bar{\mathbf{P}}^*)^T. \qquad (23.28)$$

Furthermore, for quadratic statistics it can be shown that as $n \rightarrow \infty$ and $B = cn$ for some constant $c > 0$,

$$\begin{aligned}
\widehat{\text{bias}} - \text{bias}_\infty &= O_p(n^{-3/2}) \\
\overline{\text{bias}} - \text{bias}_\infty &= O_p(n^{-1}).
\end{aligned} \qquad (23.29)$$

This means that $\widehat{\text{bias}}$ approaches the ideal variance more quickly that does $\overline{\text{bias}}$. Table 23.1 shows the results of a small simulation study to compare estimates of bias. For each row of the table a sample of size 10 was drawn from a uniform $(0,1)$ distribution. The statistic of interest is $\hat{\theta} = \log \bar{x}$. The leftmost column shows $\overline{\text{bias}}_{1000}$, which is a good estimate of the ideal bias bias_∞. In each row, 25 separate bootstrap analyses with $B = 20$ were carried out and columns 2,3,4 show the average bias over the 25 replications. Column 2 corresponds to the simple bias estimate $\overline{\text{bias}}_{20}$, while column 3 shows the improved estimate $\widehat{\text{bias}}_{20}$. In column 4, the least-squares control function, described in the next section, is used. Column 5 corresponds to the "permutation bootstrap" $\widehat{\text{bias}}_{\text{perm}}$ described in Section 23.5. Columns 6, 7 and 8 show the ratio of the variance of $\overline{\text{bias}}_{20}$ to that of $\widehat{\text{bias}}_{20}$, $\widehat{\text{bias}}_{\text{con}}$ and $\widehat{\text{bias}}_{\text{perm}}$, respectively. All are roughly unbiased; $\widehat{\text{bias}}_{20}$ has approximately 57 times less variance (on the average) than the simple estimate. Since the variance $\overline{\text{bias}}_B$ goes to zero like $1/B$, we deduce that $\widehat{\text{bias}}_{20}$ has about the same variance as $\overline{\text{bias}}_{1000}$. The estimator $\widehat{\text{bias}}_{\text{con}}$,

Table 23.1. *10 sampling experiments to compare estimates of bias. Details of column headings are given in the text. The last line shows the average over the 10 experiments.*

	$\bar{b}_{1000}$	$\bar{b}_{20}$	$\widehat{b}_{20}$	$\widehat{b}_{con}$	$\hat{b}_{p}$	$\dfrac{\mathrm{var}(\bar{b}_{20})}{\mathrm{var}(\hat{b}_{20})}$	$\dfrac{\mathrm{var}(\bar{b}_{20})}{\mathrm{var}(\hat{b}_{con})}$	$\dfrac{\mathrm{var}(\bar{b}_{20})}{\mathrm{var}(\hat{b}_{p})}$
1	-0.019	-0.018	-0.016	-0.018	-0.019	29.8	16.6	0.6
2	-0.017	-0.015	-0.014	-0.012	-0.005	88.5	44.5	2.1
3	-0.030	-0.009	-0.019	-0.016	-0.022	67.2	24.2	1.1
4	-0.012	-0.026	-0.016	-0.014	-0.020	82.6	37.8	1.1
5	-0.019	-0.021	-0.020	-0.017	-0.035	24.8	32.8	0.7
6	-0.012	-0.012	-0.016	-0.015	-0.016	62.7	25.9	0.7
7	-0.039	-0.031	-0.045	-0.041	-0.048	12.2	7.3	1.0
8	-0.014	-0.014	-0.016	-0.016	-0.009	42.9	44.8	0.8
9	-0.020	-0.010	-0.008	-0.006	-0.002	103.0	72.7	1.2
10	-0.018	-0.018	-0.016	-0.004	-0.019	53.4	34.5	1.4
Ave	-0.020	-0.017	-0.019	-0.017	-0.019	56.8	34.1	1.1

which uses a control function rather than making the approximation leading to $\widehat{\mathrm{bias}}_{20}$, has approximately 34 times less variance (on the average) than the simple estimate. Surprisingly, though, it is outperformed by the apparently cruder estimator $\widehat{\mathrm{bias}}_{20}$.

23.4 Application to bootstrap variance estimation

For estimation of variance, we consider again the use of a linear control function and write

$$T(\mathbf{P}^*) = a_0 + \mathbf{a}^T\mathbf{P}^* + T(\mathbf{P}^*) - (a_0 + \mathbf{a}^T\mathbf{P}^*). \qquad (23.30)$$

Then our estimate of variance is

$$\widehat{\mathrm{var}}(T(\mathbf{P}^*)) = \quad \mathbf{a}^T\hat{\Sigma}\mathbf{a} + \frac{1}{B}\sum_{b=1}^{B}(T(\mathbf{P}^{*b}) - a_0 - \mathbf{a}^T\mathbf{P}^{*b})^2$$
$$+ \quad \frac{2}{B}\sum_{b=1}^{B}(a_0 + \mathbf{a}^T\mathbf{P}^{*b})(T(\mathbf{P}^{*b}) - a_0 - \mathbf{a}^T\mathbf{P}^{*b}). \qquad (23.31)$$

Reasonable choices for the control function would be the jackknife or infinitesimal jackknife planes described in Chapter 20. One drawback of these is that they require the additional computation of the n jackknife (or infinitesimal jackknife) derivatives of $\hat{T}$. An alternative that avoids this is the least-squares fit of $T(\mathbf{P}^{*b})$ on $\mathbf{P}^{*b}$ for $b = 1, 2, \ldots B$. Denote the fitted least-squares plane by

$\hat{a}_0 + \hat{\mathbf{a}}^T \mathbf{P}^*$. By a proper choice of constraints, the cross-product term in (23.31) drops out and we obtain

$$
\begin{aligned}
\widehat{\text{var}}(T(\mathbf{P}^*)) &= \hat{\mathbf{a}}^T \hat{\Sigma} \hat{\mathbf{a}} + \frac{1}{B} \sum_{b=1}^{B} (T(\mathbf{P}^{*b}) - \hat{a}_0 - \hat{\mathbf{a}}^T \mathbf{P}^{*b})^2 \\
&= \sum_{i=1}^{n} a_i^2 + \frac{1}{B} \sum_{b=1}^{B} (T(\mathbf{P}^{*b}) - \hat{a}_0 - \hat{\mathbf{a}}^T \mathbf{P}^{*b})^2.
\end{aligned}
$$
(23.32)

Details are given in Problem 23.5.

Table 23.2 shows the results of a simulation study designed to compare estimates of variance. Data $y_1, y_2, \ldots y_{10}$ were generated from a uniform $(0,1)$ distribution, and $z_1, z_2, \ldots z_{10}$ generated independently from $G_1^2/2$, where G_1 denotes a standard negative exponential distribution. The statistic of interest is $\hat{\theta} = \bar{z}/\bar{y}$. For each of 10 samples, 30 bootstrap analyses were carried out with $B = 100$. The left-hand column shows the average of the simple bootstrap variance estimate with $B = 1000$; this is close to the value that we would obtain as $B \to \infty$. The next three columns show the average of the variance estimates for the simple bootstrap based on 100 replications ($\hat{v}_{100}$), control functions ($\hat{v}_{\text{con}}$) and permutation bootstrap ($\hat{v}_{\text{perm}}$). The 5th and 6th columns show the ratio of the variances of $\hat{v}_{100}$ to $\hat{v}_{\text{con}}$ and $\hat{v}_{\text{p}}$. The table shows that the control function estimate is roughly unbiased and on the average is about 5 times less variable than the simple bootstrap estimate based on 100 bootstrap replications. The control function estimate is less variable than the usual estimate in all but the last sample, where it is five times more variable! A closer examination reveals that this is largely due to just one of the 30 analyses for that sample. When that analysis is removed, the ratio becomes 1.25.

The last column gives a diagnostic to aid us in determining when control functions are likely to be advantageous. It is the estimated percentage of the variance explained by the linear approximation to $\hat{T}$:

$$
\hat{R}^2 = \frac{\sum_1^n \hat{a}_i^2}{\widehat{\text{var}}(T(\mathbf{P}^*))}.
$$
(23.33)

A linear control function will tend to be helpful when $\hat{R}^2$ is high, and we see that it is lowest ($=.89$) for the last sample. For the

Table 23.2. *10 sampling experiments to compare estimates of variance. Details of column headings are given in the text. The last line shows the average over the 10 experiments.*

	$\bar{v}_{1000}$	$\bar{v}_{100}$	$\hat{v}_{con}$	$\hat{v}_p$	$\dfrac{\text{var}(\bar{v}_{100})}{\text{var}(\hat{v}_{con})}$	$\dfrac{\text{var}(\bar{v}_{100})}{\text{var}(\hat{v}_p)}$	$\text{mean}(R^2)$
1	1.97	1.92	2.03	1.94	2.68	0.79	0.94
2	0.08	0.09	0.09	0.08	2.08	2.07	0.95
3	0.44	0.46	0.46	0.46	6.12	1.72	0.96
4	0.52	0.50	0.51	0.51	7.17	0.82	0.97
5	3.21	3.00	3.16	3.02	9.20	0.56	0.98
6	0.37	0.42	0.40	0.39	2.08	1.39	0.92
7	6.28	4.89	4.97	4.95	2.83	1.05	0.95
8	0.77	0.76	0.74	0.74	15.50	1.16	0.99
9	18.80	18.10	19.30	19.00	3.34	0.53	0.96
10	4.06	3.89	4.40	3.90	0.19	0.52	0.89
Ave	3.65	3.41	3.60	3.50	5.12	1.06	0.95

one analysis that led to the large variance ratio mentioned above, $\hat{R}^2 = .69$. While .69 is the lowest value of $\hat{R}^2$ that we observed in this study, it is not clear in general what a "dangerously low" value is. Further study is needed on this point.

23.5 Pre- and post-sampling adjustments

The methods described in this section approach the problem of efficient bootstrap computation by modification of the sampling scheme. The first method is called the *balanced bootstrap*. Consider for example the problem of estimating the bias of a linear function $T(\mathbf{P}^*) = c_0 + (\mathbf{P}^* - \mathbf{P}^0)^T \mathbf{U}$ where $\sum_1^n U_i = 0$. As we have seen in Section 23.3 the simple estimate of bias can be written as

$$\overline{\text{bias}}_B = \frac{1}{B} \sum_1^B T(\mathbf{P}^{*b}) - T(\mathbf{P}^0) = (\bar{\mathbf{P}}^* - \mathbf{P}^0)^T \mathbf{U}. \quad (23.34)$$

The true bias is zero, but $\overline{\text{bias}}_B$ is non-zero due to the difference between the average bootstrap resampling vector $\bar{\mathbf{P}}^*$ and its theoretical expectation $\mathbf{P}^0$.

One way to rectify this is to modify bootstrap sampling to ensure that $\bar{\mathbf{P}}^* = \mathbf{P}^0$. This can be achieved by arranging so that each data item appears exactly B times in the total collection of nB resampled items. Rather than sampling with replacement, we concatenate B copies of $x_1, x_2, \ldots x_n$ into a string L of length $n \cdot B$, and then take a random permutation of L into say $\tilde{L}$. Finally, we define the first bootstrap sample to be elements $1, 2, \ldots n$ of $\tilde{L}$, the second bootstrap sample to be elements $n + 1, \ldots 2n$ of $\tilde{L}$, and so on.

Since the resulting bootstrap samples are balanced with respect to the occurrences of each individual data item, this procedure is called the *first order balanced bootstrap*. Alternatively, since it can be carried out by a simple permutation as described above, it is also called the *permutation bootstrap*.

Of course estimation of the bias of a linear statistic is not of interest. But for non-linear statistics with large linear components, it is reasonable to hope that this procedure will reduce the variance of our estimate. The first order balanced bootstrap was carried out in the experiments of Tables 23.1 and 23.2. In both cases, it improved upon the simple estimate for some samples, but did worse for other samples. Overall, the average performance was about the same as the simple bootstrap estimate.

It is possible to achieve higher order balance in the set of bootstrap samples, through the use of Latin squares. For example, second order balance ensures that each data item, and each pair of data items appears the same number of times. Higher order balanced samples improve somewhat on the first order balanced bootstrap, but limited evidence suggests that they are not as effective as the other methods described in this chapter.

23.6 Importance sampling for tail probabilities

In this section we discuss a method than can provide many-fold reductions in the number of bootstrap samples needed for estimating a tail probability. We first describe the technique in general and then apply it in the bootstrap context.

Suppose we are interested in estimating

$$e = \int f(z)g(z)dz \tag{23.35}$$

for some function $f(z)$, where $g(z)$ is a probability density function. This quantity is the expectation of f with respect to g. The simple

Monte Carlo estimate of e is

$$\hat{e}_0 = \frac{1}{B} \sum_{b=1}^{B} f(z_b), \qquad (23.36)$$

where $z_1, z_2, \ldots z_B$ are random variates sampled from g. Now suppose further that we have a probability density function $h(z)$ that is roughly proportional to $f(z)g(z)$

$$h(z) \approx f(z)g(z), \qquad (23.37)$$

and we have a convenient way of sampling from $h(z)$. Then we can write (23.35) as

$$e = \int [\frac{f(z)g(z)}{h(z)}]h(z)dz. \qquad (23.38)$$

To estimate e, we can now focus on $f(z)g(z)/h(z)$ rather than $f(z)$. We draw $z_1, z_2, \ldots z_B$ from $h(z)$ and then compute

$$\begin{aligned}
\hat{e}_1 &= \frac{1}{B} \sum_{b=1}^{B} \frac{f(z_b)g(z_b)}{h(z_b)} \\
&= \frac{1}{B} \sum_{b=1}^{B} f(z_b) \cdot \frac{g(z_b)}{h(z_b)}. \qquad (23.39)
\end{aligned}$$

The second line in (23.39) is informative, as it expresses the new estimate as a simple Monte Carlo estimate for f with weights $w_b = g(z_b)/h(z_b)$, to account for the fact that samples were drawn from $h(z)$ rather than from $g(z)$.

The quantity $\hat{e}_1$ is called the *importance sampling* estimate of e. The name derives from the fact that, by sampling from $h(z)$, we sample more often in the regions where $f(z)g(z)$ is large. Clearly $\hat{e}_1$ is unbiased since

$$E(\hat{e}_1) = E_h\left[\frac{f(z_b)g(z_b)}{h(z_b)}\right] = E_g[f(z_b)]. \qquad (23.40)$$

The variance of $\hat{e}_1$ is

$$\text{var}(\hat{e}_1) = \frac{1}{B}\text{var}_h\left[\frac{f(z_b)g(z_b)}{h(z_b)}\right] \qquad (23.41)$$

as compared to

$$\text{var}(\hat{e}_0) = \frac{1}{B}\text{var}_g[f(z_b)] \qquad (23.42)$$

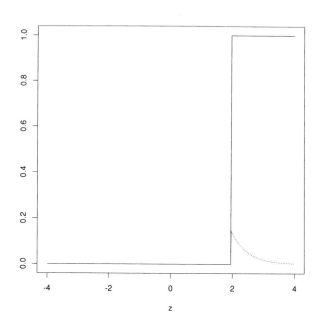

Figure 23.2. *Indicator function* $I_{\{z>1.96\}}$ *(solid line) and importance sampler* $I_{\{z>1.96\}}\phi(z)/\phi_{1.96}(z)$ *(dotted line) for estimating an upper tail probability. The two functions coincide for* $z \leq 1.96$.

for the simple Monte Carlo estimate. Since we chose $h(z) \approx f(z)g(z)$, $\mathrm{var}(\hat{e}_1)$ should be less than $\mathrm{var}(\hat{e}_0)$.

As an example, suppose Z is a standard normal variate and we want to estimate $\mathrm{Prob}\{Z > 1.96\}$. We can write this as

$$\mathrm{Prob}\{Z > 1.96\} = \int I_{\{z>1.96\}}\phi(z)dz, \qquad (23.43)$$

where $\phi(z)$ is the standard normal density function. The simple Monte Carlo estimate of $\mathrm{Prob}\{Z > 1.96\}$ is $\sum_{b=1}^{B} I_{\{z_b>1.96\}}/B$ where z_b are standard normal numbers.

A reasonable choice for the importance sampling function is $h(z) = \phi_{1.96}(z)$, the density function of $N(1.96, 1)$. Figure 23.2 shows why. The solid line is the function $I_{\{z>1.96\}}$; the broken line is $I_{\{z>1.96\}}\phi(z)/\phi_{1.96}(z)$ (the two lines coincide for $z < 1.96$). The importance sampling integrand has much less variance than $I_{\{Z>1.96\}}$, and hence is easier to integrate.

If $Z \sim N(0,1)$, $Y \sim N(1.96,1)$, then one can show that

$$\text{var}(I_{\{Z>1.96\}})/\text{var}\left(\frac{I_{\{Y>1.96\}}\phi(Y)}{\phi_{1.96}(Y)}\right) \approx 16.8. \qquad (23.44)$$

Hence the importance sampling estimate achieves a roughly 17-fold increase in efficiency over the simple Monte Carlo estimate.

As the above example shows, tail probabilities are ideal candidates for importance sampling estimates because the indicator function $I_{\{Z>c\}}$ is highly variable. Importance sampling works in this case by shifting the sampling distribution so that the mean of z is roughly c. This implies that approximately half of the samples will have $z > 1.96$, as opposed to only $100 \cdot \alpha\%$ under the original distribution.

Importance sampling can break down if some of the weights $g(z)/h(z)$ for a nonzero $f(z)$ get very large and hence dominate the sum (23.39). This occurs if a sample z_b is obtained having negligible probability $h(z_b)$, but non-negligible probability $g(z_b)$, and $f(z_b) \neq 0$. For estimation of tail probabilities $\text{Prob}\{Z > c\}$, we need to ensure that $h(z) > g(z)$ in the region $z > c$. This is the case for the example given above.

23.7 Application to bootstrap tail probabilities

Let's consider how importance sampling can be applied to the computation of a bootstrap tail probability

$$\text{Prob}\{\hat{\theta}^* > c\}. \qquad (23.45)$$

where $\hat{\theta}^* = T(\mathbf{P}^*)$, a statistic in resampling form. Of course the α-level bootstrap percentile is the value c such that $\text{Prob}\{\hat{\theta}^* > c\} = 1 - \alpha$, and hence can be derived from bootstrap tail probabilities.

The simple estimate of (23.45) is the proportion of bootstrap values larger than c:

$$\overline{\text{Prob}}\{\hat{\theta}^* > c\} = \frac{1}{B}\sum_{b=1}^{B} I_{\{T(P^{*b})>c\}}. \qquad (23.46)$$

Denote by $m_{\tilde{P}}(\mathbf{P})$ the probability mass function of the rescaled multinomial distribution $\text{Mult}(n, \mathbf{P})/n$ having mean $\tilde{\mathbf{P}}$. An obvious choice for the importance sampler $h(z)$ is a multinomial distribution with mean $\tilde{\mathbf{P}} \neq \mathbf{P}^0 = (1/n, 1/n \cdots 1/n)^T$, that is, a sampling scheme that gives unequal weights to the observations.

What weights should be used? The answer is clearest when $T = \bar{x}$, the sample mean. Intuitively, we want to choose $\tilde{\mathbf{P}}$ so that the event $\bar{x}^* > c$ has about a 50% chance of occurring under $m_{\tilde{P}}(\mathbf{P}^*)$. Suppose c is an upper percentile of $\bar{x}^*$. Then we can increase the chance that $\bar{x}^* > c$ by putting more mass on the larger values of x.

A convenient form for the weights $\tilde{\mathbf{P}} = (\tilde{P}_1, \tilde{P}_2, \ldots \tilde{P}_n)$ is

$$\tilde{P}_i(\lambda) = \frac{\exp\left[\lambda(x_i - \bar{x})\right]}{\sum_1^n \exp\left[\lambda(x_i - \bar{x})\right]}. \tag{23.47}$$

When $\lambda = 0$, $\tilde{\mathbf{P}} = \mathbf{P}^0$; for $\lambda > 0$, $\tilde{P}_i(\lambda) > \tilde{P}_j(\lambda)$ if $x_i > x_j$, and conversely for $\lambda < 0$. We choose λ so that the mean of $\bar{x}^*$ under $m_{\tilde{P}}(\mathbf{P}^*)$ is approximately c. Thus we choose λ_c to be the solution to the equation

$$c = \frac{\sum_1^n x_i \exp\left[\lambda(x_i - \bar{x})\right]}{\sum_1^n \exp\left[\lambda(x_i - \bar{x})\right]}. \tag{23.48}$$

For illustration, we generated a sample of size 100 from a $N(0,1)$ distribution. Suppose that the tail probabilities to be estimated are approximately 2.5% and 97.5%. Since the 2.5% and 97.5% points of the standard normal distribution are -1.96 and 1.96, we solve (23.48) for $c = -1.96$ and 1.96 giving $\lambda_{-1.96} = -3.89$, $\lambda_{1.96} = 2.37$. Figure 23.3 shows the weights $\tilde{\mathbf{P}}(0)$, $\tilde{\mathbf{P}}(-3.89)$, and $\tilde{\mathbf{P}}(2.37)$. The largest and smallest x value are given a weight of about .5, which is 50 times as large as the usual bootstrap weight of $1/100$.

In order to use importance sampling for statistics T other than the mean, we need to know how to shift the probability mass on the observations in order to make $T(\mathbf{P}^*)$ large or small. For statistics with a significant linear component, it is clear how to do this. If U_i denotes the ith influence component, we define a family of weights by

$$\tilde{P}_i(\lambda) = \frac{\exp\left(\lambda U_i\right)}{\sum_1^n \exp\left(\lambda U_i\right)}; \quad i = 1, 2, \cdots n. \tag{23.49}$$

The mean of $T(\mathbf{P}^*)$ under multinomial sampling with probability vector $\tilde{\mathbf{P}}(\lambda)$ is approximately

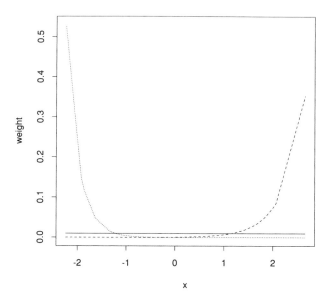

Figure 23.3. *Observation weights for estimating lower (dotted line) and upper (dashed line) tail probabilities. Solid line shows equal weights 1/100.*

$$\hat{\theta} + \frac{\sum_1^n U_i \exp\left(\lambda U_i\right)}{\sum_1^n \exp\left(\lambda U_i\right)}. \qquad (23.50)$$

To estimate $\text{Prob}\{T(\mathbf{P}^*) > c\}$, we solve for λ_c by setting this expectation equal to c, and then use resampling weights $\tilde{\mathbf{P}}(\lambda_c)$. Our estimate is

$$\widehat{\text{Prob}}\{T(\mathbf{P}^*) > c\} = \frac{1}{B} \sum_{b=1}^{B} I_{\{T(P^{*b}) > c\}} \frac{m_{\tilde{P}(\lambda)}(\mathbf{P}^{*b})}{m_{P^0}(\mathbf{P}^{*b})}. \qquad (23.51)$$

If T is a location and scale equivariant functional, a careful analysis of the choice of λ to minimize the variance of the estimate is possible. Details are in Johns (1988). As Johns notes, however, the performance of the importance sampler does not seem to be very

Table 23.3. *10 sampling experiments to compare estimates of an upper (97.5%) tail probability. Details of column headings are given in the text. The last line shows the average over the 10 experiments.*

	$\overline{\mathrm{Prob}_{100}}$	$\widehat{\mathrm{Prob}_{100}}$	$\dfrac{\mathrm{var}\left(\overline{\mathrm{Prob}_{100}}\right)}{\mathrm{var}\left(\widehat{\mathrm{Prob}_{100}}\right)}$
1	0.021	0.018	12.6
2	0.025	0.029	8.5
3	0.031	0.026	5.6
4	0.026	0.026	8.1
5	0.024	0.025	0.3
6	0.026	0.020	7.5
7	0.035	0.031	6.3
8	0.022	0.025	8.7
9	0.021	0.025	7.3
10	0.033	0.030	6.1
Ave	0.026	0.025	7.1

sensitive to the choice of λ.

As an example, we consider again the problem of Section 23.3 and Table 23.2. Data $y_1, y_2, \ldots y_{10}$ were generated from a uniform $(0,1)$ distribution, and $z_1, z_2, \ldots z_{10}$ from $G_1^2/2$, where G_1 denotes a standard negative exponential distribution. The statistic of interest is $\hat{\theta} = \bar{z}/\bar{y}$, and we wish to estimate $\mathrm{Prob}\{\hat{\theta}^* > c\}$. For each of 10 samples, we used the simple bootstrap estimate (23.46) based on $B = 1000$ bootstrap samples to find the value of c such that $\mathrm{Prob}\{\hat{\theta}^* > c\} \approx 0.025$. For each of the 10 samples, 30 bootstrap analyses were carried out, each with $B = 100$ bootstrap samples. Column 1 gives the average of the simple estimates based on $B = 100$, while column 2 gives the average of the importance sampling estimates based on $B = 100$. Both estimates are roughly unbiased. Column 3 shows the ratio of the variances of the two estimates. On the average the importance sampling estimate achieves a 7-fold reduction in variance. We note that in the fifth experiment, however, the importance sampling estimate had a variance roughly three times *greater* than the simple estimate.

The performance of the importance sampling estimate in the preceding example is encouraging. Remember however that the statistic of interest has a large linear component (an $\hat{R}^2$ of at least 90% according to Table 23.2) and hence the shifted distribution (23.49) was successful in generating larger values of $\hat{\theta}^*$. For statistics without a large linear component, the shifted distribution will not work nearly as well. Note also that the importance sampling method requires a separate simulation for lower and upper tail probabilities, whereas the simple bootstrap method uses a single simulation for both points. Hence a 7-fold savings is actually a 3.5-fold savings if both the lower and upper percentiles are required.

23.8 Bibliographic notes

Hammersley and Handscomb (1964) is a standard reference for Monte Carlo variance reduction techniques. Thisted (1986) has some discussion relevant for statistical applications. Therneau (1983) studies a number of different Monte Carlo methods for bias and variance estimation in the bootstrap context, including control functions, antithetic variables, conditioning and stratification. He finds that control functions are a clear winner. Oldford (1985) studies the benefit of approximations prior to bootstrap sampling. Davison, Hinkley, and Schechtman (1986) propose the permutation or first order balanced bootstrap. Gleason (1988), Graham, Hinkley, John, and Shi (1990), and Hall (1990) investigate balanced bootstrap sampling in depth. Latin hypercube sampling (McKay, Beckman, and Conover 1979, Stein 1987) is a closely related research area. Johns (1988) studies importance sampling for percentiles, with a particular emphasis on location-scale equivariant functionals. Davison (1988) suggest similar ideas and Hinkley and Shi (1989) propose importance sampling for nested bootstrap computations. Hesterberg (1988) gives some new variance reduction techniques, including a generalization of importance sampling; Hesterberg (1992) proposes some modifications of control variates and importance sampling in the bootstrap context. Other approaches are given in Hinkley and Shi (1989), and Do and Hall (1991). Efron (1990) describes the estimate bias studied in section 23.2 and the least-squares control function of section 23.4. He also applies the least-squares control function to percentile estimation by a cumulant matching approach; this is related to similar ideas in the Davison *et al.* (1986) paper. Hall (1991) describes balanced importance resampling, while Hall (1989b) investigates antithetic variates for the bootstrap. Further details on computational methods may be found in Hall (1989a) and Appendix II of Hall (1992).

23.9 Problems

23.1 Verify the variance reduction expression (23.12).

23.2 Show that if $\hat{\theta} = c_0 + (\mathbf{P}^* - \mathbf{P}^0)^T \mathbf{U}$ is a linear statistic, $\widehat{\text{bias}} = \text{bias}_\infty = 0$, but $\overline{\text{bias}} = (\bar{\mathbf{P}}^* - \mathbf{P}^0)^T \mathbf{U}$.

23.3 Verify equations (23.25) — (23.27) for the true and estimated bias of a quadratic functional.

23.4 Establish relations (23.29).   [Section 6 of Efron, 1990].

23.5 *Control functions for variance estimation.*

(a) In the notation of section 23.4, let $\mathbf{R} = (R_1, \cdots R_B)^T$ be the centered version of $T(\mathbf{P}^{*b})$:

$$R_b = T(\mathbf{P}^{*b}) - \overline{T(\mathbf{P}^{*b})}; \quad b = 1, 2, \ldots B. \qquad (23.52)$$

Let $\mathbf{Q}$ be the $B \times n$ centered matrix of resampling vectors:

$$\mathbf{Q} = (\mathbf{P}^{*1} - \bar{\mathbf{P}}^*, \mathbf{P}^{*2} - \bar{\mathbf{P}}^*, \cdots \mathbf{P}^{*B} - \bar{\mathbf{P}}^*)^T. \quad (23.53)$$

Show that the least-squares regression coefficient $\mathbf{a}$ of $\mathbf{R}$ on $\mathbf{Q}$, constrained to satisfy $\mathbf{1}^T \hat{\mathbf{a}} = 0$ is

$$\hat{\mathbf{a}} = (\mathbf{Q}^T \mathbf{Q} + \mathbf{1}^T \mathbf{1})^{-1} \mathbf{Q}^T \mathbf{R}, \qquad (23.54)$$

where $\mathbf{1}$ is a vector of ones.

(b) For the choice $\hat{\mathbf{a}}$ given in part (a), derive the decomposition (23.32).

23.6 Consider the problem of estimating the largest eigenvalue of the covariance matrix of a set of multivariate normal data. Take the sample size to be 40, and dimension to 4 and let the eigenvalues of the true covariance matrix be $2.7, 0.7, 0.5$, and 0.1. Carry out experiments like those of Tables 23.2 and 23.3 to estimate the variance and upper 95% point of the largest eigenvalue. In the variance estimation experiment, compare the simple bootstrap estimate, permutation bootstrap and least-squares control function. In the tail probability estimation experiment, compare the simple bootstrap estimate and the importance sampling estimator. Discuss the results.

## CHAPTER 24

Approximate likelihoods

24.1 Introduction

The likelihood plays a central role in model-based statistical inference. Likelihoods are usually derived from parametric sampling models for the data. It is natural to ask whether a likelihood can be formulated in situations like those discussed in this book in which a parametric sampling model is not specified. A number of proposals have been put forth to answer this question, and we describe and illustrate some of them in this chapter.

Suppose that we have data $\mathbf{x} = (x_1, x_2, \ldots x_n)$, independent and identically distributed according to a distribution F. Our statistic $\hat{\theta} = \theta(\hat{F}) = s(\mathbf{x})$ estimates the parameter of interest $\theta = \theta(F)$, and we seek an approximate likelihood function for θ. There are several reasons why we might want a likelihood in addition to the point estimate $\hat{\theta}$, or confidence intervals for θ. First, the likelihood is a natural device for combining information across experiments: in particular, the likelihood for two independent experiments is just the product of the individual experiment likelihoods. Second, prior information for θ may be combined with the likelihood to produce a Bayesian posterior distribution for inference.

To begin our discussion, suppose first that we have a parametric sampling model for $\mathbf{x}$ given by the density function $p(x|\theta)$. By definition the likelihood is proportional to the density of the sample, thought of as a function of θ:

$$L(\theta) = c \cdot \prod_1^n p(x_i|\theta). \tag{24.1}$$

Here c is any positive constant. For convenience we will choose c so that the maximum value of $L(\theta)$ is equal to 1.

In most situations $p(x|\cdot)$ depends on additional "nuisance" parameters λ, besides the parameter of interest θ. The full likelihood then has the form $L(\theta, \lambda)$. The main objective of the methods described here is to get rid of the nuisance parameters in order to have a likelihood for θ alone.

One popular tool is the *profile likelihood*

$$L_{\mathrm{pro}}(\theta) = L(\theta, \hat{\lambda}_\theta). \qquad (24.2)$$

Here $\hat{\lambda}_\theta$ is the restricted maximum likelihood estimate for λ when θ is fixed. Another approach is to find some function of the data, say $v = v(\mathbf{x})$ whose density function $q_v(v|\theta)$ involves only θ not λ. Then the *marginal likelihood* for θ is defined to be

$$L_{\mathrm{mar}}(\theta) = q_v(v|\theta). \qquad (24.3)$$

A major difficulty in the nonparametric setting is that the form of $p(x|\theta, \lambda)$ is not given. To overcome this, the *empirical likelihood* focuses on the empirical distribution of the data $\mathbf{x}$: it uses the profile likelihood for the data-based multinomial distribution. Some other methods discussed in this chapter instead focus directly on $\hat{\theta}$ and seek an approximate marginal likelihood for θ. The goal is to estimate the sampling density

$$p(\hat{\theta}|\theta). \qquad (24.4)$$

In other words for each θ we need an estimate of the sampling distribution of $\hat{\theta}$ when the true parameter is θ. The *approximate pivot method* assumes that some function of $\hat{\theta}$ and θ is pivotal. That is, it has a distribution not depending on any unknown parameters. This allows estimation of $p(\hat{\theta}|\theta)$ from the bootstrap distribution of $\hat{\theta}$. The *bootstrap partial likelihood* approach does not assume the existence of a pivot but estimates $p(\hat{\theta}|\theta)$ directly from the data using a nested bootstrap computation. The *implied likelihood* is a somewhat different approach: it derives an approximate likelihood from a set of nonparametric confidence intervals.

It is important to note that, in general, none of these methods produces a true likelihood: a function that is proportional to the probability of a fixed event in the sample space. The marginal likelihood is a likelihood in ideal cases. However in the nonparametric problem treated here typically $v(\mathbf{x})$ is not completely free of the nuisance parameters and furthermore, the form of $v(\mathbf{x})$ may be estimated from the data.

24.2 Empirical likelihood

Suppose once again that we have independent and identically distributed observations $\mathbf{x} = (x_1, x_2, \cdots x_n)$ from a distribution F, and a parameter of interest $\theta = t(F)$. As in section 21.7, we may define the (nonparametric) likelihood for F by

$$L(F) = \prod_1^n F(\{x_i\}). \tag{24.5}$$

where $F(\{x_i\})$ is the probability of the set $\{x_i\}$ under F. The profile likelihood for θ is

$$L_{\mathrm{pro}}(\theta) = \sup_{F:t(F)=\theta} L(F). \tag{24.6}$$

Computation of $L_{\mathrm{pro}}(\theta)$ requires, for each θ, maximization of $L(F)$ over all distributions satisfying $t(F) = \theta$. This is a difficult task. An important simplification is obtained by restricting attention to distributions having support entirely on $x_1, x_2, \ldots x_n$. Let $\mathbf{w} = (w_1, w_2, \ldots w_n)$ and define F_w to be the discrete distribution putting probability mass w_i on x_i, $i = 1, 2, \cdots n$. The probability of obtaining our sample $\mathbf{x}$ under F_w is $\prod_1^n w_i$. Hence we define the *empirical likelihood* by

$$L_{\mathrm{emp}}(\theta) = \sup_{w:t(F_w)=\theta} \prod_1^n w_i. \tag{24.7}$$

The empirical likelihood is just the profile likelihood for the data-based multinomial distribution having support on $x_1, x_2, \ldots x_n$. Note that there are n parameters in this distribution, with $n - 1$ of the dimensions representing nuisance parameters. Often it is unwise to maximize a likelihood over a large number of nuisance parameters, as this can lead to inconsistent or inefficient estimates. However, this does not seem to be a problem with the empirical likelihood. It is possible to show that in many suitably smooth problems, the likelihood ratio statistic derived from the empirical likelihood, namely $-2\log\{L_{\mathrm{emp}}(\theta)/L_{\mathrm{emp}}(\hat{\theta})\}$, has a χ_1^2 distribution asymptotically just as in parametric problems. Empirical likelihood has also been extended to regression problems and generalized linear models.

Consider the problem of estimating the pth quantile of F, defined by $\theta = \inf\{x; F(x) \geq p\}$. The quantity $s = \#\{x_i \leq \theta\}$ has a

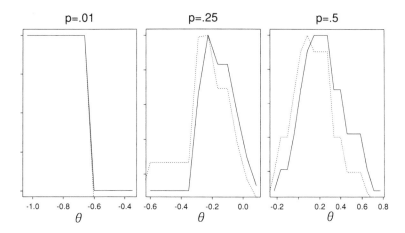

Figure 24.1. *Nonparametric likelihood (solid curve) and empirical likelihood (dotted curve) for estimating the pth quantile from a standard normal sample of size 20. The panels show (from left to right), p = .01, .25, .5. Note the different ranges on the horizontal axes.*

binomial distribution $\text{Bi}(n, p)$ if θ is the pth quantile of F. Thus an approximate nonparametric likelihood for θ is

$$L(\theta) = \binom{n}{s} p(\theta)^s (1 - p(\theta))^{n-s}. \qquad (24.8)$$

where $p(\theta)$ satisfies $\theta = \inf\{x; F(x) \geq p(\theta)\}$. It is not clear whether this is a likelihood in the strict sense, but it does seem a reasonable function on which to base inference. It turns out that the empirical likelihood can be computed exactly for this problem (Problem 24.4). It has a similar form, but is not the same, as the nonparametric likelihood (24.8). For a standard normal sample of size 20, Figure 24.1 shows the nonparametric likelihood (solid curve) and the empirical likelihood (dotted curve) for the $p = .01, .25, .5$ quantiles. The two are very similar.

Figure 24.2 shows a more complex example. The data are a random sample of 22 of the 88 test scores given in Table 7.1. The statistic of interest $\hat{\theta}$ is the maximum eigenvalue of the covariance matrix. The solid curve in Figure 24.2 is the empirical likelihood.

The dotted curve is the standard normal approximate likelihood

$$L_{\text{nor}}(\theta) = \exp\left\{-\frac{(\hat{\theta} - \theta)^2}{2\hat{\sigma}^2}\right\} \tag{24.9}$$

based on the normal theory approximation $\hat{\theta} \sim N(\theta, \hat{\sigma}^2)$, where $\hat{\sigma}$ is the bootstrap estimate of the standard deviation of $\hat{\theta}$. By definition both achieve their maximum at $\theta = \hat{\theta}$; however, the empirical likelihood is shifted to the right compared to normal theory curve.

Computation of the empirical likelihood is quite difficult in general. For statistics derived as the solution to a set of estimating equations, the constrained optimization problem may be recast into an unconstrained problem via convex duality. Newton or other multidimensional minimization procedures can then be applied. In the maximum eigenvalue problem, we were unable to compute the empirical likelihood exactly; the solid curve that appears in Figure 24.2 is actually the likelihood evaluated over the least favorable family in the parameter space. This can be viewed as a Taylor series approximation to the empirical likelihood.

Attractive properties of the empirical likelihood include: a) it transforms as a likelihood should [the empirical likelihood of $g(\theta)$ is $L_{\text{emp}}(g(\theta))$], and b) it is defined only for permissible values of θ, (for example $[-1, 1]$ for a correlation coefficient). A further advantage of empirical likelihood is its simple extension to multiple parameters of interest. This feature is not shared by most of the other techniques described in this chapter.

24.3 Approximate pivot methods

Suppose we assume that $\hat{\theta} - \theta$ is a pivotal quantity; that is, if θ is the true value then

$$\hat{\theta} - \theta \sim H \tag{24.10}$$

with the distribution function H not involving θ. If this is the case, we can estimate H for the single value $\theta = \hat{\theta}$ and then infer its value for all θ. Let

$$\hat{\theta}^* - \hat{\theta} \sim \hat{H}; \tag{24.11}$$

$\hat{H}$ is the cumulative distribution function of $\hat{\theta}^* - \hat{\theta}$ under bootstrap sampling. Usually $\hat{H}$ cannot be given in closed form, but rather is

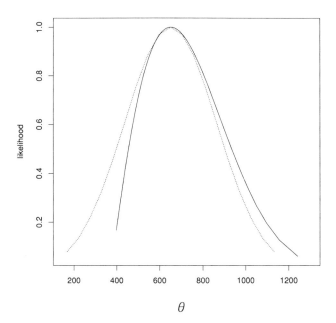

Figure 24.2. *Empirical likelihood (solid curve) and standard normal likelihood (dotted curve) for the maximum eigenvalue problem.*

estimated by generating B bootstrap samples $\mathbf{x}^{*b}, b = 1, 2, \ldots B$, computing $\hat{\theta}^*(b) = s(\mathbf{x}^{*b})$ for each sample, and defining

$$\rho^*(b) = \hat{\theta}^*(b) - \hat{\theta}; \quad b = 1, 2, \ldots B \qquad (24.12)$$

Our estimate $\hat{H}$ is the empirical distribution of the B values $\rho^*(b)$. Note, however, that the empirical distribution does not possess a density function, and for construction of an approximate likelihood, a density function is required. Let $\hat{h}(\rho)$ be a kernel density estimate of the distribution of $\hat{\rho} = \hat{\theta} - \theta$ based on the values $\rho^*(b)$:

$$\hat{h}(\rho) = \frac{1}{Bs} \sum_{b=1}^{B} k\left(\frac{\rho - \rho^*(b)}{s}\right). \qquad (24.13)$$

Here $k(\cdot)$ is any kernel function, for example, the standard normal density function, with window width s (see Section 16.5 for details). Using the pivot assumption (24.10), an estimate of the density function of $\hat{\theta}$ when the true value is θ is given by

$$
\begin{aligned}
p(\hat{\theta}|\theta) &= \hat{h}(\hat{\theta} - \theta) \\
&= \frac{1}{Bs} \sum_{1}^{B} k\left[(2\hat{\theta} - \theta - \hat{\theta}^*(b))/s\right]
\end{aligned}
\qquad (24.14)
$$

This gives the approximate likelihood $L(\theta) = \hat{h}(\hat{\theta} - \theta)$ thought of as a function of θ. The success of this approach will depend on how close $\hat{\theta} - \theta$ is to being pivotal. Alternatively, one might use the studentized approximate pivot $(\hat{\theta} - \theta)/\hat{\sigma}$ or the variance stabilized approximate pivot $g(\hat{\theta}) - g(\theta)$ discussed in Chapter 12. Some care must be used, however, when defining a likelihood from a pivot. For example, notice that if $\hat{\theta} \sim N(\theta, \theta)$, then $Z = (\hat{\theta} - \theta)/\theta$ has a standard normal distribution but the likelihood of θ is not $\exp(-Z^2/2)$. In general, to form a likelihood from the distribution of a pivot $Z = g(\hat{\theta}, \theta)$, the Jacobian of the mapping $\hat{\theta} \to Z$ should not involve θ. If it doesn't, then the density function of $\hat{\theta}$ is proportional to the density of Z.

Figure 24.3 shows an approximate likelihood for the maximum eigenvalue problem. The solid curve is the approximate likelihood computed using (24.14) with $B = 100$ bootstrap samples and a Gaussian kernel with a manually chosen window width. The dotted curve is the standard normal theory likelihood (24.9). The pivot-based curve is shifted to the right compared to normal curve, as was the empirical likelihood.

24.4 Bootstrap partial likelihood

The bootstrap partial likelihood approach estimates the distribution $p(\hat{\theta}|\theta)$ using a nested bootstrap procedure. The method proceeds as follows. We generate B_1 bootstrap samples $\mathbf{x}^{*1}, \cdots \mathbf{x}^{*B_1}$ giving bootstrap replications $\hat{\theta}_1^*, \cdots \hat{\theta}_{B_1}^*$. Then from each of the bootstrap samples, $\mathbf{x}^{*b}$, we generate B_2 second stage bootstrap samples,[1] giving second stage bootstrap replicates $\hat{\theta}_{b1}^{**}, \cdots \hat{\theta}_{bB_2}^{**}$. We

---

[1] A second stage bootstrap sample consists of n draws with replacement from a bootstrap sample $\mathbf{x}^{*b}$.

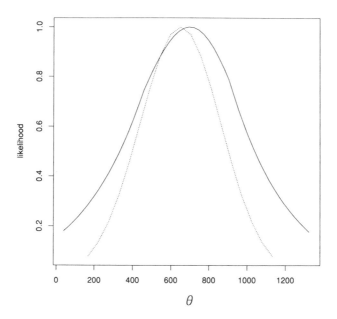

Figure 24.3. *Approximate likelihood based on* $\hat{\theta} - \theta$ *(solid curve) and standard normal likelihood (dotted curve) for the maximum eigenvalue problem.*

form the kernel density estimates

$$\hat{p}(t|\hat{\theta}_b^*) = \frac{1}{B_2 s} \sum_{j=1}^{B_2} k\left(\frac{t - \theta_{bj}^{**}}{s}\right) \tag{24.15}$$

for $b = 1, 2, \cdots B_1$. As in the previous section $k(\cdot)$ is any kernel function, for example the standard normal density function, with window width s (see Section 16.5 for details). We then evaluate $\hat{p}(t|\hat{\theta}_b^*)$ for $t = \hat{\theta}$. Since the values $\hat{\theta}_{bj}^{**}$ were generated from a distribution governed by parameter value $\hat{\theta}_b^*$, $\hat{p}(\hat{\theta}|\hat{\theta}_b^*)$ provides an estimate of the likelihood of θ for parameter value $\theta = \hat{\theta}_b^*$.

A smooth estimate of the likelihood is then obtained by applying a scatterplot smoother to the pairs $[\hat{\theta}_b^*, \hat{p}(\hat{\theta}|\hat{\theta}_b^*)]$, $b = 1, 2, \ldots B_1$.

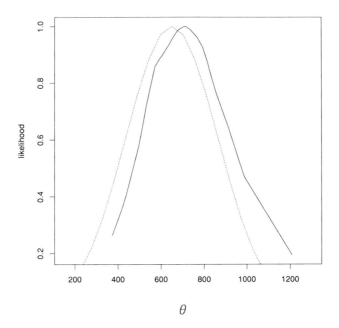

Figure 24.4. *Bootstrap partial likelihood (solid curve) and standard normal likelihood (dotted curve) for the maximum eigenvalue problem.*

This construction is called *bootstrap partial likelihood* because it estimates the likelihood based on $\hat{\theta}$ rather than the full data **x**. Further details of the implementation may be found in Davison *et al.* (1992).

Figure 24.4 shows the bootstrap partial likelihood and normal theory likelihood for the maximum eigenvalue problem. We used 40 bootstrap replications at each level, for a total of 1600 bootstrap samples. The window sizes for the kernel density estimate and the scatterplot smoother (a local least-squares fit) were chosen manually to make the final estimate look smooth. The bootstrap partial likelihood is similar to the previous empirical and pivot-based likelihoods.

24.5 Implied likelihood

Suppose we have a set of α-level confidence points $\theta_{\mathbf{X}}(\alpha)$ for a parameter θ based on a data set $\mathbf{x}$. The *implied likelihood* approach deduces a likelihood for θ from $\theta_{\mathbf{X}}(\alpha)$. The idea is as follows. Let $\alpha_{\mathbf{X}}(\theta)$ be the inverse of $\theta_{\mathbf{X}}(\alpha)$, that is, the coverage level corresponding to endpoint θ. Now define a density $\pi_{\mathbf{X}}^{\text{imp}}(\theta)$ by

$$\pi_{\mathbf{X}}^{\text{imp}}(\theta) = d\alpha_{\mathbf{X}}(\theta)/d\theta \qquad (24.16)$$

or $\alpha_{\mathbf{X}}(\theta) = \int^{\theta} \pi_{\mathbf{X}}^{\text{imp}}(\theta)d\theta$. We think of $\pi_{\mathbf{X}}^{\text{imp}}(\theta)$ as the *implied posterior distribution* whose α percentage points are given by $\alpha_{\mathbf{X}}(\theta)$; $\pi_{\mathbf{X}}^{\text{imp}}(\theta)$ is sometimes called the confidence distribution for θ. The implied likelihood for θ is defined by

$$L_{\text{imp}}(\theta) = \frac{\pi_{\mathbf{XX}}^{\text{imp}}(\theta)}{\pi_{\mathbf{X}}^{\text{imp}}(\theta)}, \qquad (24.17)$$

where $\mathbf{xx}$ denotes the data set consisting of two independent, identical copies of $\mathbf{x}$. The motivation for $L_{\text{imp}}(\theta)$ is the following. Suppose $\pi_{\mathbf{X}}(\theta)$ is an actual posterior distribution corresponding to a prior $\pi_0(\theta)$ and a likelihood $L_{\mathbf{X}}(\theta)$. Then

$$\pi_{\mathbf{X}}(\theta) = \pi_0(\theta)L_{\mathbf{X}}(\theta) \qquad (24.18)$$

$$\pi_{\mathbf{XX}}(\theta) = \pi_0(\theta)L_{\mathbf{X}}^2(\theta) \qquad (24.19)$$

and therefore

$$\frac{\pi_{\mathbf{XX}}(\theta)}{\pi_{\mathbf{X}}(\theta)} = L_{\mathbf{X}}(\theta) \qquad (24.20)$$

so that the ratio $\pi_{\mathbf{XX}}/\pi_{\mathbf{X}}$ recovers the likelihood $L_{\mathbf{X}}(\theta)$. It is known (Lindley, 1958) that the use of the confidence distribution $\pi_{\mathbf{X}}^{\text{imp}}(\theta)$ as a likelihood can lead to inconsistencies. The reason is that the confidence distribution contains an implied prior distribution that must be removed in order to obtain a quantity with likelihood properties. The definition (24.17) removes this prior in the correct manner. We can also obtain an expression for the *implied (noninformative) prior*

$$\pi_0^{\text{imp}}(\theta) = \frac{[\pi_{\mathbf{X}}^{\text{imp}}(\theta)]^2}{\pi_{\mathbf{XX}}^{\text{imp}}(\theta)}. \qquad (24.21)$$

The term "noninformative" means that Bayesian intervals based on $\pi_0^{\mathrm{imp}}(\theta)$ will have accurate frequentist coverage properties.

A simple example helps to illustrate this. Consider inference for the success probability θ in a binomial experiment with s successes out of n trials. If $\theta_{\mathbf{X}}(\alpha)$ is the exact α-level confidence point then

$$\pi_{\mathbf{X}}^{\mathrm{imp}}(\theta) = c\theta^s(1-\theta)^{n-s}\left[\frac{1-\hat{\theta}}{1-\theta} + \frac{\hat{\theta}}{\theta}\right] \qquad (24.22)$$

while $L_{\mathrm{imp}}(\theta)$ is the actual binomial likelihood $c\theta^s(1-\theta)^{n-s}$ (Problem 24.1). The confidence distribution contains an implied prior distribution, in square brackets, that must be removed in order to obtain the likelihood function. Suppose instead that we used negative binomial sampling, that is, we ran the experiment until a fixed number s of successes. Then the confidence distribution changes but the implied likelihood still equals $c\theta^s(1-\theta)^{n-s}$ as it should (Problem 24.1).

Computation of the implied likelihood requires a set of confidence intervals for θ. Suppose we assume that $\hat{\theta} - \theta$ is an approximate pivotal quantity and base the confidence intervals on the bootstrap distribution of $\hat{\theta}^* - \hat{\theta}$. Then a simple calculation shows that $L_{\mathrm{imp}}(\theta)$ is equal to the approximate likelihood derived in section 24.3 (Problem 24.2). More generally, one might base the implied likelihood on the BC_a intervals. However, the ABC intervals of Chapter 22 turn out to be more convenient computationally. Define the series of transformations

$$\theta \rightarrow \quad \xi \equiv \frac{\theta-\hat{\theta}}{\hat{\sigma}} \rightarrow \quad \lambda \equiv \frac{2\xi}{1+[1+4\hat{c}_q\xi]^{1/2}}$$

$$\rightarrow \qquad w \equiv \frac{2\lambda}{(1+2\hat{a}\lambda)+(1+4\hat{a}\lambda)^{1/2}}, \qquad (24.23)$$

where $\hat{a}$ and $\hat{c}_q$ are the acceleration and curvature constants that are defined in Section 22.6. Then the implied likelihood is simply

$$\exp\{-\frac{1}{2}w(\theta)^2\}. \qquad (24.24)$$

Figure 24.5 shows the implied likelihood (solid curve) for the maximum eigenvalue problem, obtained using (24.24). It shifts the normal theory likelihood (dotted curve) to the right, although the position of the maxima coincides at $\theta = \hat{\theta}$, since $w(\hat{\theta}) = 0$.

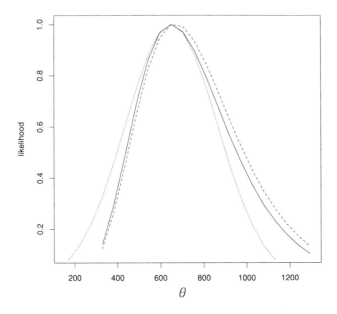

Figure 24.5. *Implied likelihood (solid curve), modified version of implied likelihood (dashed curve), and standard normal likelihood (dotted curve) for the maximum eigenvalue problem.*

In exponential family problems, it is possible to show that the implied likelihood based on the ABC intervals agrees with the profile likelihood up to second order. One can also make a more refined adjustment that can move the position of the maximum and make the implied likelihood close to the conditional profile likelihood of Cox and Reid (1987). The modification has the form

$$\exp\{-\frac{1}{2}w(\theta)^2\}\exp\{-(\hat{\gamma}/\hat{\sigma})\theta\} \qquad (24.25)$$

where $\hat{\gamma}$ is the total curvature of $\hat{\theta}$ as defined in Section 22.6. The dashed curve in Figure 24.5 shows that the modification shifts the likelihood a short distance to the right.

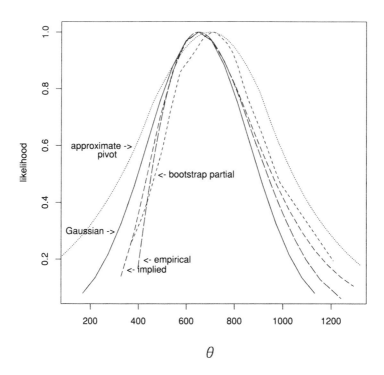

Figure 24.6. *Likelihoods for the maximum eigenvalue problem.*

24.6 Discussion

We have described a number of different methods for obtaining
approximate non-parametric likelihoods. Figure 24.6 summarizes
the results for the maximum eigenvalue problem. Some theoretical
results suggest that the bootstrap partial likelihood and the implied likelihood will agree closely with the profile likelihood, and
the latter is the empirical likelihood in the nonparametric setting.
Figure 24.5 seems to confirm this for our example. Finally, it is
important to note that the techniques described here are relatively
new. More research and experience is needed to understand their
properties.

24.7 Bibliographic notes

Nonparametric likelihood is discussed by Kiefer and Wolfowitz (1956), Scholz (1980), and in the special case of quantiles by Jeffreys (1961) and Wasserman (1990). Empirical likelihood is studied by Owen (1988, 1990), Kolaczyk (1993), Qin and Lawless (1991), and Hall and La Scala (1990). Theoretical adjustments to empirical likelihood are given by DiCiccio, Hall and Romano (1989b). Boos and Monahan (1986) and Hall (1987) discuss pivot methods for approximate likelihoods. A related procedure for contrast parameters is given by Ogbonmwan and Wynn (1988). Bootstrap partial likelihood is described in Davison, Hinkley and Worton (1992). Efron (1992c) proposes the implied likelihood. A brief overview of approximate likelihoods appears in Hinkley (1988).

24.8 Problems

24.1 (a) Derive equation (24.22) for the confidence distribution in the binomial experiment.

 (b) Show that the implied likelihood for the binomial experiment equals $c\theta^s(1-\theta)^{n-s}$.

 (c) Under negative binomial sampling, show that the confidence distribution changes but the implied likelihood is still equal to $c\theta^s(1-\theta)^{n-s}$.

24.2 Show that if $\hat{\theta} - \theta$ is a pivotal quantity, then the implied likelihood, based on the confidence intervals from the pivot, is equal to the marginal likelihood.

24.3 Derive expression (24.24) for the implied likelihood based on the ABC intervals.

24.4 Show that the empirical likelihood for the pth quantile is given by

$$L_{\text{emp}}(\theta) = cp^s(1-p)^{n-s}s^{-s}(n-s)^{n-s} \qquad (24.26)$$

where $s = \#\{x_i \leq \theta\}$ if $\theta < \hat{\theta}$, $s = np$ if $\theta = \hat{\theta}$ and $s = \#\{x_i < \theta\}$ if $\theta > \hat{\theta}$. Compare this to the nonparametric likelihood (24.8) in a numerical example [Wasserman, 1990].

## CHAPTER 25

Bootstrap bioequivalence and power calculations: a case history

25.1 Introduction

A small data set often requires proportionately greater amounts of statistical analysis. This chapter concerns a bioequivalence study involving only eight patients. The bioequivalence problem is a good one for understanding the advantages and limitations of bootstrap confidence intervals. Power calculations give us a chance to see bootstrap prediction methods in action. We begin by describing the problem, and then give solutions based on the simplest bootstrap ideas. An improved analysis based on more advanced bootstrap methods completes the chapter.

25.2 A bioequivalence problem

A drug company has separately applied each of three hormone supplement medicinal patches to eight patients who suffer from a hormone deficiency. One of the three patches is "Approved", meaning that it has received approval from the Food and Drug Administration (FDA). Another of the three patches is a "Placebo", which contains no hormone. The third patch is "New", meaning that it is manufactured at a new facility but is otherwise intended to be identical to "Approved." The three wearings occur in random order. Each patient's blood level of the hormone is measured after each patch wearing, with the results shown in Table 25.1. Notice that both the Approved and New patches raise the blood level of the hormone above that for the Placebo in all eight patients.

The FDA requires proof of *bioequivalence* before it will approve for sale a previously approved product manufactured at a new facility. Bioequivalence has a technical definition: let x indicate the

Table 25.1. *A small bioequivalence study; $n = 8$ patients each received three patches; measurements are blood levels of a hormone that these patients are deficient in. "Approved" is the blood level after wearing hormone supplement patch approved by the FDA; "New" patches come from a new manufacturing facility, but otherwise are supposed to be identical to the approved patches; "Placebo" patches contain no active ingredients. Are the new patches bioequivalent to the old according to the FDA's definition?*

Patient	Placebo	Approved	New	App-Pla	New-App.
1.	9243	17649	16449	8406	-1200
2.	9671	12013	14614	2342	2601
3.	11792	19979	17274	8187	-2705
4.	13357	21816	23798	8459	1982
5.	9055	13850	12560	4795	-1290
6.	6290	9806	10157	3516	351
7.	12412	17208	16570	4796	-638
8.	18806	29044	26325	10238	-2719
mean	11328	17671	17218	6342	-452

difference between Approved and Placebo measurements on the same patient, and let y indicate the difference between New and Approved,

$$x = \text{Approved} - \text{Placebo} \qquad y = \text{New} - \text{Approved}. \qquad (25.1)$$

Let μ and ν be the expectations of x and y,

$$\mu = \text{E}(x), \qquad \nu = \text{E}(y), \qquad (25.2)$$

and define ρ to be the ratio of ν to μ

$$\rho = \nu/\mu. \qquad (25.3)$$

The FDA bioequivalence requirement is that a .90 central confidence interval for ρ lie within the range $[-.2, .2]$. If the .90 confidence interval is expressed in terms of a lower .05 limit and an upper .95 limit,

$$\rho \in (\hat{\rho}[.05], \ \hat{\rho}[.95]), \qquad (25.4)$$

then the requirement is that

$$-.2 < \hat{\rho}[.05] \quad \text{and} \quad \hat{\rho}[.95] < .2. \qquad (25.5)$$

In other words, the FDA requires the New patches to have the same efficacy as the Approved patches, within an error tolerance of 20%.

Table 25.1 shows the (x_i, y_i) pairs for the $n = 8$ patients. These are the data that we analyze in this chapter,

$$\text{data}_n = \{(x_i, y_i), i = 1, 2, \cdots, n\}. \tag{25.6}$$

Figure 25.1 plots the data. The "+" symbol indicates the mean of the vectors (x_i, y_i),

$$(\bar{x}, \bar{y}) = (6342, -452) = (\hat{\mu}, \hat{\nu}). \tag{25.7}$$

This means $\bar{x}$ and $\bar{y}$ estimate the expectations μ and ν in (25.2) and provide a natural estimate of the ratio ρ,

$$\hat{\rho} = \hat{\nu}/\hat{\mu} = \bar{y}/\bar{x} = -.071. \tag{25.8}$$

The fact that ρ lies well inside the range $(-.2, .2)$ does not necessarily imply that the bioequivalence criteria (25.5) are satisfied.

The drug company wishes to answer two related questions:

Question 1   Are the FDA bioequivalence criteria satisfied by the data in Table 25.1?

Question 2   If not, how many patients should be measured in a future experiment so that the FDA requirements will have a good chance of being satisfied?

The second question relates to what is usually called a *power* or *sample size* question.

25.3 Bootstrap confidence intervals

The left panel of Figure 25.2 shows the histogram of $B = 4000$ nonparametric bootstrap replications of $\hat{\rho}$. The original data set can be thought of as a sample of size $n = 8$ from an unknown bivariate probability distribution F for the pairs (x, y),

$$F \rightarrow \text{data}_n = \{(x_1, y_1), (x_2, y_2), \cdots, (x_8, y_8)\}. \tag{25.9}$$

A nonparametric bootstrap sample is a random sample of size $n = 8$ from the empirical distribution $\hat{F}$, as in Chapter 6,

$$\hat{F} \rightarrow \text{data}_n^* = \{(x_1^*, y_1^*), (x_2^*, y_2^*), \cdots, (x_8^*, y_8^*)\}. \tag{25.10}$$

In this case $\hat{F}$ is the distribution putting probability $1/8$ on each original data point (x_i, y_i), $i = 1, 2, \cdots, 8$. In other words, data_n^*

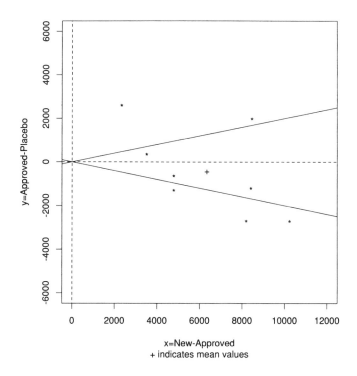

Figure 25.1. *A plot of the eight patch data points (x_i, y_i) from Table 25.1;* $+$ *indicates observed mean* $(\bar{x}, \bar{y}) = (\hat{\mu}, \hat{\nu})$; *wedge indicates the FDA bioequivalence region for expectations* (μ, ν). *The observed mean is in the wedge, but does the confidence interval for* ρ *pass the bioequivalence criteria (25.5)?*

is a random sample of size 8 drawn with replacement from data$_n$, (25.6).

Each bootstrap sample data$_n^*$ gives a bootstrap replication of $\hat{\rho}$,

$$\hat{\rho}^* = \bar{y}^*/\bar{x}^* = (\sum_{i=1}^{8} y_i^*/8)/(\sum_{i=1}^{8} x_i^*/8). \qquad (25.11)$$

$B = 4000$ independent bootstrap samples gave 4000 bootstrap

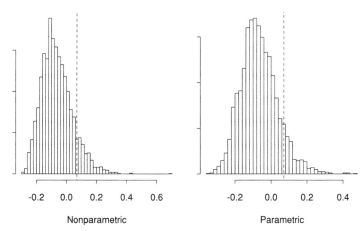

Figure 25.2. *Left panel: 4000 nonparametric bootstrap replications of* $\hat{\rho} =$ $\bar{y}/\bar{x}$. *Right panel: 4000 normal theory bootstrap replications of* $\hat{\rho}$. *The two histograms look similar and have nearly the same means and standard deviations. They give moderately different bootstrap confidence intervals though, as seen in Table 25.2.*

replications

$$\hat{\rho}^*(b) \quad \text{for} \quad b = 1, 2, \cdots, B = 4000. \tag{25.12}$$

These had mean $\hat{\rho}^*(\cdot) = -.063$ and standard deviation

$$\widehat{\text{se}}_{4000} = \{\sum_{b=1}^{B}[\hat{\rho}^*(b) - \hat{\rho}^*(\cdot)]^2/(B-1)\}^{1/2} = .103, \tag{25.13}$$

this being the bootstrap estimate of standard error for $\hat{\rho}$. The histogram is notably long-tailed toward the right. $B = 4000$ is twenty times as big as necessary for a reasonable standard error estimate, but it is only twice as big as necessary for computing bootstrap confidence intervals.

The right panel of Figure 25.2 is the histogram of 4000 normal-theory parametric bootstrap replications of $\hat{\rho}$. The only difference from the nonparametric theory comes in the choice of $\hat{F}$ in (25.9). Instead of using the empirical distribution, we take $\hat{F}$ equal to the best-fitting bivariate normal distribution to the (x, y) data in Table 25.1. "Best-fitting" refers to choosing the expectation vector λ and covariance matrix $\mathbf{\Sigma}$ for the bivariate normal distribution according to maximum likelihood theory,

$$\hat{\lambda} = \begin{pmatrix} \bar{x} \\ \bar{y} \end{pmatrix} \quad \text{and} \quad \hat{\mathbf{\Sigma}} = \frac{1}{8} \begin{pmatrix} \sum_1^8 (x_i - \bar{x})^2 & \sum_1^8 (x_i - \bar{x})(y_i - \bar{y}) \\ \sum_1^8 (x_i - \bar{x})(y_i - \bar{y}) & \sum_1^8 (y_i - \bar{y})^2 \end{pmatrix} \tag{25.14}$$

Instead of (25.9) we generate the bootstrap data according to

$$\hat{F}_{\text{norm}} \to \text{data}_n^* = \{(x_1^*, y_1^*), (x_2^*, y_2^*), \cdots, (x_8^*, y_8^*) \tag{25.15}$$

where $\hat{F}_{\text{norm}} = N_2(\hat{\lambda}, \hat{\mathbf{\Sigma}})$. In other words, data_n^* is a random sample of size 8 from $\hat{F}_{\text{norm}}$.

Figure 25.3 shows the ellipses of constant density for $\hat{F}_{\text{norm}}$. $\hat{F}_{\text{norm}}$ is much smoother than the empirical distribution $\hat{F}$, but that doesn't seem to make much difference to the bootstrap results. The two histograms in Figure 25.2 have similar shapes, and nearly the same means and standard deviations. Closer inspection reveals that the parametric histogram is shifted a little to the left of the nonparametric histogram. This shift shows up in the bootstrap confidence intervals.

Table 25.2 shows the BC_a confidence intervals based on the percentiles of the histograms in Figure 25.2. The central .90 nonparametric BC_a interval is

$$\rho \in (-.204, .146), \tag{25.16}$$

which comes close to satisfying the FDA bioequivalence criteria (25.5). The values of $\hat{a}$ and $\hat{z}_0$ required for the BC_a intervals were $(\hat{a}, \hat{z}_0) = (.028, .021)$, calculated from (14.14) and (14.15). By following the BC_a definitions (14.9) and (14.10), the reader can calculate that $-.204$ and $.146$ are, respectively, the 6.25th and 96.14th percentiles of the left-hand histogram in Figure 25.2.

The normal theory BC_a .90 interval is

$$\rho \in (-.221, .112), \tag{25.17}$$

shifted downward from (25.16). The difference isn't large, but it

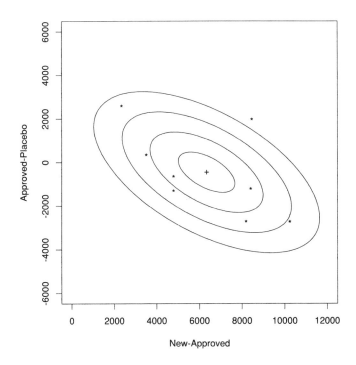

Figure 25.3. *The ellipses indicate curves of constant density for $\hat{F}_{\mathrm{norm}}$, the best-fitting bivariate normal distribution to the eight (x, y) pairs in Figure 25.1. $\hat{F}_{\mathrm{norm}}$ is much smoother than the empirical distribution $\hat{F}$, which is concentrated in the eight starred points.*

moves the results further into violation of criteria (25.5). Later we will see that the bootstrap intervals are somewhat too short in this case, for reasons having to do with the small sample size 8.

Table 25.2 also shows the ABC confidence interval endpoints, computed from the algorithm **abcnon** as described in the Appendix. These are quite close to the BC_a endpoints, and required only 1% as much computational effort.

Table 25.2. *Bootstrap approximate confidence intervals for the ratio* ρ; *left: nonparametric* BC_a *and ABC intervals; right: normal theory parametric* BC_a *and ABC intervals. The nonparametric* BC_a *.90 central interval* $(-.204, .146)$ *nearly satisfies the bioequivalence requirement (25.5) the corresponding parametric interval is* $(-.221, .112)$. *The constants required for the intervals appear in the bottom row.*

| | Nonparametric | | Parametric | |
α	BC_a	ABC	BC_a	ABC
.025	-.226	-.222	-.249	-.239
.05	-.204	-.202	-.225	-.215
.10	-.178	-.177	-.193	-.186
.16	-.158	-.155	-.168	-.162
.84	.041	.043	.031	.033
.90	.085	.082	.065	.067
.95	.146	.136	.112	.110
.975	.192	.188	.150	.152
$(\hat{a}_1, \hat{z}_0, \hat{c}_q)$:	(.028,.021,—)	(.028,.028,.073)	(0,-.010,—)	(0,0,.073)

25.4 Bootstrap power calculations

The drug company decided to run a larger study in order to definitively verify bioequivalence. This meant answering Question 2: how many patients should be enrolled in the new study to give it good power, i.e., a good probability of satisfying the bioequivalence criteria? Bootstrap methods are well-suited to answering power and sample size questions.

We can imagine drawing a future sample of size say N from the distribution F that yielded the original data (25.9):

$$F \to \text{data}_N = \{(X_j, Y_j), \ j = 1, 2, \cdots, N\}. \qquad (25.18)$$

The capital letters (X_j, Y_j) are intended to avoid confusion with the actual data set $\{(x_i, y_i), i = 1, 2, \cdots, n\}$. Having obtained data$_N$, we will use it to calculate a confidence interval for ρ, say

$$\rho \in (\hat{\rho}_N[.05], \ \hat{\rho}_N[.95]), \qquad (25.19)$$

and hope that the bioequivalence criteria are satisfied,

$$-.2 < \hat{\rho}_N[.05] \quad \text{and} \quad \hat{\rho}_N[.95] < .2. \qquad (25.20)$$

The power, or sample size, calculation consists of choosing N so that (25.20) is likely to occur. (Usually it is not allowed to use the original n points in the new study, which is why the sample size in (25.19) is N rather than $N + n$.)

Let $\pi_N(\text{lo})$ and $\pi_N(\text{up})$ be the probabilities of violating the lower and upper bioequivalence criterion,

$$\begin{aligned} \pi_N(\text{lo}) &= \text{Prob}_{\hat{F}}\{\hat{\rho}_N[.05] < -.2\} \quad \text{and} \\ \pi_N(\text{up}) &= \text{Prob}_{\hat{F}}\{\hat{\rho}_N[.95] > .2\}. \end{aligned}$$

(25.21)

We can estimate $\pi_N(\text{lo})$ and $\pi_N(\text{up})$ by the plug-in principle of Chapter 4,

$$\begin{aligned} \hat{\pi}_N(\text{lo}) &= \text{Prob}_{\hat{F}}\{\hat{\rho}_N[.05]^* < .2\} \quad \text{and} \\ \hat{\pi}_N(\text{up}) &= \text{Prob}_{\hat{F}}\{\hat{\rho}_N[.95]^* > .2\}. \end{aligned}$$

(25.22)

The calculation of $\hat{\pi}_N(\text{lo})$ and $\hat{\pi}_N(\text{up})$ is done by the bootstrap. Let data_N^* be a bootstrap sample of size N,

$$\hat{F} \rightarrow \text{data}_N^* = \{(X_j^*, Y_j^*), \ j = 1, 2, \cdots, N\}. \tag{25.23}$$

Since $\hat{F}$ still is the empirical distribution of the original data, data_N^* is a random sample of N pairs drawn with replacement from the original n pairs $\text{data}_n = \{(x_i, y_i), i = 1, 2, \cdots, n\}$. We calculate the confidence limits for ρ based on data_N^*, and check to see if the bioequivalence criteria are violated. The proportion of violations in a large number of bootstrap replications gives estimates of $\pi_N(\text{lo})$ and $\pi_N(\text{up})$.

Table 25.3 gives estimates of $\pi_N(\text{lo})$ and $\pi_N(\text{up})$ based on $B = 100$ bootstrap replications for $N = 12, 24, 36$. The endpoints $(\hat{\rho}_N[.05]^*, \hat{\rho}_N[.95]^*)$ in (25.22) were obtained by applying the nonparametric ABC algorithm for the ratio statistic $\hat{\rho}$ to the bootstrap data sets data_N^*. We see that $N = 12$ is too small, giving large estimated probabilities of violating the bioequivalence criteria, but $N = 24$ is much better, and $N = 36$ is almost perfect.

The computation in Table 25.3 uses modern computer-intensive bootstrap methodology, but it is identical in spirit to traditional sample size calculations: a preliminary data set, data_n, is used to estimate a probability distribution, in this case $\hat{F}$. Then the desired power or sample size calculations are carried out as if $\hat{F}$ were the true distribution. This is just another way of describing the plug-in principle.

The last column in Table 25.3 shows $\hat{\pi}_N(\text{lo})$ and $\hat{\pi}_N(\text{up})$ going to zero as N goes to infinity. This has to be true given the definitions

Table 25.3. *Estimated probabilities of violating the bioequivalency criteria for future sample sizes* N, *(25.22); based on* $B = 100$ *bootstrap replications for each sample size; confidence limits obtained from the nonparametric ABC algorithm abcnon.*

N:	12	24	36	∞
$\hat{\pi}_N(\text{lo})$:	.43	.15	.04	.00
$\hat{\pi}_N(\text{up})$:	.16	.01	.00	.00

we have used. Let $\hat{\rho}_N^*$ be the estimate of ρ based on data_N^*, (25.23)

$$\hat{\rho}_N^* = \bar{Y}^*/\bar{X}^* = (\sum_{j=1}^{N} Y_j^*/N)/(\sum_{j=1}^{N} X_j^*/N). \qquad (25.24)$$

It is easy to see that if N is large then $\hat{\rho}_N^*$ must be close to $\hat{\rho} = -.071$, the original estimate of ρ, (25.8). In fact $\hat{\rho}_N^*$ will have bootstrap expectation approximately $\hat{\rho}$, and bootstrap standard deviation approximately

$$\hat{\sigma}_N = [\frac{\hat{\rho}^2 \widehat{\Sigma}_{11} - 2\hat{\rho}\widehat{\Sigma}_{12} + \widehat{\Sigma}_{22}}{N\bar{x}^2}]^{1/2}, \qquad (25.25)$$

this being the delta-method estimate of standard error for $\hat{\rho}_N^*$. Here $\widehat{\Sigma}_{11}, \widehat{\Sigma}_{12}, \widehat{\Sigma}_{22}$ are the elements of $\widehat{\Sigma}$, (25.14). As N gets large, the bootstrap confidence limits $(\hat{\rho}_N[.05]^*, \hat{\rho}_N[.95]^*)$ will approach the standard limits

$$(\hat{\rho} - 1.645\hat{\sigma}_N, \ \hat{\rho} + 1.645\hat{\sigma}_N). \qquad (25.26)$$

Since $\hat{\sigma}_N \to 0$ as $N \to \infty$, and $\hat{\rho}$ is well inside the range $(-.2, .2)$, this means that $\hat{\pi}_N(\text{lo})$ and $\hat{\pi}_N(\text{up})$ both go to zero.

25.5 A more careful power calculation

The last column in Table 25.3 is worrisome. It implies that the drug company would certainly satisfy the bioequivalency criteria if the future sample size N got very large. But this can only be so if in fact the true value of ρ lies in the range $(-.2, .2)$. Otherwise we will certainly *disprove* bioequivalency with a large N. The trouble comes from the straightforward use of the plug-in principle. In assuming that F is $\hat{F}$ we are assuming that ρ equals $\hat{\rho} = -.071$, that is, that bioequivalency is true. This kind of assumption is standard in power calculations, and usually acceptable for the rough purposes and sample size planning, but it is possible to use the plug-in principle more carefully.

To do so we will use a method like that of the bootstrap-t of Chapter 12. Define

$$T = \frac{\hat{\rho}_n - \hat{\rho}_N[.05]}{\hat{\sigma}_n}, \tag{25.27}$$

where $\hat{\rho}_n$ is the estimate based on data$_n$, $\hat{\rho}_n = \hat{\rho} = -.071$. The denominator $\hat{\sigma}_n$ is the delta-method estimate of standard error for $\hat{\rho}_n$, formula (25.25) with $N = n$; $\hat{\sigma}_n = .097$ based on data$_n$. The statistic T measures how many standard error units it is from the original point estimate $\hat{\rho}_n$ to the confidence interval lower limit $\hat{\rho}_N[.05]$ based on a future sample data$_N$. The value $-.2$ is 1.33 such units from $\hat{\rho}_n = .071$,

$$\frac{\hat{\rho}_n - (-.2)}{\hat{\sigma}_n} = 1.33. \tag{25.28}$$

We see that the statement "$\hat{\rho}_N[.05] < -.2$" is equivalent to the statement "$T > 1.33$", given the observed values of $\hat{\rho}_n$ and $\hat{\sigma}_n$. We can estimate $\pi_N(\text{lo})$, (25.21), by using the bootstrap to estimate the probability that T exceeds 1.33,

$$\hat{\pi}_N(\text{lo}) = \text{Prob}_{\hat{F}}\{T^* > 1.33\}. \tag{25.29}$$

A bootstrap replication of T is of the form

$$T^* = \frac{\hat{\rho}_n^* - \hat{\rho}_N^*[.05]}{\hat{\sigma}_n^*}. \tag{25.30}$$

Here $\hat{\rho}_n^*$ and $\hat{\sigma}_n^*$ are the parameter estimate and standard error estimate for a bootstrap sample of size $n = 8$ as in (25.11), while $\hat{\rho}_N^*[.05]$ is the lower confidence limit based on a *separate* bootstrap sample of size N, as in (25.23).

The numbers in Table 25.4 are each based on $B = 100$ bootstrap replications of T^*, using (25.29). These results are much less optimistic than those in Table 25.3. This is because (25.30) takes into account the variability in the original sample of size $n = 8$, as well as in the future sample of size N, in estimating the probabilities $\pi_n(\text{lo})$ and $\pi_n(\text{up})$. Table 25.4 which is more realistic than Table 25.3, suggests that the drug company might well consider enrolling $N = 48$ patients in the new study.

Table 25.4. *Using the bootstrap-t method (25.29) to estimate the probabilities of violating the bioequivalence criteria. This analysis gives less optimistic results than those in Table 25.3.*

N:	12	24	36	48	∞
$\hat{\pi}_N(\text{lo})$:	.38	.33	.27	.18	.07
$\hat{\pi}_N(\text{up})$:	.35	.16	.18	.11	.07

It is possible to study more closely the relative merits of the two power calculations. The observed value of $\hat{\rho}$ was -0.071. What would have happened if we had observed $\hat{\rho}$ to be one standard deviation lower, that is, $\hat{\rho} = -.071 - .097 = -.168$? We perturbed our original data set by adding a value Δ to each y_i and subtracting the same Δ from each x_i, choosing Δ so that $\hat{\rho}$ for the perturbed data set was -0.168. Then we repeated each of the two power calculations. Then we found the value of Δ so that $\hat{\rho}$ was one standard deviation *larger* than -0.071 ($-0.071 + .097 = .26$), and again repeated the two power calculations. For brevity, we only tried $N = 24$ and did computations for the lower endpoint. The left panel of Figure 25.4 shows the 100 simulated values of $\hat{\rho}_N^*[.05]$ corresponding to data sets with $\hat{\rho} = -0.168, -0.071$, and .26. The right panel shows the boxplots for $(\hat{\rho}_n^* - \hat{\rho}_N^*[.05])/\hat{\sigma}_n^*$ for the same three data sets. Notice how the distributions shift dramatically in the left panel, but are quite stable in the right panel. The estimates of $\text{Prob}\{\hat{\rho}_N[.05]\}$ are .79, .15, and 0 from the left panel, while the estimates of $\text{Prob}\{(\hat{\rho}_n - \hat{\rho}_N[.05])/\hat{\sigma}_n > 1.33\}$ are .34, .33, and .31. from the right panel. This illustrates that the power calculation based on $(\hat{\rho}_n^* - \hat{\rho}_N^*[.05])/\hat{\sigma}_n^*$ is more reliable.

Both Tables 25.3 and 25.4 are based on technically correct applications of the bootstrap. The bootstrap is not a single technique, but rather a general method of solving statistical inference problems. In complicated situations like the one here, alternate bootstrap methods may coexist, requiring sensible decisions from the statistician. It never hurts to do more than one analysis.

The calculations in this section are related to the construction of a prediction interval using the bootstrap: see Problem 25.8.

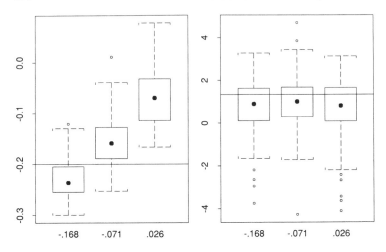

Figure 25.4. *The left figure shows boxplots of $\hat{\rho}_N^*[.05]$ for data sets with $\hat{\rho} = -0.071 - \hat{\sigma}_n, -0.071,$ and $-0.071 + \hat{\sigma}_n,$ respectively (from left to right). Each boxplot corresponds to 100 simulations; a horizontal line is drawn at -0.2. The right figure shows boxplots for the quantity $(\hat{\rho}_n^* - \hat{\rho}_N^*[.05])/\hat{\sigma}_n^*$ for the same three cases. A horizontal line is drawn at the value 1.33.*

25.6 Fieller's intervals

Suppose we are willing to assume that the true probability distribution F giving the $n = 8$ data pairs (x_i, y_i) in (25.9) is bivariate normal,

$$F = N_2(\lambda, \mathfrak{L}), \qquad \lambda = \begin{pmatrix} \mu \\ \nu \end{pmatrix}. \qquad (25.31)$$

In this case there exist exact confidence limits for $\rho = \nu/\mu$, called *Fieller's intervals*. We will use the Fieller intervals as a gold standard to check the parametric bootstrap confidence intervals in Table 25.2. Some discrepancies will become apparent, due to the small sample size of the data set. Finally, we will use a calibration approach to improve the bootstrap results.

Fieller's method begins by defining a function $T(\rho)$ depending on the statistics $\hat{\lambda}$ and $\widehat{\Sigma}$ in (25.14),

$$T(\rho) = \frac{\sqrt{n}[\bar{y} - \rho\bar{x}]}{[\rho^2\widehat{\Sigma}_{11} - 2\rho\widehat{\Sigma}_{12} + \widehat{\Sigma}_{22}]^{1/2}}, \qquad (25.32)$$

where $\widehat{\Sigma}_{11}$ is the upper left-hand element of $\widehat{\Sigma}$, $\widehat{\Sigma}_{11} = \sum_{i=1}^{n}(x_i - \bar{x})^2/n$, etc. Given data$_n$, we can calculate $T(\rho)$ for all possible values of ρ. From the true value of ρ, $T(\rho)$ has a rescaled Student's t distribution with $n - 1$ degrees of freedom,

$$T(\rho) \sim \sqrt{\frac{n}{n-1}}\, t_{n-1} \qquad \text{(for } \rho \text{ the true value).} \qquad (25.33)$$

The Fieller central .90 confidence interval for ρ consists of all values of ρ that put $T(\rho)$ between the 5th and 95th percentile of a $\sqrt{n/(n-1)}\, t_{n-1}$ distribution. We can express this as

$$-\sqrt{\frac{n}{n-1}}\, t_{n-1}^{(.95)} < T(\rho) < \sqrt{\frac{n}{n-1}}\, t_{n-1}^{(.95)}. \qquad (25.34)$$

There is a simple formula for the Fieller limits as asked for in Problem 25.4.

The top row of Table 25.5 shows that the central .90 Fieller interval for ρ is $(\hat{\rho}[.05], \hat{\rho}[.95]) = (-.249, .170)$. Two descriptors of the intervals are given,

$$\text{Length} = \hat{\rho}[.95] - \hat{\rho}[.05] \quad \text{and} \quad \text{Asymmetry} = \frac{\hat{\rho}[.95] - \hat{\rho}}{\hat{\rho} - \hat{\rho}[.05]}. \qquad (25.35)$$

Asymmetry describes how much further right than left the interval extends from the point estimate $\hat{\rho}$. The standard intervals $\hat{\rho} \pm 1.645\, \hat{\sigma}_n$ always have Asymmetry $= 1.00$, compared to the gold standard asymmetry 1.36 here. The length of the standard intervals is also considerably too small, .32 compared to .42.

The BC$_a$ and ABC intervals, taken from the parametric side of Table 25.2, have almost the correct asymmetry, but only slightly better length than the standard interval. This is not an accident. Theoretical results show that the bootstrap achieves the second order accuracy described in Chapter 22 by correcting the 1.00 asymmetry of the standard intervals. This corrects the leading source of coverage errors for the standard method.

The length deficiency seen in Table 25.5 is a third order effect, lying below the correction abilities of the BC$_a$ and ABC methods.

Table 25.5. *Parametric normal-theory confidence limits for ρ based on the $n = 8$ data pairs (x_i, y_i) in Table 25.1. The exact Fieller limits, top row, are compared with various approximate intervals; BC_a and ABC intervals have about the correct asymmetry, but are no better than the standard intervals in terms of length. Calibrating the ABC intervals gives nearly exact results.*

	Limits		Length	Asymmetry
	$\hat{\rho}[.05]$	$\hat{\rho}[.95]$	$\hat{\rho}[.95] - \hat{\rho}[.05]$	$\frac{(\hat{\rho}[.95]-\hat{\rho})}{(\hat{\rho}-[.05])}$
1. Fieller:	-.249	.170	.42	1.36
2. Standard: $\hat{\rho} \pm 1.645\hat{\sigma}_n$	-.232	.089	.32	1.00
3. BC_a:	-.212	.115	.33	1.32
4. ABC:	-.215	.111	.33	1.27
5. Fieller: "n"$= \infty$	-.217	.119	.34	1.31
6. ABC: calibrated	-.257	.175	.43	1.33

Theoretically the third order effects become negligible, compared to second order effects, as sample size gets large. In this case, however, the small sample size allows for a big third order effect on length.

In this problem we can specifically isolate the third order effect. It relates to the constant $\sqrt{n/(n-1)} \, t_{n-1}^{(.95)} = 2.08$ in (25.34). As the degrees of freedom $n - 1$ goes to infinity, this constant approaches $z^{(.95)} = 1.645$, the normal percentile point. Using 1.645 instead of 2.08 in calculating the Fieller limits (25.34) gives interval $(-.217, .119)$ for ρ, row 5 of Table 25.5. It is no coincidence that this nearly matches the BC_a and ABC intervals.

Of course what we really want is a bootstrap method that gives the actual Fieller limits of row 1. This requires improving the second order BC_a or ABC accuracy to third order. We conclude by using calibration, as described in Chapter 18, to achieve this improvement.

A confidence limit $\hat{\rho}[\alpha]$ is supposed to have probability α of covering (exceeding) the true value ρ,

$$\text{Prob}_F\{\rho < \hat{\rho}[\alpha]\} = \alpha. \qquad (25.36)$$

Thus ρ is supposed to be less than $\hat{\rho}[.95]$ 95% of the time, and less than $\hat{\rho}[.05]$ 5% of the time, so $\text{Prob}_F\{\hat{\rho}[.05] < \rho < \hat{\rho}[.95]\} = .90$. For an approximate confidence limit like ABC there is a true probability β that ρ is less than $\hat{\rho}[\alpha]$, say

$$\beta(\alpha) = \text{Prob}_F\{\rho < \hat{\rho}[\alpha]\}. \qquad (25.37)$$

An exact confidence interval method is one that has $\beta(\alpha) = \alpha$ for all α, but we are interested in inexact methods here.

If we knew the function $\beta(\alpha)$ then we could *calibrate*, or adjust, an approximate confidence interval to give exact coverage. Suppose we know that $\beta(.03) = .05$ and $\beta(.96) = .95$. Then instead of $(\hat{\rho}[.05], \hat{\rho}[.95])$ we would use $(\hat{\rho}[.03], \hat{\rho}[.96])$ to get a central .90 interval with correct coverage probabilities.

In practice we usually don't know the calibration function $\beta(\alpha)$. However we can use the bootstrap to estimate $\beta(\alpha)$. The bootstrap estimate of $\beta(\alpha)$ is

$$\hat{\beta}(\alpha) = \text{Prob}_{\hat{F}}\{\hat{\rho} < \hat{\rho}[\alpha]^*\}. \qquad (25.38)$$

In this definition $\hat{F}$ and $\hat{\rho}$ are fixed, nonrandom quantities, while $\hat{\rho}[\alpha]^*$ is the αth confidence limit based on a bootstrap data set from $\hat{F}$. (We are working in a parametric normal-theory mode in this section, so $\hat{F}$ is $\hat{F}_{\text{norm}}$, (25.15).) The estimate $\hat{\beta}(\alpha)$ is obtained by taking B bootstrap data sets, and seeing what proportion of them have $\hat{\rho} < \hat{\rho}[\alpha]^*$. See Problem 25.6 for an efficient way to do this calculation.

This calculation method was applied to the normal-theory ABC limits for ρ based on the patch data of Table 25.1. $B = 1000$ normal-theory data sets were drawn as in (25.15), and for each one the parametric ABC endpoint $\hat{\rho}[\alpha]^*$ based on data$_n^*$ was evaluated for α between 0 and 1. The value $\hat{\beta}(\alpha)$ is the proportion of the B endpoints exceeding $\hat{\rho} = -.071$. The curve $\hat{\beta}(\alpha)$ is shown in Figure 25.5.

The calibration tells us to widen the ABC limits. In particular

$$\hat{\beta}(.0137) = .05 \quad \text{and} \quad \hat{\beta}(.9834) = .95. \qquad (25.39)$$

This suggests replacing the ABC interval

$$(\hat{\rho}[.05], \hat{\rho}[.95]) = (-.215, .111) \qquad (25.40)$$

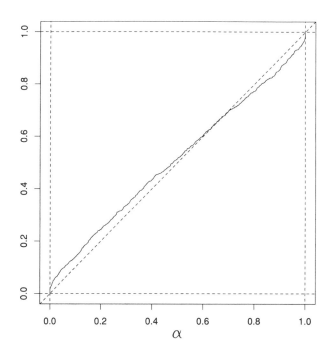

Figure 25.5. *Bootstrap calibration curve* $\hat{\beta}(\alpha)$; *based on* $B = 100$ *normal-theory bootstrap replications of the ABC endpoints. Important values:* $\hat{\beta}(.0137) = .05$ *and* $\hat{\beta}(.9834) = .95$.

with the calibrated ABC endpoints

$$(\hat{\rho}[.0137], \hat{\rho}[.9834]) = (-.257, .175). \qquad (25.41)$$

Row 6 of Table 25.5 shows that (25.41) has nearly the gold standard Length as well as Asymmetry.

   This data set was chosen deliberately to be one in which third order effects were large so that calibration gave substantial improvements. The calibration effects would have been noticeably less dramatic if n equaled 16 instead of 8. Nevertheless, it is nice to be able to check bootstrap results like those in Table 25.2, especially since the calibration requires no new assumptions or data. It does require a lot more computation, essentially a second level of bootstrapping. The computational efficiency of the ABC method, compared to BC_a was important here, since it is the entire confidence interval process that is bootstrapped in a calibration.

25.7 Bibliographic notes

References for bootstrap confidence intervals are given in the bibliographic notes at the end of Chapter 22. The paper of Efron (1986) looks specifically at confidence intervals for functions of a multivariate normal mean; the parametric analysis in the right half of Table 25.2 is a special case. He also discusses Fieller's interval of section 25.6 (Fieller, 1954). Power calculations based on normal theory are described in most applied statistics texts, for example Snedecor and Cochran (1980). The more careful power calculation of section 25.5 is an example of the use of bootstrap for *prediction*. This topic is discussed in Stine (1985), Bai and Olshen (1988), and Bai, Bickel, and Olshen (1990); see also Problem 25.8.

25.8 Problems

25.1 Give an explicit description of the calculation of $\hat{\pi}_N(\text{up})$ in Table 25.4.

25.2 Draw a diagram like Figure 8.1 showing the logic of the bootstrap method leading to Table 25.4.

25.3 How were the numbers corresponding to $N = \infty$ calculated in Table 25.4? Why aren't they zero, as in Table 25.3?

25.4 Derive a closed-form expression for the Fieller limits (25.33).

25.5 Suppose that in definition (25.32) of $T(\rho)$ we replaced the $\widehat{\Sigma}_{ij}$ with the usual unbiased estimates that divide by $n-1$ instead of n, $\widehat{\Sigma}_{11} = \Sigma(x_i - \bar{x})^2/(n-1)$, etc. How would that change (25.34)?

25.6 Figure 25.5 was actually calculated as follows: for each bootstrap data set data_n^*, the value $\hat{\alpha}^*$ such that $\hat{\rho}[\hat{\alpha}^*]^* = \hat{\rho}$ was calculated. (In other words, the ABC level $\hat{\alpha}^*$ limit for data_n^* exactly equaled $\hat{\rho} = -.007$.) Let $\hat{\alpha}^*(i)$ be the ith ordered value of the 1000 $\hat{\alpha}^*$'s. The plotted points in Figure 25.5 are

$$(\hat{\alpha}^*(i), (i - .5)/1000). \tag{25.42}$$

Explain why this is almost the same as carrying out the calibration calculation described in Section 25.5, if we assume that $\hat{\rho}[\alpha]^*$ is always an increasing function of α.

25.7 Suppose we wished to calibrate the parametric BC_a intervals, rather than ABC intervals, in Table 25.2. Describe how the calculations would be done.

25.8 *Prediction intervals from the bootstrap.*

Suppose we are in the one-sample situation $F \to \mathbf{x} = (x_1, x_2, \ldots x_n)$ and require a $1 - 2\alpha$ prediction interval for a new observation $Z \sim F$. That is, we would like random variables $a(\mathbf{x})$ and $b(\mathbf{x})$ so that

$$\text{Prob}_F\{a(\mathbf{x}) \leq Z \leq b(\mathbf{x})\} = 1 - 2\alpha. \tag{25.43}$$

It is important to note that the probability in (25.43) refers to the randomness in both $x_i \sim F$, $i = 1, 2, \ldots n$ and $Z \sim F$. To proceed, we find a value $\hat{t}^{(\alpha)}$ so that

$$\text{Prob}_F\{\frac{\bar{x} - Z}{s} \leq \hat{t}^{(\alpha)}\} = \alpha, \tag{25.44}$$

where $s = \sum_1^n (x_i - \bar{x})^2/(n - 1)$. Our prediction interval is then obtained by pivoting expression (25.44) giving

$$(\bar{x} - \hat{t}^{(1-\alpha)}s, \ \bar{x} - \hat{t}^{(\alpha)}s). \tag{25.45}$$

If we assume that F is standard normal with mean μ and unknown variance σ^2, we obtain

$$\hat{t}^{(\alpha)} = t_{n-1}^{(\alpha)}\sqrt{1 + 1/n}, \tag{25.46}$$

where $t_{n-1}^{(\alpha)}$ is the α-percentile of the t distribution with $n-1$ degrees of freedom. This differs from a confidence interval for $\mu = \text{E}(Y)$ in that the factor $\sqrt{1 + 1/n}$ appears rather than $\sqrt{1/n}$. The extra "1" accounts for the variance of the new observation Z.

The bootstrap approach resamples $\hat{F} \to \mathbf{x}^*$ and $\hat{F} \to Z^*$ independently, and then estimates $\hat{t}^{(\alpha)}$ by the empirical αth quantile of the values

$$\frac{\bar{x}^* - Z^*}{s^*} \tag{25.47}$$

where $\bar{x}^*$ and s^* are the mean and sample standard deviation of the bootstrap sample $\mathbf{x}^*$.

This is closely related to the bootstrap-t method of constructing confidence intervals, but there is an interesting difference. In the confidence interval setting, the bootstrap-t method produce second-order correct and accurate intervals (Chapter 22). It turns out that bootstrap-t prediction intervals are "almost" second-order accurate but are only first order correct. For details, see Bai and Olshen (1988) and Bai, Bickel, and Olshen (1990).

(a) Explain carefully how the calculations of section 25.5 fit into the framework described above.

(b) For the control group times of the mouse data of Table 2.1, compute prediction intervals for a new observation, using both normal theory and the bootstrap-t approach. Compare these to the corresponding confidence intervals for the mean.

(c) We can write

$$\text{Prob}\{\frac{\bar{x}^* - Z^*}{s^*} \le \hat{t}^{(\alpha)}\} \;\; = \;\; \text{E}(\text{Prob}\{\frac{\bar{x}^* - Z^*}{s^*} \le \hat{t}^{(\alpha)}\}|\mathbf{x}^*). \tag{25.48}$$

Use this to suggest a more computationally efficient way of approximating $\text{Prob}\{(\bar{x}^* - Z^*)/s^* \le t\}$.

CHAPTER 26

Discussion and further topics

26.1 Discussion

Statisticians work at the interface between science, mathematics, and philosophy. A statistical analysis of a data set, for example the one in Chapter 25, is supposed to conclude what the data say, and how far these conclusions can be trusted. This is an ambitious program, an attempt to quantify "learning from experience." A correspondingly ambitious theory of statistical inference was developed to carry out this program, mainly in the first half of the twentieth century. The work of Pearson, Fisher, Neyman, Wald and others coalesced statistical inference around a small set of powerful theoretical ideas: likelihood, sufficiency, power, risk, confidence, etc. These ideas have continued to dominate statistical thinking in the post-war era.

What has changed in the past forty years is how these ideas are implemented. Modern electronic computation, ten million times faster than the pre-war variety, has vastly increased the scope and power of statistical reasoning. This is not just a matter of working faster or on bigger data sets. Computational power has freed statisticians from the grip of mathematical tractability. We can now answer the questions scientists are really interested in, rather than choosing from a very small catalogue of mathematically solvable cases. As a result, powerful new statistical methodologies are being developed to take advantage of electronic computation in the practical business of statistical inference.

The bootstrap is one such methodology. It aims to carry out familiar statistical calculations, standard errors, biases, confidence intervals, etc., in an unfamiliar way: by purely computational means, rather than through the use of mathematical formulas. In fact a lot of mathematical theory goes into the development of bootstrap methods, in order to make these methods fully compatible with traditional theories of statistical inference.

The biggest difference between pre- and post-war statistical practice is the degree of automation. The theory of the bootstrap is "pre-loaded" into an algorithm and carried out entirely by the computer for any particular application. This doesn't free the statistician from thinking, of course, but it does allow the thinking to concern inferential questions of direct interest to the scientist, rather than a host of small mathematical difficulties.

One can describe the ideal computer-based statistical inference machine of the future. The statistician enters the data, the questions of interest, and the class of allowable probability models (for instance, the one-sample model of Chapter 4). Without further intervention, the machine answers the questions, in a way that is optimal according to statistical theory.

This book concerns how closely current bootstrap theory approaches this ideal. For standard errors and confidence intervals, the ideal is in sight if not in hand. The programs bcanon and abcnon compress a large fraction of nonparametric confidence interval theory into a surprisingly short algorithm. The inferences are not perfect yet, as we have seen, but they are substantially better than most of the traditional approximate methods.

The current era does not mark the first attempt at statistical automation. Fisher's theory of maximum likelihood estimation, developed in the 1920's, was notably successful in automating statistical estimation. In fact it fits our picture of the ideal statistical inference machine, at least within the framework of parametric estimation. The bootstrap is closely related to Fisher's way of thinking. The plug-in principle, which leads directly to the bootstrap in Chapter 6, could just as well be called nonparametric maximum likelihood. Bootstrap methods can be thought of as maximum likelihood theory applied via the computer to a more complicated class of estimation problems.

Fisher's theory produces reasonably good statistical estimates on a routine basis. Interestingly enough, this theory fell into disuse in the post-war period, at least in the United States. Decision theory, which aims for optimal solutions and not merely good ones, monopolized theoretical interest from 1945 to 1965. Decision theory remains with us in hypothesis testing, but Fisher's ideas have reclaimed the center stage in estimation.

One can at least conceive of a decision-theory bootstrap that would automatically produce optimal inferences for arbitrarily complicated testing situations. The fact is that bootstrap ideas have been least successful in hypothesis testing problems, the statistical area where exact inferences are most highly prized. The calibration calculations of Section 25.6 are an attempt to raise the accuracy of bootstrap methods to an acceptable level for hypothesis testing.

Bootstrap methods, and other computationally intensive statistical techniques, continue to develop at a robust pace. Areas not much discussed in this book, such as Bayesian methods, discriminant analysis, data-based selection of regression models, prediction problems, etc., are in various stages of the automation process. The twenty-first century may or may not use different theories of statistical inference, but it will certainly be a different, better world for statistical practitioners.

The remainder of this chapter discusses some general questions about the bootstrap, and a brief list of related topics not covered in this book.

26.2 Some questions about the bootstrap

In order to illuminate some general points concerning the bootstrap and its role in statistical inference, we provide answers to some specific questions.

1. *What are the attractive features of the bootstrap?*

The bootstrap allows the data analyst to assess the statistical accuracy of complicated procedures, by exploiting the power of the computer. The use of the bootstrap either relieves the analyst from having to do complex mathematical derivations, or in some instances provides an answer where no analytical answer can be obtained.

The bootstrap can be used either nonparametrically, or parametrically. In nonparametric mode, it avoids restrictive and sometimes dangerous parametric assumptions about the form of the underlying populations. In parametric mode, it can provide more accurate estimates of error than traditional Fisher information-based methods.

2. *Isn't the bootstrap just another form of simulation?*

Yes, approximation of bootstrap quantities usually involves some type of simulation, either sampling with replacement from the data or sampling from a parametric model. But it is a special kind of simulation, namely, a *data-based simulation*. That is, we simulate from a data-based estimate of the population. Hence the bootstrap is used not to learn about the general properties of a statistical procedure, as in most statistical simulations, but rather to assess its properties for *the data at hand*. It is interesting to see how far this simple idea of data-based simulation can be pushed, as evidenced by the broad scope of the topics covered in this book.

3. *When should the bootstrap be used, and when should other methods be used instead?*

This question is difficult, with many different factors to consider. The bootstrap is an approach to frequentist or Fisherian inference[1]. Therefore, at one level, a discussion of the merits of the bootstrap involves the question of Bayesian versus frequentist and Fisherian inference. We do not intend to give a discussion of this issue here, but refer the reader to Efron (1986) and the accompanying commentary for some opposing points of view.

It may be more productive to compare the nonparametric bootstrap to parametric modeling for frequentist or Fisherian inference. The bootstrap is a fairly crude form of inference, that can be used when the data analyst is either unable or unwilling to carry out more extensive modeling. Nonparametric bootstrap inferences are asymptotically efficient. That is, for large samples they give accurate answers no matter what the underlying population. Unlike methods such as permutation tests, they do not enjoy exact finite sample nonparametric properties. However the scope of their application is much greater than the scope of permutation tests.

In place of the nonparametric bootstrap, there are situations where one can instead use flexible parametric modeling. One might start with a tentative model for the data, draw inferences based on the model, and then perturb the model in various ways and check the sensitivity of the inferences. Some discussion of this approach is given Cox and Snell (1981). We might try this approach, for example, in the stamp problem of Chapter 16. We could start by fitting normal mixtures to the data, and make inferences about the number of subpopulations. Then we could try other models and see how much our inference changes. This approach would be successful if our inference was stable over our choice of models and we were satisfied that the family of models considered was large enough in some sense. It would also give a stronger inference than the

---

[1] This isn't strictly true, as there is a Bayesian form of the bootstrap. References on the "Bayesian bootstrap" are given in the next section.

bootstrap approach because it would tell us about the number of subpopulations (rather than just about the number of modes) and it would tell us an approximate form for the distribution of each subpopulation. However, in the stamp problem and many other situations, it may not be clear what constitutes a "large enough" family of models.

There is no clear choice that can be made between these different approaches. Often it can be informative to carry out more than one form of analysis. We hope that future research in statistics will shed more light on these important issues.

4. *How does the bootstrap deal with problems of dependence?*

Independence between observational units is often an important assumption in data analysis and is usually present in bootstrap-based inferences. Lack of independence can reduce the accuracy of inferences: see Hampel *et al.* (1986, chapter 8) for a discussion of this. There is no easy solution to problems of dependence: one approach is to model the dependence in some way, and then draw inferences from the model. The use of the bootstrap in the auto-regressive time series model of Chapter 8 is an example of this, although in that example the dependence is in fact the main quantity of interest. The moving blocks bootstrap of Chapter 8 represents a more model-free approach to handling dependence and looks to be a promising tool. However, problems of dependence do not appear to be well understood and are an important area for further research.

26.3 References on further topics

In recent years the bootstrap has been an active and broad topic for research. We have not attempted to give a complete survey of this research here, and as a result, a number of important topics have been omitted. In this chapter we provide some references on a number of these topics.

The bootstrap (and jackknife) have potential for use in survey sampling, but cannot be simply applied without modification. This is due to the fact that sampling without replacement is commonly used in surveys, and the sampling design is often stratified in one more more stages. Kish and Frankel (1974) give a review of inference problems in survey sampling. McCarthy (1969) described half-sampling and balanced repeated replications, a systematic sampling method for providing unbiased estimates of variance in stratified designs. The topic is discussed in Efron (1982, chapter 8). Sitter (1993) extends this concept to more complex designs through the use of orthogonal arrays. The "without replacement bootstrap" is proposed in Gross (1980) and Bickel and Freedman (1984). Krewski and Rao (1981), and Rao and Wu (1985, 1988) propose linearization methods to adjust the jackknife and bootstrap for complex designs. Linearization methods are applicable only to statistics that are smooth functions of sample means. Sitter (1992) gives an excellent overview of current research in bootstrap methods for sample surveys.

The connection of the bootstrap with Bayesian inference was pointed out by Rubin (1981) and Efron (1982, chapter 10). Laird and Louis (1987) developed related ideas in the context of empirical Bayes inference. Newton and Raftery (1992) propose the "weighted likelihood bootstrap", a method for simulating from the posterior distribution in nonparametric Bayesian inference.

Saddlepoint methods are a potentially useful tool for more efficient computation of bootstrap quantities, as shown by Davison and Hinkley (1988), Feuerverger (1989) and Wang (1992). An overview of the use of saddlepoint approximations in statistics was given by Reid (1988). At the present time, these approximations can only be derived easily for smooth functions of sample means, and this limits their applicability to bootstrap problems. DiCiccio, Martin and Young (1992) discuss saddlepoint methods for nonlinear statistics.

Inference through estimating equations (Godambe 1960, Godambe and Thompson, 1984) is an increasingly active research area, as evidenced by the edited volume of Godambe (1991). Application of the bootstrap to estimating equations was studied by Lele (1991).

Bootstrap analysis of directional data was studied by Ducharme *et al.* (1985) and Fisher and Hall (1989).

Appendix: software for bootstrap computations

Introduction

As indicated in Chapter 6, simple bootstrapping of a statistic $\hat{\theta} = s(\mathbf{x})$ consists of the following steps:

1. B samples are drawn with replacement from the original data set $\mathbf{x}$, with each sample the same size as the original data set. Call these bootstrap samples $\mathbf{x}^{*1}, \mathbf{x}^{*2} \ldots \mathbf{x}^{*B}$.

2. The statistic of interest $\hat{\theta}$ is computed for each bootstrap sample, that is $\hat{\theta}^*(b) = s(\mathbf{x}^{*b})$ for $b = 1, 2, \ldots B$. The mean, standard deviation and percentiles of these B values form the basis for the bootstrap approach to inference, as described in Chapter 6.

Implementation of these steps in a computer language is not difficult. A necessary ingredient for any bootstrap program is a high quality uniform number generator. Most packages have built-in generators, but their quality can vary greatly. See Knuth (1969) or Thisted (1988) for more details about uniform random number generators.

Bootstrapping can be performed in most computer languages, for example, Fortran, C, Pascal, APL, Gauss, Matlab, Lisp, or XLISP-Stat. An elementary bootstrap program in Fortran is given in Efron and Tibshirani, (1985). However, it is important to remember that the bootstrap (and associated methods) are not tools that are used in isolation but rather are applied to other statistical techniques. For this reason, they are most effectively used in an integrated environment for data analysis. In such an environment, a bootstrap procedure has the ability to call other procedures with different sets of inputs (data) and then collect them together and analyze the results. The S, S-PLUS, XLISP-Stat, Gauss and Matlab packages are examples of integrated environments. The ability of a package or language to deal with complicated data structures is also important. For example, S and S-PLUS have built-in facilities for vectors, matrices, high-dimensional arrays, time series and lists.

Some available software

- *"Resampling stats."* This is an MS-DOS package for resampling and randomization tests. Details can be obtained from Resampling Stats, 612 N. Jackson St., Arlington, Va. 22201.

- *SAS language.* Tibshirani (1985) describes some programs for bootstrapping in SAS. These programs are not particularly efficient and better approaches surely exist.

- *S or S-PLUS.* We describe a collection of functions for this language below.

S language functions

The following function **bootstrap** performs bootstrap sampling of an S function **theta**. It works for the one-sample problem but can also be applied to more complicated data situations. The function is defined by

```
"bootstrap" < - function(x,nboot,theta,...){
     data < - matrix(sample(x,size=length(x)*nboot,
         replace=T),nrow=nboot)
return(apply(data,1,theta,...))
}
```

The following pages contain documentation for a more powerful version of **bootstrap** that has an option for jackknife-after-bootstrap computations, as well as a number of other S functions for confidence interval construction, prediction error estimation, the jackknife and cross-validation. In order to use these functions, the S or S-PLUS statistical language is required. Becker, Chambers and Wilks (1988) describe the S language. S is currently available from AT&T Software Sales, P.O Pox 25000, Greensboro, North Carolina 27420. S-PLUS is an enhancement of S, and is available from StatSci, 1700 Westlake Ave. N., Suite 500, Seattle, Washington 98109.

The functions described here are available from the statistics archive at Carnegie-Mellon University, by sending electronic mail to statlib@lib.stat.cmu.edu with the one-line mail message send bootstrap.funs from S. Alternatively, you can retrieve the software by ftp access to lib.stat.cmu.edu: login with the username statlib and look for a shar file named bootstrap.funs in the directory S. If neither of these options are available to you, you can request a diskette from the second author.

---

abcnon   Nonparametric ABC confidence limits   **abcnon**

```
abcnon(x, tt, epsilon=0.001,
   alpha=c(0.025, 0.05, 0.1, 0.16,
       0.84, 0.9, 0.95, 0.975))
```

ARGUMENTS

x   the data. Must be either a vector, or a matrix whose rows are the observations

tt   function defining the parameter in the resampling form tt(p,x), where p is the vector of proportions and x is the data

epsilon   optional argument specifying step size for finite difference calculations

alpha   optional argument specifying confidence levels desired

VALUE   list with following components

limits   The estimated confidence points, from the ABC and standard normal methods

stats   list consisting of t0=observed value of tt, sighat=infinitesimal jackknife estimate of standard error of tt, bhat= estimated bias

constants   list consisting of a=acceleration constant, z0=bias adjustment, cq=curvature component

tt.inf   (approximate) influence components of tt

pp   matrix whose rows are the resampling points in the least favourable family . The abc confidence points are the function tt evaluated at these points

REFERENCES   Efron, B, and DiCiccio, T. (1992) More accurate confidence intervals in exponential families. Biometrika 79, pages 231-245.

EXAMPLE

```
# compute abc intervals for the mean
x <- rnorm(10)
theta <- function(p,x) {sum(p*x)/sum(p)}
results <- abcnon(x, theta)
# compute abc intervals for the correlation
x <- matrix(rnorm(20),ncol=2)
theta <- function(p, x)
{
    x1m <- sum(p * x[, 1])/sum(p)
    x2m <- sum(p * x[, 2])/sum(p)
    num <- sum(p * (x[, 1] - x1m) * (x[, 2] - x2m))
    den <- sqrt(sum(p * (x[, 2] - x1m)^2) *
        sum(p * (x[, 2] - x1m)^2))
    return(num/den)
}
results <- abcnon(x, theta)
```

---

abcpar　　　Parametric ABC confidence limits　　　**abcpar**

```
abcpar(x, tt, S, etahat, mu, n=rep(1,length(x)),
  lambda=0.001,  alpha=c(0.025, 0.05, 0.1, 0.16))
```

ARGUMENTS

x	vector of data
tt	function of expectation parameter mu defining the parameter of interest
S	maximum likelihood estimate of the covariance matrix of x
etahat	maximum likelihood estimate of the natural parameter eta
mu	function giving expectation of x in terms of eta
n	optional argument containing denominators for binomial (vector of length len(x))
lambda	optional argument specifying step size for finite difference calculation
alpha	optional argument specifying confidence levels desired
VALUE	list with the following components

call the call to abcpar

limits The nominal confidence level, ABC point, quadratic ABC point, and standard (normal) point.

stats list consisting of observed value of tt, estimated standard error and estimated bias

constants list consisting of a=acceleration constant, z0=bias adjustment, cq=curvature component

REFERENCES   Efron, B, and DiCiccio, T. (1992) More accurate confidence intervals in exponential families. Biometrika 79, pages 231-245.

EXAMPLE

```
# binomial random variables
# x is a p-vector of successes, n is a p-vector of
#   number of trials

S <- matrix(0,nrow=p,ncol=p)
S[row(S)==col(S)] <- x*(1-x/n)
mu <- function(eta,n){n/(1+exp(eta))}
etahat <- log(x/(n-x))

#suppose p=2 and we are interested in mu2-mu1

tt <- function(mu){mu[2]-mu[1]}
x <- c(2,4); n <- c(12,12)
a <- abcpar(x, tt, S, etahat,n)
```

---

bcanon   Nonparametric BCa confidence limits   **bcanon**

```
bcanon(x, nboot, theta, ...,
   alpha=c(0.025, 0.05, 0.1, 0.16,
       0.84, 0.9, 0.95, 0.975))
```

ARGUMENTS

x a vector containing the data. To bootstrap more complex data structures (e.g bivariate data) see the last example below.

nboot number of bootstrap replications

theta function defining the estimator used in constructing the confidence points

... additional arguments for theta

alpha optional argument specifying confidence levels desired

VALUE   a list consisting of

confpoint estimated bca confidence limits

z0 estimated bias correction

acc estimated acceleration constant

u jackknife influence values

REFERENCES   Efron, B. and Tibshirani, R. (1986). The Bootstrap Method for standard errors, confidence intervals, and other measures of statistical accuracy. Statistical Science, Vol 1., No. 1, pp 1-35.

Efron, B. (1987). Better bootstrap confidence intervals (with discussion). J. Amer. Stat. Assoc. vol 82, pg 171

EXAMPLE

```
# bca limits for the  mean
# (this is for illustration;
#   since "mean" is a built in function,
#   bcanon(x,100,mean) would be simpler)

x <- rnorm(20)
theta <- function(x){mean(x)}
results <- bcanon(x,100,theta)

# To obtain bca limits for functions of more
# complex data structures, write theta so that
# its argument x is the set of observation numbers
# and simply pass as data to bcanon the vector 1..n.
# For example, find bca limits for the
# correlation coefficient based on 15 data pairs:

xdata <- matrix(rnorm(30),ncol=2)
n <- 15
theta <- function(x,xdata)
  { cor(xdata[x,1],xdata[x,2]) }
results <- bcanon(1:n,100,theta,xdata)
```

| **bootstrap** | Non-parametric bootstrapping | **bootstrap** |

```
bootstrap(x,nboot,theta,..., func=NULL)
```

ARGUMENTS

    x a vector containing the data. To bootstrap more complex data structures (e.g bivariate data) see the last example below.

  nboot The number of bootstrap samples desired.

  theta function to be bootstrapped. Takes x as an argument, and may take additional arguments (see below and last example).

  ... any additional arguments to be passed to theta

  func (optional) argument specifying the functional of the distribution of thetahat that is desired. If func is specified, the jackknife-after-bootstrap estimate of its standard error is also returned. See example below.

  VALUE   list with the following components:

thetastar the nboot bootstrap values of theta

func.thetastar the functional func of the bootstrap distribution of thetastar, if func was specified

jack.boot.val the jackknife-after-bootstrap values for func, if func was specified

jack.boot.se the jackknife-after-bootstrap standard error estimate of func, if func was specified

REFERENCES   Efron, B. and Tibshirani, R. (1986). The bootstrap method for standard errors, confidence intervals, and other measures of statistical accuracy. Statistical Science, Vol 1., No. 1, pp 1-35.

Efron, B. (1992) Jackknife-after-bootstrap standard errors and influence functions. J. Roy. Stat. Soc. B, vol 54, pages 83-127

EXAMPLE

```
# 100 bootstraps of the sample mean
# (this is for illustration;  since "mean" is  a
# built in function, bootstrap(x,100,mean)
# would be simpler!)

x <- rnorm(20)
theta <- function(x){mean(x)}

results <- bootstrap(x,100,theta)

# as above, but also estimate the 95th percentile
# of the bootstrap dist'n of the mean, and
# its jackknife-after-bootstrap  standard error

perc95 <- function(x){quantile(x, .95)}

results <-  bootstrap(x,100,theta, func=perc95)

# To bootstrap functions of more  complex data
# structures, write theta so that its argument x
# is the set of observation numbers
# and simply  pass as data to bootstrap
#   the vector 1,2,..n.
# For example, to bootstrap the
# correlation coefficient based on 15 data pairs:

xdata <- matrix(rnorm(30),ncol=2)
n <- 15
theta <- function(x,xdata)
  { cor(xdata[x,1],xdata[x,2]) }
results <- bootstrap(1:n,20,theta,xdata)
```

---

bootpred   bootstrap estimates of prediction error   **bootpred**

```
bootpred(x,y,nboot,theta.fit,theta.predict,
    err.meas,...)
```

ARGUMENTS

    x a matrix containing the predictor (regressor) values. Each row corresponds to an observation.

y a vector containing the response values

nboot the number of bootstrap replications

theta.fit function to be cross-validated. Takes x and y as an argument. See example below.

theta.predict function producing predicted values for theta.fit. Arguments are a matrix x of predictors and fit object produced by theta.fit. See example below.

err.meas function specifying error measure for a single response y and prediction yhat. See examples below.

... any additional arguments to be passed to theta.fit

VALUE   list with the following components

app.err the apparent error rate- that is, the mean value of err.meas when theta.fit is applied to x and y, and then used to predict y.

optim the bootstrap estimate of optimism in app.err. A useful estimate of prediction error is app.err+optim

err.632 the ".632" bootstrap estimate of prediction error.

REFERENCES   Efron, B. (1983). Estimating the error rate of a prediction rule: improvements on cross-validation. J. Amer. Stat. Assoc, vol 78. pages 316-31.

EXAMPLE

```
# bootstrap prediction error estimation in least
#  squares regression

x <- rnorm(85)
y <- 2*x +.5*rnorm(85)
theta.fit <- function(x,y){lsfit(x,y)}
theta.predict <- function(fit,x){
            cbind(1,x)%*%fit$coef
            }
sq.err_function(y,yhat) { (y-yhat)^2}
results <- bootpred(x,y,20,theta.fit,theta.predict,
            err.meas=sq.err)

# for a classification problem, a standard choice
# for err.meas would simply count up the
#  classification errors:
```

```
miss.clas <- function(y,yhat){ 1*(yhat!=y)}

# with this specification,  bootpred estimates
#   misclassification rate
```

---

boott	Bootstrap-*t* confidence limits	**boott**

```
boott(x,theta, ..., sdfun=MISSING,nbootsd=25,
   nboott=200, VS=F,
      v.nbootg=100,v.nbootsd=25,v.nboott=200,
         perc=c(.001,.01,.025,.05,.10,.50,.90,.95,
            .975,.99,.999))
```

ARGUMENTS

 x a vector containing the data. Nonparametric bootstrap sampling is used. To bootstrap from more complex data structures (e.g bivariate data) see the last example below.

 theta function to be bootstrapped. Takes x as an argument, and may take additional arguments (see below and last example).

 ... any additional arguments to be passed to theta

 sdfun optional name of function for computing standard deviation of theta based on data x. Should be of the form: sdmean $< -$ function(x,nbootsd,theta,...) where nbootsd is a dummy argument that is not used. If theta is the mean, for example,

 sdmean $< -$ function(x,nbootsd,theta,...)

 {sqrt(var(x)/length(x))}.

 If sdfun is missing, then boott uses an inner bootstrap loop to estimate the standard deviation of theta(x)

 nbootsd The number of bootstrap samples used to estimate the standard deviation of theta(x)

 nboott The number of bootstrap samples used to estimate the distribution of the bootstrap T statistic. 200 is a bare minimum and 1000 or more is needed for reliable alpha % confidence points, alpha $< .05$ or $> .95$ say. Total number of bootstrap samples is nboott*nbootsd.

VS If true, a variance stabilizing transformation is estimated, and the interval is constructed on the transformed scale, and then is mapped back to the original theta scale. This can improve both the statistical properties of the intervals and speed up the computation. See the reference Tibshirani (1988) given below. If false, variance stabilization is not performed.

v.nbootg The number of bootstrap samples used to estimate the variance stabilizing transformation g. Only used if VS=T.

v.nbootsd The number of bootstrap samples used to estimate the standard deviation of theta(x). Only used if VS=T.

v.nboott Number of bootstrap samples used in estimation of percentiles of g(thetahat)-g(theta) (final stage). Only used if VS=T. Total number of bootstrap samples is v.nbootg*v.nbootsd + v.nboott

perc Confidence points desired.

VALUE   list with the following components:

confpoints Estimated confidence points

theta

g theta and g are only returned if VS=T was specified. (theta[i],g[i]), i=1,length(theta) represents the estimate of the variance stabilizing transformation g at the points theta[i].

REFERENCES   Tibshirani, R. (1988) Variance stabilization and the bootstrap. Biometrika , vol 75, pages 433-44.

Hall, P. (1988) Theoretical comparison of bootstrap confidence intervals. Ann. Statist. 16, 1-50.

EXAMPLE


```
#  estimated confidence points for the mean

x <- rchisq(20,1)
theta <- function(x){mean(x)}
results <- boott(x,theta)
```

```
# using variance-stabilization bootstrap-T method

results <-  boott(x,theta,VS=T)
results$confpoints        # gives confidence points

# plot the estimated var stabilizing transformation
plot(results$theta,results$g)


# use standard formula for stand dev of mean
# rather than an inner bootstrap loop

sdmean <- function(x,nbootsd,theta)
          {sqrt(var(x)/length(x))}

results <-  boott(x,theta,sdfun=sdmean)

# To bootstrap functions of more  complex data
#   structures, write theta so that its argument x
#   is the set of observation numbers
#   and simply  pass as data to boot
#   the vector 1,2,..n.
# For example, to bootstrap the
#   correlation coefficient based on 15 data pairs:

xdata <- matrix(rnorm(30),ncol=2)
n <- 15
theta <- function(x, xdata)
   { cor(xdata[x,1],xdata[x,2]) }
results <- boott(1:n,theta, xdata)
```

crossval	K-fold cross-validation	**crossval**

```
crossval(x,y,theta.fit,theta.predict,...,ngroup=n)
```

ARGUMENTS

> x a matrix containing the predictor (regressor) values. Each row corresponds to an observation.
>
> y a vector containing the response values
>
> theta.fit function to be cross-validated. Takes x and y as an argument. See example below.

theta.predict function producing predicted values for theta.fit. Arguments are a matrix x of predictors and fit object produced by theta.fit. See example below.

... any additional arguments to be passed to theta.fit

ngroup optional argument specifying the number of groups formed. Default is ngroup=sample size, corresponding to leave-one out cross-validation.

VALUE   list with the following components

cv.fit The cross-validated fit for each observation. The numbers 1 to n (the sample size) are partitioned into ngroup mutually disjoint groups of size "leave.out". leave.out, the number of observations in each group, is the integer part of n/ngroup. The groups are chosen at random if ngroup < n. (If n/leave.out is not an integer, the last group will contain > leave.out observations). Then theta.fit is applied with the kth group of observations deleted, for k=1, 2, ngroup. Finally, the fitted value is computed for the kth group using theta.predict.

ngroup The number of groups

leave.out The number of observations in each group

groups A list of length ngroup containing the indices of the observations in each group. Only returned if leave.out > 1.

REFERENCES   Stone, M. (1974). Cross-validation choice and assessment of statistical predictions. Journal of the Royal Statistical Society, B-36, 111–147.

EXAMPLE

```
# cross-validation of least squares regression
# note that crossval is not very efficient, and
#  being a general purpose function, it does not
#  use the Sherman-Morrison identity
x <- rnorm(85); y <- 2*x +.5*rnorm(85)
theta.fit <- function(x,y){lsfit(x,y)}
theta.predict <- function(fit,x){
                cbind(1,x)%*%fit$coef
                }
results <- crossval(x,y,theta.fit,theta.predict,
                ngroup=6)
```

jackknife	jackknife estimation	**jackknife**

```
jackknife(x,theta,...)
```

ARGUMENTS

  x   a vector containing the data. To jackknife more complex data structures (e.g bivariate data) see the last example below.

  theta   function to be jackknifed. Takes x as an argument, and may take additional arguments (see below and last example).

  ...   any additional arguments to be passed to theta

  VALUE   list with the following components

jack.se   The jackknife estimate of standard error of theta. The leave-one out jackknife is used.

jack.bias   The jackknife estimate of bias of theta. The leave-one out jackknife is used.

jack.values   The n leave-one-out values of theta, where n is the number of observations. That is, theta applied to x with the 1st observation deleted, theta applied to x with the 2nd observation deleted, etc.

REFERENCES   Efron, B. and Tibshirani, R. (1986). The Bootstrap Method for standard errors, confidence intervals, and other measures of statistical accuracy. Statistical Science, Vol 1., No. 1, pp 1-35.

EXAMPLE

```
# jackknife values for the sample mean
# (this is for illustration;   since "mean" is  a
#  built in function,  jackknife(x,mean)
#   would be simpler!)

x <- rnorm(20)
```

```
theta <- function(x){mean(x)}

results <- jackknife(x,theta)


# To jackknife functions of more  complex data
#  structures, write theta so that its argument x
#  is the set of observation numbers
#  and simply  pass as data to jackknife
#   the vector 1,2,..n.
# For example, to jackknife the
#  correlation coefficient based on 15 data pairs:

xdata <- matrix(rnorm(30),ncol=2)
n <- 15
theta <- function(x,xdata)
    { cor(xdata[x,1],xdata[x,2]) }
results <- jackknife(1:n,theta,xdata)
```

References

Abramovitch, L. and Singh, K. (1985) Edgeworth corrected pivotal statistics and the bootstrap. *Ann. Statist.* **13**, 116–132.

Akaike, H. (1973) Information theory and an extension of the maximum likelihood principle. In *Second International Symposium on Information Theory*, (eds. B.N. Petrov and F. Czáki) Akademiai Kiadó, Budapest, 267–81.

Allen, D.M. (1974) The relationship between variable selection and data augmentation and a method of prediction. *Technometrics* **16**, 125–7.

Anderson, T.W. (1958) *An introduction to multivariate statistical analysis*. Wiley, New York.

Bai, C., Bickel, P.J., and Olshen, R.A. (1990) Hyperaccuracy of bootstrap based prediction. *Probability in Banach Spaces VII Proceedings of the Seventh International Conference* (edited by E. Eberlein, J. Kuelbs, and M.B. Marcus), Birkhauser Boston, Cambridge, Massachusetts, 31–42.

Bai. C, and Olshen, R.A. (1988) Discussion of "Theoretical comparison of bootstrap confidence intervals" by P. Hall, *Ann. Statist.* **16**, 953–956.

Barnard, G.A. (1963) Contribution to discussion. *J. Royal. Statist. Soc. B* **25**, 294.

Barndorff-Neilson, O.E. and Cox, D.R. (1989) *Asymptotic techniques for use in statistics*. Chapman and Hall, London, New York.

Becker, R., Chambers, J. and Wilks. A. (1988) *The S language.* Wadsworth, Belmont CA.

Behrens, B.-U. (1929) Ein Beitrag zur Fehlen-Berechnung bei wenigen Beobachtungen. *Landwirtsch. Jb.* **68**, 807–837.

Beran, R. (1984) Bootstrap methods in statistics. *Jber. d. Dt. Math. Verein.* **86**, 14–30.

Beran, R. (1987) Prepivoting to reduce level error of confidence sets. *Biometrika* **74**, 457–468.

Beran, R. (1988) Prepivoting test statistics: a bootstrap view of asymptotic refinements, *J. Amer. Statist. Assoc.* **83**, 687–697.

Beran, R. and Ducharme, G. (1991) *Asymptotic theory for bootstrap methods in statistics.* Centre de Reserches Mathematiques, Univ. of Montreal.

Beran, R. and Millar, P.W. (1987) Stochastic estimation and testing. *Ann. Statist.* **15**, 1131–1154.

Beran, R. and Srivastava, M.S. (1985) Bootstrap tests and confidence regions for functions of a covariance matrix. *Ann. Statist.* **13**, 95–115.

Besag, J. and Clifford, P. (1989) Generalized Monte Carlo significance tests. *Biometrika* **76**, 633-642.

Bickel, P.J. and Freedman, D.A. (1981) Some asymptotic theory for the bootstrap. *Ann. Statist.* **9**, 1196–1217.

Bickel, P.J. and Freedman, D.A. (1983) Bootstrapping regression models with many parameters, *A Festschrift for Erich L. Lehmann*, P.J. Bickel, K. Doksum and J.L. Hodges eds., Wadsworth, Belmont, CA.

Bickel, P.J. and Freedman, D.A. (1984) Asymptotic normality and the bootstrap in stratified sampling, *Ann. Statist.* **12**, 470–482.

Boos, D. and Monahan, J. (1986) Bootstrap methods using prior information. *Biometrika* **73**, 77–83

Box, G.E. and Jenkins, G.M (1970) *Times series analysis, forecasting, and control.* Holden-Day, San Francisco.

Breiman, L. (1992) The little bootstrap and other methods for dimensionality selection in regression: X-fixed prediction error. *J. Amer. Statist. Assoc.* **87**, 738-754.

Breiman, L., Friedman, J., Olshen, R., and Stone, C. (1984) *Classification and Regression Trees.* Wadsworth, Belmont, CA.

Breiman, L. and Spector, P. (1992) Submodel selection and evaluation in Regression. The X-Random Case. *Int. Stat. Rev.* **60**, 291–319.

Buckland, S.T. (1983) Monte Carlo methods for confidence interval estimation using the bootstrap technique. *Bull. Appl. Statist.* **10**, 194–212.

Buckland, S.T. (1984) Monte Carlo confidence intervals. *Biometrics* **40**, 811–817.

Buckland, S.T. (1985) Calculation of Monte Carlo confidence intervals. *Appl. Statist.* **34**, 296–301.

Carlstein, E. (1986) The use of subseries values for estimating the variance of a general statistic from a stationary sequence. *Ann. Statist.* **14**, 1171–1179.

Chambers, J. and Hastie. T.J. (eds) (1991) *Statistical models in S.* Wadsworth, Belmont, CA.

Chatfield, C. (1980) *The analysis of time series: an introduction* (second edition). Chapman and Hall, London.

Cleveland, W.S. (1979) Robust locally-weighted regression and smoothing scatterplots. *J. Amer. Statist. Assoc.* **74**, 829–36.

Cox, D.R. and Hinkley, D.V. (1974) *Theoretical Statistics.* Chapman and Hall, London.

Cox, D.R. and Reid, N. (1987) Orthogonal parameters and approximate conditional inference (with discussion) *J. Royal. Statist. Soc. B* **49**, 1–39.

Cox, D. R. and Snell. E.J. (1981) *Applied statistics. Principles and examples.* Chapman and Hall, London.

Crawford, S. (1989) Extensions to the CART algorithm. *Int. J. of Machine Studies* **31**, 197-217.

Davison, A.C. (1988) Discussion of papers by D.V. Hinkley and by T.J. DiCiccio and J.P. Romano. *J. Royal. Statist. Soc. B* **50**, 356–357.

Davison A.C. and Hinkley, D.V. (1988) Saddlepoint approximations in resampling methods. *Biometrika* **75**, 417–431.

Davison A.C., Hinkley, D.V. and Schechtman, E. (1986) and Efficient bootstrap simulations, *Biometrika* **73**. 555–566.

Davison A.C., Hinkley, D.V. and Worton, B.J. (1992) Bootstrap likelihoods, *Biometrika* **79**, 113–130.

Devroye, L. (1986) *Non-uniform random variate generation.* Springer, Berlin, New York.

Diaconis, P. and Efron, B. (1983) Computer-intensive methods in statistics. *Sci. Amer.* **248**, 116–130.

DiCiccio, T.J. Hall, P. and Romano, J.P. (1989b) Comparison of parametric and empirical likelihood functions. *Biometrika* **76**, 465–476.

DiCiccio, T.J., Martin, M. and Young, A. (1992) Analytic approximations to bootstrap distribution functions using saddlepoint methods. Tech. report, Stanford University.

DiCiccio, T.J. and Romano, J.P. (1988) A review of bootstrap confidence intervals (with discussion). *J. Royal. Statist. Soc. B* **50**, 338–370.

DiCiccio, T.J. and Romano, J.P. (1989) The automatic percentile method: Accurate confidence limits in parametric models, *Can. J. Statist.* **17**, 155–169.

DiCiccio, T.J. and Romano, J.P. (1990) Nonparametric confidence limits by resampling methods and least favorable families, *Inter. Statist. Review* **58**, 59–76.

DiCiccio, T.J. and Tibshirani, R. (1987) Bootstrap confidence intervals and bootstrap approximations. *J. Amer. Statist. Assoc.* **82**, 163–170.

Diggle, P. (1990) *Time Series, a Biostatistical Introduction.* Clarendon Press, Oxford.

Do, K.A. and Hall, P. (1991) On importance resampling for the bootstrap, *Biometrika* **78**, 161–167.

Draper, N. and Smith, H. (1981) *Applied regression analysis*, (second edition). Wiley, New York.

Ducharme, G.R., Jhun, M., Romano, J.P., and Truong, K.N. (1985) Bootstrap confidence cones for directional data. *Biometrika* **72**, 637–645.

Edgington, E.S. (1987) *Randomization Tests*, 2nd ed. Dekker, New York.

Efron, B. (1979a) Bootstrap methods: another look at the jackknife. *Ann. Statist.* **7**, 1–26.

Efron, B. (1979b) Computers and the theory of statistics: thinking the unthinkable. *SIAM Review* **21**, 460–480.

Efron, B. (1981a) Nonparametric standard errors and confidence intervals. (With discussion.) *Can. J. Statist.* **9**, 139–172.

Efron, B. (1981b) Nonparametric estimates of standard error: the jackknife, the bootstrap, and other methods. *Biometrika* **68**, 589–599.

Efron, B. (1982) *The jackknife, the bootstrap and other resampling plans.* Volume 38 of *CBMS-NSF Regional Conference Series in Applied Mathematics.* SIAM.

Efron, B. (1983) Estimating the error rate of a prediction rule: improvements on cross-validation. *J. Amer. Statist. Assoc.* **78**, 316–331.

Efron, B. (1985) Bootstrap confidence intervals for a class of parametric problems. *Biometrika* **72**, 45–58.

Efron, B. (1986) How biased is the apparent error rate of a prediction rule? *J. Amer. Statist. Assoc.* **81**, 461–70.

Efron, B. (1987) Better bootstrap confidence intervals. (with discussion.) *J. Amer. Statist. Assoc.* **82**, 171–200.

Efron, B. (1988) Bootstrap confidence intervals: good or bad? (With discussion.) *Psychol. Bull.* **104**, 293–296.

Efron, B. (1990) More efficient bootstrap computations. *J. Amer. Statist. Assoc.* **85**, 79–89.

Efron, B. (1991) Regression percentiles using asymmetric squared error loss. *Statistica Sinica* **1**, 93-125.

Efron, B. (1992a). Six questions raised by the bootstrap. *Exploring the Limits of Bootstrap* (Eds. R. LePage and L. Billard), John Wiley and Sons, 99–126, New York.

Efron, B. (1992b) Jacknife-after-bootstrap standard errors and influence functions. *J. Royal. Statist. Soc. B* **54**, 83–127.

Efron, B. (1992c) Bayes and likelihood calculations from confidence intervals. Tech, rep., Dept. of Statistics, Stanford Univ.

Efron, B. and Feldman, D. (1991) Compliance as an explanatory variable in clinical trials. *J. Amer. Statist. Assoc.* **86**, 9-26.

Efron, B. and Gong, G. (1983) A leisurely look at the bootstrap, the jackknife and cross-validation. *Amer. Statistician* **37**, 36–48.

Efron, B. and Stein, C. (1981) The jackknife estimate of variance, *Ann. Statist.* **9**, 586–596.

Efron, B. and Tibshirani, R. (1985) The bootstrap method for assessing statistical accuracy. *Behaviormetrika* **17**, 1–35.

Efron, B. and Tibshirani, R. (1986) Bootstrap measures for standard errors, confidence intervals, and other measures of statistical accuracy. *Statistical Science* **1**, 54–77.

Efron, B. and Tibshirani, R. (1991) Statistical data analysis in the computer age. *Science* **253**, 390–395.

Eubank, R.L. (1988) *Smoothing Splines and Nonparametric Regression.* Marcel Dekker, New York and Basel.

Faraway, J. and Jhun, M. (1990) Bootstrap choice of bandwidth for density estimation. *J. Amer. Statist. Assoc.* **85**, 1119–1122.

Fernholz, L.T. (1983). *Von Mises calculus for statistical functionals.* Lecture Notes in Statistics, **19**, Springer, New York.

Feuerverger, A. (1989) On the empirical saddlepoint approximation. *Biometrika* **76**, 457–464.

Fisher, N.I. and Hall, P. (1989) Bootstrap confidence regions for directional data. *J. Amer. Statist. Assoc.* **84**, 996–1002.

Fisher, N.I. and Hall, P. (1990) On bootstrap hypothesis testing. *Austral. J. Statist.* **32**, 177–190.

Freedman, D.A. (1981) Bootstrapping regression models. *Ann. Statist.* **9**, 1218–1228.

Freedman, D.A. and Peters, S.C. (1984) Bootstrapping a regression equation: Some empirical results. *J. Amer. Statist. Assoc.* **79**, 97–106.

Giampaolo, C., Gray, A., Olshen, R. and Szabo, S. (1988) Predicting induced duodenal ulcer and adrenal necrosis with classification trees. Technical Report 125, Division of Biostatistics, Stanford University.

Gine, E. and Zinn, J. (1989) Necessary conditions for the bootstrap of the mean, *Ann. Statist.* **17**, 684–691.

Gine, E. and Zinn, J. (1990) Bootstrap general empirical measures, *Ann. Prob.* **18**, 851–869.

Gleason, J.R. (1988) Algorithms for balanced bootstrap simulations, *Amer. Statist.* **42**, 263–266.

Godambe, V.P. (1960) An optimum property of regular maximum likelihood estimation. *Ann. Math. Statist.* **31**, 1208–1211.

Godambe, V.P. (ed.) (1991) *Estimating functions.* Clarendon Press, Oxford.

Godambe, V.P. and Thompson, M.W. (1984) Robust estimation through estimating equations. *Biometrika* **71**, 115–125.

Golub, G., Heath, M., and Wahba, G. (1979) Generalized cross validation as a method for choosing a good ridge parameter. *Technometrics* **21**, 215–224.

Golub, G. and Van Loan, C. V. (1983) *Matrix computations.* Johns Hopkins University Press.

Graham, R.L., Hinkley, D.V., John, P.W.M. and Shi, S. (1990) Balanced design of bootstrap simulations. *J. Royal. Statist. Soc. B* **52**, 185–202.

Gray, H.L. and Schucany, W.R. (1972) *The Generalized Jackknife Statistics,* Marcel Dekker, New York.

Gross, S. (1980) Median estimation in sample surveys. *Proc. Sect. Survey Res. Methods,* Amer. Stat. Assoc., 181–184.

Hall, P. (1986a) On the bootstrap and confidence intervals. *Ann. Statist.* **14**, 1431–1452.

Hall, P. (1986b) On the number of bootstrap simulations required to construct a confidence interval. *Ann. Statist.* **14**, 1453–1462.

Hall, P. (1987) On the bootstrap and likelihood-based confidence intervals. *Biometrika* **74**, 481–493.

Hall, P. (1988a) Theoretical comparison of bootstrap confidence intervals. (with discussion.) *Ann. Statist.* **16**, 927–953.

Hall, P. (1988b) On symmetric bootstrap confidence intervals. *J. Royal. Statist. Soc. B* **50**, 35–45.

Hall, P. (1989a) On efficient bootstrap simulation. *Biometrika* **76**, 613–617.

Hall, P. (1989b) Antithetic resampling for the bootstrap. *Biometrika* **76**, 713–724.

Hall, P. (1990) Performance of bootstrap balanced resampling in distribution function and quantile problems. *Prob. Th. Rel. Fields* **85**, 239–267.

Hall, P. (1991) Bahadur representations for uniform resampling and importance resampling, with applications to asymptotic relative efficiency. *Ann. Statist.* **19**, 1062–1072.

Hall, P. (1992) *The Bootstrap and Edgeworth Expansion.* Springer-Verlag, New York, Berlin, Heidelberg, London, Paris, Toyko, Hong Kong, Barcelona, Budapest.

Hall, P., DiCiccio, T.J. and Romano, J.P. (1989) On smoothing and the bootstrap. *Ann. Statist.* **17**, 692–704.

Hall, P. and La Scala, B. (1990) Methodology and algorithms of empirical likelihood. *Internat. Statist. Rev.* **58**, 109–127.

Hall, P. and Martin, M.A. (1988) On bootstrap resampling and iteration. *Biometrika* **75**, 661–671.

Hall, P. and Titterington, D. (1987) Common structure of techniques for choosing smoothing parameters in regression problems. *J. Royal. Statist. Soc. B* **49**, 184–198.

Hall, P. and Titterington, M. (1988) On confidence bands in nonparametric density estimation and regression. *J. Mult. Anal.* **27**, 228–254.

Hall, P. and Titterington, M. (1989) The effect of simulation order on level accuracy and power of Monte Carlo tests. *J. Royal. Statist. Soc. B* **51**, 459–467.

Hall, P. and Wilson, S.R. (1991) Two guidelines for bootstrap hypothesis testing. *Biometrics* **47**, 757-762.

Hammersley, J.M. and Handscomb, D.C. (1964) *Monte Carlo Methods.* Methuen, London.

Hampel, F.R. (1974) The influence curve and its role in robust estimation. *J. Amer. Statist. Assoc.* **69**, 383–393.

Hampel, F.R., Ronchetti, E. M., Rousseeuw, P.J. and Stahel, W. A. (1986) *Robust statistics: The approach based on influence functions.* Wiley, New York.

Härdle, W. (1990) *Applied Non-parametric Regression.* Oxford University Press.

Härdle, W. and Bowman, A. (1988) Bootstrapping in nonparametric regression: local adaptive smoothing and confidence bands. *J. Amer. Statist. Assoc.* **83**, 102–110

Härdle, W., Hall, P., and Marron, S. (1988) How far are automatically chosen smoothing parameters from their optimum? *J. Amer. Statist. Assoc.* **83**, 86–95, Rejoinder 100–101.

Hartigan, J.A. (1969) Using subsample values as typical values. *J. Amer. Statist. Assoc.* **64**, 1303–1317.

Hartigan, J.A. (1971) Error analysis by replaced samples. *J. Royal. Statist. Soc. B* **33**, 98–110.

Hartigan, J.A. (1975) Necessary and sufficient conditions for asymptotic joint normality of a statistic and its subsample values. *Ann. Statist.* **3**, 573–580.

Hartigan, J.A. (1986) Discussion of Efron and Tibshirani (1986). *Statistical Science* **1**, 75–77.

Hastie, T. and Tibshirani, R. (1990) *Generalized additive models.* Chapman and Hall, London.

Hesterberg, T. (1988) Advances in importance sampling. Ph.D. dissertation, Dept. of Statistics, Stanford University.

Hesterberg, T. (1992) Efficient bootstrap simulations I: Importance sampling and control variates. Unpublished.

Hinkley, D.V. (1977) Jackknifing in unbalanced situations. *Technometics* **19**, 285–292.

Hinkley, D.V. (1988) Bootstrap methods (with discussion). *J. Royal. Statist. Soc. B* **50**, 321–337.

Hinkley, D.V. (1989) Bootstrap significance tests, *Proceedings of the 47th session of International Statistical Institute.* Paris, 65–74.

Hinkley, D.V. and Shi, S. (1989) Importance sampling and the nested bootstrap. *Biometrika* **76**, 435–446.

Hinkley, D.V. and Wei, B.C. (1984) Improvement of jackknife confidence limit methods. *Biometrika* **71**, 331–339.

Huber, P. J. (1981) *Robust Statistics.* Wiley, New York.

Hope, A.C.A. (1968) A simple Monte Carlo test procedure. *J. Royal. Statist. Soc. B* **30**, 582–598.

Izenman, A.J. and Sommer, C.J. (1988) Philatelic mixtures and multimodal densities. *J. Amer. Statist. Assoc.* **83**, 941–953.

Jaeckel, L. (1972) The infinitesimal jackknife. *Memorandum*, MM 72-1215-11, Bell Lab. Murray Hill, N.J.

Jeffreys, H. (1961) *Theory of probability*, 3rd Edition. Oxford, Clarendon Press.

Johns, M.V. Jr. (1988) Importance sampling for bootstrap. confidence intervals. *J. Amer. Statist. Assoc.* **83**, 709–714.

Kendall, M.G. and Stuart, A. (1977) *The Advanced Theory of Statistics,* 4th Edition. Griffin, London.

Kent, T. J. (1982) Robust Properties of Likelihood Ratio Tests. *Biometrika* **69**, 19–27.

Kiefer, J. and Wolfowitz, J. (1956) Consistency of the maximum likelihood estimator in the presence of infinitely many incidental parameters. *Ann. Math. Statist.* **27**, 887-906,

Kish, L. and Frankel, M.R. (1974) Inference from complex samples (with discussion). *J. Royal. Statist. Soc. B* **36**, 1–37.

Knuth, D. (1969) *The art of computer programming.* Addison-Wesley, Reading, Mass.

Knuth, D. (1986) *The TEX book.* Addison-Wesley, Reading, Mass.

Kolaczyk, E. (1993) Empirical likelihood for generalized linear models. To appear, *Statist. Sinica.*

Konishi, S. (1991) Normalizing transformations and bootstrap confidence intervals, *Ann. Statist.* **19**, 2209–2225.

Krewski, D. and Rao, J.N.K. (1981) Inference from stratified samples: properties of the linearization, jackknife and balanced repeated replication methods, *Ann. Statist.* **9**, 1010–1019.

Künsch, H. (1989) The jackknife and the bootstrap for general stationary observations. *Ann. Statist.* **17**, 1217-1241.

Laird, N. and Louis, T.A. (1987) Empirical Bayesian confidence intervals based on bootstrap sampling. *J. Amer. Statist. Assoc.* **82**, 739–750.

Lamport, L. (1986) LATEX: *a document preparation system.* Addison-Wesley. Reading, Mass.

Leger, C., Politis, D. and Romano, J. (1992) Bootstrap technology and applications. *Technometrics* **34**, 378–398.

Lehmann, E.L. (1983) *Theory of Point Estimation .* Wiley, New York.

Lele, S. (1991) Resampling using estimating equations. In *Estimating functions,* 295–304. V. Godambe ed. Clarendon Press, Oxford.

Lin, D. Y. and Wei, L. J. (1989) Robust Inference for the Cox Proportional Hazards Model. *J. Amer. Statist. Ass.* **84**, 1074-8.

Lindley, D.V. (1958) Fiducial distributions and Bayes theorem. *J. Royal. Statist. Soc. B* **20**, 102–107.

Linhart, H. and Zucchini, W. (1986) *Model selection.* Wiley, New York.

Liu, R. Y. and Singh, K. (1992) Moving blocks jackknife and bootstrap capture weak dependence. In *Exploring the limits of bootstrap,* ed. by LePage and Billard. John Wiley, New York.

Loh, W.-Y. (1987) Calibrating confidence coefficients. *J. Amer. Statist. Assoc.* **82**, 155–162.

Loh, W.-Y. (1991) Bootstrap calibration for confidence interval construction and selection, *Statist. Sinica* **1**, 479–495.

Lunneborg, C.E. (1985) Estimating the correlation coefficient: The bootstrap approach. *Psychol. Bull.* **98**, 209–215.

Mallows, C. (1973) Some comments on C_p. *Technometrics* **15**, 661–675.

Mardia, K.V., Kent, J.T. and Bibby, J.M. (1979) *Multivariate Analysis.* Academic Press.

Martin, M.A. (1990) On bootstrap iteration for converge correction in confidence intervals. *J. Amer. Statist. Assoc.* **85**, 1105–1108.

Marriot, F.H.C. (1979) Barnard's Monte Carlo test: how many simulations? *Appl. Statist.* **28**, 75–77.

McCarthy, P.J. (1969) Pseudo-replication: half samples. *Rev. Internat. Statist. Inst.* **37**, 239–264.

McKay, M.D. Beckman, R.M. and Conover, W.J. (1979) A comparison of three methods for selecting values of input variables in the analysis of output from a computer code. *Technometrics* **21**, 239-245.

Miller, R.G. (1964) A trustworthy jackknife. *Ann. Math. Statist.* **39**, 1594–1605.

Miller, R.G. (1974) The jackknife – a review. *Biometrika* **61**, 1–17.

Morrison, D.F. (1976) *Multivariate statistical methods, 2nd Ed.* McGraw-Hill, New York.

Newton, M., and Raftery, A. (1993) Approximate Bayesian inference via the weighted likelihood bootstrap (with discussion) To appear, *J. Royal. Statist. Soc. B.*

Noreen, E.W. (1989) *Computer Intensive Methods for Testing Hypotheses: An Introduction.* Wiley, New York.

Ogbonmwan, S.M. and Wynn, H.P. (1988) Resampling generated likelihoods. *Statistical Decision Theory and Related Topics IV.* S.S. Gupta and J.O. Berger, eds.), Springer-Verlag, New York, pp. 133–147.

Oldford, R.W. (1985) Bootstrapping by Monte Carlo versus approximating the estimator and bootstrapping exactly: Cost and performance. *Comm. Statist. Ser. B* **14**, 395–242.

Owen, A.B. (1988) Empirical likelihood ratio confidence intervals for a single functional. *Biometrika* **75**, 237–249.

Owen, A.B. (1990) Empirical likelihood confidence regions. *Ann. Statist.* **18**, 90–120.

Parr, W.C. (1983) A note on the jackknife, the bootstrap and delta method estimators of bias and variance. *Biometrika* **70**, 719–722.

Parr, W.C. (1985) Jackknifing differentiable statistical functionals. *J. Royal. Statist. Soc. B* **47**, 56–66.

Peters, S.C. and Freedman, D.A. (1984a) Bootstrapping an econometric model: Some empirical results. *J. Bus. Econ. Studies* **2**, 150–158.

Peters, S.C. and Freedman, D.A. (1984b) Some notes on the bootstrap in regression problems. *J. Bus. Econ. Studies* **2**, 406–409.

Peters, S.C. and Freedman, D.A. (1987) Balm for bootstrap confidence intervals. *J. Amer. Statist. Assoc.* **82**, 186–187.

Politis, D. and Romano, J. (1992) The stationary bootstrap. Unpublished.

Qin, J. and Lawless, J. (1993) Empirical likelihood and general estimating equations. To appear, *Ann. Statist.*

Quenouille, M. (1949) Approximate tests of correlation in time series. *J. Royal. Statist. Soc. B* **11**, 18–44.

Rao, J.N.K. and Wu, C.F.J. (1985) Inference from stratified samples: second-order analysis of three methods for nonlinear statistics. *J. Amer. Statist. Assoc.* **80**, 620–630.

Rao, J.N.K. and Wu, C.F.J. (1988) Resampling inference with complex survey data. *J. Amer. Statist. Assoc.* **83**, 231–241.

Rasmussen, J. (1987) Estimating correlation coefficients: bootstrap and parametric approaches. *Psych. Bull.* **101**, 136–139.

Reeds, J.A. (1978) Jackknifing maximum likelihood estimates. *Ann. Statist.* **6**, 727–739.

Reid, N. (1988) Saddlepoint methods and statistical inference (with discussion). *Statistical Science* **3**, 213–238.

Rice, J. (1984) Bandwidth choice for nonparametric regression. *Ann. Statist.* **12**, 1215–1230.

Robinson, G.K. (1982) Behrens-Fisher Problem. In *Encyclopedia of Statistical Science*, Vol. 1. Ed. S. Kotz and N.L. Johnson. Wiley, New York. pp. 205–209.

Robinson, J.A. (1986) Bootstrap and randomization confidence intervals. *Proceedings of the Pacific Statistical Congress, 20-24 May 1985, Auckland* (I.S. Francis, B.F.J. Manly, and F.C. Lam, eds.), North-Holland, Groningen, pages 49–50.

Robinson, J.A. (1987) Nonparametric confidence intervals in regression: The bootstrap and randomization methods. *New Perspectives in Theoretical and Applied Statistics* (M.L. Puri, J.P. Vilaplana, and W. Wertz, eds.), Wiley, New York, pages 243-256.

Romano, J.P. (1988) A bootstrap revival of some nonparametric distance tests. *J. Amer. Statist. Assoc.* **83**, 698–708.

Romano, J.P. (1989) Bootstrap and randomization tests of some nonparametric hypotheses. *Ann. Statist.* **17**, 141–159.

Rousseeuw, P. (1984) Least median of squares regression. *J. Amer. Statist. Assoc.* **79**, 871–80.

Royall, R. M. (1986) Model robust confidence intervals using maximum likelihood estimators. *Int. Statist. Rev.* **54**, 221–6.

Rubin, D.B. (1981) The Bayesian bootstrap. *Ann. Statist.* **9**, 130–134.

Schenker, N. (1985) Qualms about bootstrap confidence intervals. *J. Amer. Statist. Assoc.* **80**, 360–361.

Scholz, F.W. (1980) Towards a unified definition of maximum likelihood. *Can. J. Statist.* **8**, 193- 203.

Schwarz, G. (1978) Estimating the dimension of a model. *Ann. Statist.* **6**, 461–464.

Scott. D. (1992) *Multivariate density estimation: theory, practice and visualization.* Wiley, New York.

Sen, P.K. (1988) Functional jackknifing: rationality and general asymptotics. *Ann. Statist.* **16**, 450–469.

Shao, J. (1988) On resampling methods for bias and variance in linear models. *Ann. Statist.* **16**, 986–988.

Shao, J. (1991) Consistency of jackknife variance estimators. *Statistics* **22**, 49–57.

Shao, J. (1993) Linear model selection via cross-validation. To appear *J. Amer. Statist. Assoc.*.

Shao, J. and Wu, C.F.J. (1989) A general theory for jackknife variance estimation. *Ann. Statist.* **17**, 1176–1197.

Sheather, S.J. (1987) Assessing the accuracy of the sample median: estimated standard errors versus interpolated confidence intervals. In *Statistical Data Analysis Based on the L_1-Norm.* Ed. Y. Dodge, pages 203–215. North Holland, Amsterdam.

Shorack, G.P. (1982) Bootstrapping robust regression. *Comm. Statist. A.* **11**, 961–972.

Silverman, B.W. (1981) Using kernel density estimates to investigate multimodality. *J. Royal. Statist. Soc. B* **43**, 97–99.

Silverman, B.W. (1983) Some properties of a test for multimodality based on kernel density estimates. In *Probability, Statistics, and Analysis* Eds. J.F.C. Kingman, and G.E.H. Reuter, 248–260. Cambridge Univ. Press, Cambridge, U.K.

Silverman, B.W. (1985) Some aspects of the spline smoothing approach to non-parametric regression curve fitting. *J. Royal. Statist. Soc. B* **36**, 1–52.

Silverman, B.W. (1986) *Density Estimation for Statistics and Data Analysis.* Chapman and Hall, London.

Silverman, B.W. and Young, G.A. (1987) The bootstrap: to smooth or not to smooth? *Biometrika* **74**, 469–479.

Silvey, S.D. (1975) *Statistical inference.* Chapman and Hall, London.

Simon, J.L. and Bruce, P (1991) Resampling: a tool for everyday statistical work. *Chance* **4**, 22-32.

Singh, K. (1981) On the asymptotic accuracy of Efron's bootstrap. *Ann. Statist.* **9**, 1187–1195.

Sitter R. V. (1992) Bootstrap methods for survey data. *Can. J. Statist.,* **20**, 135–154.

Sitter, R. V. (1993) Balanced repeated replications based on orthogonal multi-arrays. To appear, Biometrika, *80*

Snedecor, G.W. and Cochran, W.G. (1980) *Statistical methods.* Seventh edition. Iowa State University.

Stein, C. (1956) Efficient nonparametric testing and estimation, *Proceedings of the Third Berkeley Symposium,* University of California Press, 187–196, Berkeley.

Stein, M. (1987) Large sample properties of simulations using Latin hypercube sampling. *Technometrics* **29**, 143–151.

Stine, R.A. (1985) Bootstrap prediction intervals for regression. *J. Amer. Statist. Assoc.* **80**, 1026–1031.

Stone, M. (1974) Cross-validation choice and assessment of statistical predictions. *J. Royal. Statist. Soc. B* **36**, 111–147.

Stone, M. (1977) An asymptotic equivalence of choice of model by cross-validation and Akaike's criterion. *J. Royal. Statist. Soc. B* **39**, 44-7.

Swanepoel, J.W.H., Van Wyk, J.W.J., and Venter, J.H. (1983) Fixed width confidence intervals based on bootstrap procedures. *Sequential Anal.* **2**, 289–310.

Taylor, C.C. (1989) Bootstrap choice of the smoothing parameter in kernel density estimation. *Biometrika* **76**, 705–712.

Therneau, T. M. (1983) Variance reduction techniques for the bootstrap. Ph.D. thesis, Department of Statistics, Stanford University.

Thisted, R.A. (1986) *Elements of statistical computing.* Chapman and Hall, London.

Tibshirani, R. (1985) Bootstrapping computations. Proc. of the SAS Users group conference, Reno, Nevada.

Tibshirani, R. (1986) Bootstrap confidence intervals. *Computer Science and Statistics: Proceedings of the 18th Symposium on the Interface* (J.T. Boardman, ed.), Amer. Stat. Assoc. Washington, DC, pages 267–273.

Tibshirani, R. (1988) Variance stabilization and the bootstrap. *Biometrika* **75**, 433–444.

Tibshirani, R. (1992) Comment on "Two guidelines for bootstrap hypothesis testing", by Hall, P. and Wilson, S.R. *Biometrics* **48**, 969–970.

Tukey, J.W. (1958) Bias and confidence in not quite large samples. (Abstract.) *Ann. Math. Statist.* **29**, 614.

Wahba, G. (1980) Spline bases, regularization, and generalized cross validation for solving approximation problems with large quantities of noisy data. In Cheney, W., editor, *Approximation Theory III*, pages 905–912. Academic Press.

Wahba, G. (1990) *Spline Functions for Observational Data.* CBMS-NSF Regional Conference series, SIAM. Philadelphia.

Wasserman, L. (1990) Belief functions and statistical inference. *Can. J. Statist.* **18**, 193–196.

Wang, S.J. (1992) General saddlepoint approximations in the bootstrap. *Statist. Prob. Letters* **13**, 61–66.

Weber, N.C. (1984) On resampling techniques for regression models. *Proceedings of the Pacific Statistical Congress, 20-24 May 1985, Auckland* (I.S. Francis, B.F.J. Manly, and F.c. Lam, eds.), North-Holland, Groningen, pp. 51–55.

Weisberg, S. (1980) *Applied linear regression.* Wiley, New York.

Welch, B.L. (1947) The generalization of "Student's" problem when several different population variances are involved. *Biometrika* **34**, 28–35.

White, H. (1981) Consequences and detection of misspecified nonlinear regression Models. *J. Amer. Statist. Assoc.* **76**, 419–33.

White, H. (1982) Maximum likelihood estimation of misspecified models. *Econometrica* **82**, 1–25.

Wu, C.F.J. (1986) Jackknife, bootstrap and other resampling plans in regression analysis (with discussion.) *Ann. Statist.* **14**, 1261–1350.

Young, G.A. (1986) Conditioned data-based simulations: some examples from geometrical statistics. *Internat. Statist. Rev.* **54**, 1–13.

Young, G.A. (1988a) A note on bootstrapping the correlation coefficient. *Biometrika* **75**, 370–373.

Young, G.A. (1988b) Resampling tests of statistical hypotheses. *Proceedings of the Eighth Biannual Symposium on Computational Statistics*, D. Edwards and N.E. Raun eds., Physica- Verlag, Heidelberg.

Zhang, P. (1992) Model selection via multifold cross-validation. Tech.Report 257, Dept. of Stat., Univ. of California, Berkeley.

Author index

Subject index